Methods for Meta-analysis in Medical Research

Methods for Meta-analysis in Medical Research

Alex J. Sutton
University of Leicester, UK

Keith R. Abrams
University of Leicester, UK

David R. Jones
University of Leicester, UK

Trevor A. Sheldon
University of York, UK

Fujian Song
University of York, UK

JOHN WILEY & SONS, LTD

Chichester · New York · Weinheim · Brisbane · Singapore · Toronto

The work is based on an original NHS Health Technology Assessment funded Project (93/51/2). Adapted with kind permission of the National Coordinating Centre for Health Technology Assessment.

Contents

Preface   xv

Acknowledgements   xvii

Part A: Meta-Analysis Methodology: The Basics   1

1. Introduction – Meta-analysis: Its Development and Uses 3
    1.1 Evidence-based health care   3
    1.2 Evidence-based everything!   4
    1.3 Pulling together the evidence – systematic reviews   5
    1.4 Why meta-analysis?   8
    1.5 Aim of this book   12
    1.6 Concluding remarks   13
    References   13

2. Defining Outcome Measures used for Combining via Meta-analysis 17
    2.1 Introduction   17
    2.2 Non-comparative binary outcomes   18
         2.2.1 Odds   18
         2.2.2 Incidence rates   19
    2.3 Comparative binary outcomes   20
         2.3.1 The Odds ratio   20
         2.3.2 Relative risk (or rate ratio/relative rate)   23
         2.3.3 Risk differences between proportions
                (or the absolute risk reduction)   25
         2.3.4 The number needed to treat   27
         2.3.5 Comparisons of rates   28
         2.3.6 Other scales of measurement used in summarizing
                binary data   28
         2.3.7 Which scale to use?   28

2.4   Continuous data   28
      2.4.1   Outcomes defined on their original metric
              (mean difference)   29
      2.4.2   Outcomes defined using standardized mean differences   31
2.4   Ordinal outcomes   33
2.5   Summary/Discussion   33
References   34

3. Assessing Between Study Heterogeneity **37**
3.1   Introduction   37
3.2   Hypothesis tests for presence of heterogeneity   38
      3.2.1   Standard χ^2 test   38
      3.2.2   Extensions/alternative tests   39
      3.2.3   Example: Testing for heterogeneity in the
              cholesterol lowering trial dataset   40
3.3   Graphical informal tests/explorations of heterogeneity   41
      3.3.1   Plot of normalized (z) scores   41
      3.3.2   Forest plot   42
      3.3.3   Radial plot (Galbraith diagram)   46
      3.3.4   L'Abbé plot   47
3.4   Possible causes of heterogeneity   48
      3.4.1   Specific factors that may cause heterogeneity in RCTs   49
3.5   Methods for investigating and dealing with sources of
      heterogeneity   50
      3.5.1   Change scale of outcome variable   51
      3.5.2   Include covariates in a regression model
              (*meta-regression*)   51
      3.5.3   Exclude studies   52
      3.5.4   Analyse groups of studies separately   52
      3.5.5   Use of random effects models   52
      3.5.6   Use of mixed-effect models   53
3.6   The validity of pooling studies with heterogeneous outcomes   53
3.7   Summary/Discussion   53
References   54

4. Fixed Effects Methods for Combining Study Estimates **57**
4.1   Introduction   57
4.2   General fixed effect model – the inverse variance-weighted
      method   58
      4.2.1   Example: Combining odds ratios using the inverse
              variance-weighted method   59
      4.2.2   Example: Combining standardized mean differences
              using a continuous outcome scale   62

4.3 Specific methods for combining odds ratios   63
    4.3.1 Mantel–Haenszel method for combining odds ratios   64
    4.3.2 Peto's method for combining odds ratios   66
    4.3.3 Combining odds ratios via maximum-likelihood
        techniques   68
    4.3.4 Exact methods of interval estimation   69
    4.3.5 Discussion of the relative merits of each method   69
4.4 Summary/Discussion   70
References   71

5. Random Effects Models for Combining Study Estimates **73**
5.1 Introduction   73
5.2 Algebraic derivation for random effects models by the
    weighted method   74
5.3 Maximum likelihood and restricted maximum likelihood
    estimate solutions   75
5.4 Comparison of estimation methods   76
5.5 Example: Combining the cholesterol lowering trials using
    a random effects model   76
5.6 Extensions to the random effects model   80
    5.6.1 Including uncertainty induced by estimating the
        between study variance   80
    5.6.2 Exact approach to random effects meta-analysis of
        binary data   81
    5.6.3 Miscellaneous extensions to the random effects model   82
5.7 Comparison of random with fixed effect models   83
5.8 Summary/Discussion   84
References   84

6. Exploring Between Study Heterogeneity **87**
6.1 Introduction   87
6.2 Subgroup analyses   88
    6.2.1 Example: Stratification by study characteristics   89
    6.2.2 Example: Stratification by patient characteristics   89
6.3 Regression models for meta-analysis   93
    6.3.1 Meta-regression models (fixed-effects regression)   93
    6.3.2 Meta-regression example: a meta-analysis of Bacillus
        Calmette-Guérin (BCG) vaccine for the prevention of
        tuberculosis (TB)   95
    6.3.3 Mixed effect models (random-effects regression)   97
    6.3.4 Mixed model example: A re-analysis of Bacillus
        Calmette-Guérin (BCG) vaccine for the prevention of
        tuberculosis (TB) trials   99

	6.3.5	Mixed modelling extensions	99
6.4	Summary/Discussion		104
References			105

7. Publication Bias. **109**
	7.1	Introduction	109
7.2	Evidence of publication and related biases		110
	7.2.1	Survey of authors	110
	7.2.2	Published versus registered trials in a meta-analysis	110
	7.2.3	Follow-up of cohorts of registered studies	111
	7.2.4	Non-empirical evidence	111
	7.2.5	Evidence of language bias	111
7.3	The seriousness and consequences of publication bias for meta-analysis		112
7.4	Predictors of publication bias (factors effecting the probability a study will get published)		112
7.5	Identifying publication bias in a meta-analysis		112
	7.5.1	The funnel plot	113
	7.5.2	Rank correlation test	116
	7.5.3	Linear regression test	117
	7.5.4	Other methods to detect publication bias	119
	7.5.5	Practical advice on methods for detecting publication bias	119
7.6	Taking into account publication bias or adjusting the results of a meta-analysis in the presence of publication bias		119
	7.6.1	Analysing only the largest studies	120
	7.6.2	Rosenthal's 'file drawer' method	120
	7.6.3	Models which estimate the number of unpublished studies, but do not adjust	122
	7.6.4	Selection models using weighted distribution theory	123
	7.6.5	The 'Trim and Fill' method	123
	7.6.6	The sensitivity approach of Copas	125
7.7	Broader perspective solutions to publication bias		126
	7.7.1	Prospective registration of trials	126
	7.7.2	Changes in publication process and journals	126
7.8	Including unpublished information		127
7.9	Summary/Discussion		127
References			128

8. Study Quality **133**
	8.1	Introduction	133
8.2	Methodological factors that may affect the quality of studies		134
	8.2.1	Experimental studies	135

		8.2.2	Observational Studies	136
	8.3		Incorporating study quality into a meta-analysis	137
		8.3.1	Graphical plot	137
		8.3.2	Cumulative methods	138
		8.3.3	Regression model	138
		8.3.4	Weighting	140
		8.3.5	Excluding studies	142
		8.3.6	Sensitivity analysis	143
	8.4		Practical implementation	143
	8.5		Summary/Discussion	144
	References			144

9. Sensitivity Analysis **147**

	9.1		Introduction	147
	9.2		Sensitivity of results to inclusion criteria	147
	9.3		Sensitivity of results to meta-analytic methods	150
		9.3.1	Assessing the impact of choice of study weighting	150
	9.4		Summary/Discussion	151
	References			151

10. Reporting the Results of a Meta-analysis **153**

	10.1		Introduction	153
	10.2		Overview and structure of a report	154
	10.3		Graphical displays used for reporting the findings of a meta-analysis	155
		10.3.1	Forest plots	155
		10.3.2	Radial plots	157
		10.3.3	Funnel plots	157
		10.3.4	Displaying the distribution of effect size estimates	158
		10.3.5	Graphs investigating length of follow-up	158
	10.4		Summary/Discussion	158
	References			158

Part B: Advanced and Specialized Meta-analysis Topics   **161**

11. Bayesian Methods in Meta-analysis. **163**

	11.1		Introduction	163
	11.2		Bayesian methods in health research	163
		11.2.1	General introduction	163
		11.2.2	General advantages/disadvantages of Bayesian methods	166
		11.2.3	Example: Bayesian analysis of a single trial using a normal conjugate model	167

11.3   Bayesian meta-analysis of normally distributed data   169
       11.3.1   Example: Combining trials with continuous
                outcome measures using Bayesian methods   171
11.4   Bayesian meta-analysis of binary data   171
       11.4.1   Example: Combining binary outcome measures
                using Bayesian methods   173
11.5   Empirical Bayes methods in meta-analysis   175
11.6   Advantages/disadvantages of Bayesian methods in
       meta-analysis   176
       11.6.1   Advantages   176
       11.6.2   Disadvantages   178
11.7   Extensions and specific areas of application   179
       11.7.1   Incorporating study quality   179
       11.7.2   Inclusion of covariates   180
       11.7.3   Model selection   180
       11.7.4   Hierarchical models   181
       11.7.5   Sensitivity analysis   181
       11.7.6   Comprehensive modelling   182
       11.7.7   Other developments   183
11.8   Summary/Discussion   183
References   183

12.   **Meta-analysis of Individual Patient Data** **191**
       12.1   Introduction   191
       12.2   Procedural methodology   193
              12.2.1   Data collection   193
              12.2.2   Checking data   193
       12.3   Issues involved in carrying out IPD meta-analyses   193
       12.4   Comparing meta-analysis using IPD or summary data?   194
       12.5   Combining individual patient and summary data   195
       12.6   Summary/Discussion   196
       References   196

13.   **Missing Data** **199**
       13.1   Introduction   199
       13.2   Reasons for missing data   200
       13.3   Categories of missing data at the study level   200
       13.4   Analytic methods for dealing with missing data   201
              13.4.1   General missing data methods which can be
                       applied in the meta-analysis context   201
              13.4.2   Missing data methods specific to meta-analysis 202
              13.4.3   Example: Dealing with missing standard
                       deviations of estimates in a meta-analysis   202

13.5	Bayesian methods for missing data	203
13.6	Summary/Discussion	203
	References	204

14. Meta-analysis of Different Types of Data...................... 205

14.1	Introduction	205
14.2	Combining ordinal data	205
14.3	Issues concerning scales of measurement when combining data	206
	14.3.1 Transforming scales, maintaining same data type	207
	14.3.2 Binary outcome data reported on different scales	207
	14.3.3 Combining studies whose outcomes are reported using different data types	208
	14.3.4 Combining summaries of binary outcomes with those of continuous outcomes	208
	14.3.5 Non-parametric method of combining different data type effect measures	208
14.4	Meta-analysis of diagnostic test accuracy	209
	14.4.1 Combining binary test results	209
	14.4.2 Combining ordered categorical test results	215
	14.4.3 Combining continuous test results	215
14.5	Meta-analysis using surrogate markers	215
14.6	Combining a number of cross-over trials using the patient preference outcome	216
14.7	Vote-counting methods	217
14.8	Combining p-values/significance levels	218
	14.8.1 Minimum p method	219
	14.8.2 Sum of z's method	220
	14.8.3 Sum of logs method	220
	14.8.4 Logit method	220
	14.8.5 Other methods of combining significance levels	220
	14.8.6 Appraisal of the methods	221
	14.8.7 Example of combining p-values	221
14.9	Novel applications of meta-analysis using non-standard methods or data	223
14.10	Summary/Discussion	223
	References	223

15. Meta-analysis of Multiple and Correlated Outcome Measures....... 229

15.1	Introduction	229
15.2	Combining multiple p-values	230
15.3	Method for reducing multiple outcomes to a single measure for each study	231

15.4   Development of a multivariate model   231
        15.4.1   Model of Raudenbush *et al.*   231
        15.4.2   Model of Gleser and Olkin   232
        15.4.3   Multiple outcome model for clinical trials   232
        15.4.4   Random effect multiple outcome regression
                  model   232
        15.4.5   DuMouchel's extended model for multiple
                  outcomes   233
        15.4.6   Illustration of the use of multiple outcome models 233
15.5   Summary/Discussion   236
References   236

16. Meta-analysis of Epidemiological and Other Observational Studies . . . 239
16.1   Introduction   239
16.2   Extraction and derivation of study estimates   240
        16.2.1   Scales of measurement used to report and
                  combine observational studies   243
        16.2.2   Data manipulation for data extraction   243
        16.2.3   Methods for transforming and adjusting reported
                  results   244
16.3   Analysis of summary data   246
        16.3.1   Heterogeneity of observational studies   246
        16.3.2   Fixed or random effects?   247
        16.3.3   Weighting of observational studies   247
        16.3.4   Methods for combining estimates of
                  observational studies   247
        16.3.5   Dealing with heterogeneity and combining the
                  OC and breast cancer studies   248
16.4   Reporting the results of meta-analysis of observational
        studies   248
16.5   Use of sensitivity and influence analysis   248
16.6   Study quality considerations for observational studies   249
16.7   Other issues concerning meta-analysis of observational
        studies   250
        16.7.1   Analysing individual patient data from
                  observational studies   250
        16.7.2   Combining dose-response data   251
        16.7.3   Meta-analysis of single case research   253
16.8   Unresolved issues concerning the meta-analysis of
        observational studies   254
16.9   Summary/Discussion   255
References   255

17. Generalized Synthesis of Evidence – Combining Different
    Sources of Evidence.. **259**
    17.1   Introduction   259
    17.2   Incorporating single-arm studies: models for
           incorporating historical controls   259
           17.2.1   Example   260
    17.3   Combining matched and unmatched data   262
    17.4   Approaches for combining studies containing multiple
           and/or different treatment arms   263
           17.4.1   Approach of Gleser and Olkin   264
           17.4.2   Models of Berkey *et al.*   264
           17.4.3   Method of Higgins   264
           17.4.4   Mixed model of DuMouchel   264
    17.5   The confidence profile method   265
    17.6   Cross-design synthesis   266
           17.6.1   Beginnings   267
           17.6.2   Bayesian hierarchical models   267
           17.6.3   Grouped random effects models of
                    Larose and Dey   271
           17.6.4   Synthesizing studies with disparate designs to
                    assess the exposure effects on the incidence of a
                    rare adverse event   271
           17.6.5   Combining the results of cancer studies in
                    humans and other species   272
           17.6.6   Combining biochemical and epidemiological
                    evidence   272
           17.6.7   Combining information from disparate
                    toxicological studies using stratified ordinal
                    regression   272
    17.7   Summary/Discussion   273
    References   273

18. Meta-analysis of Survival Data **277**
    18.1   Introduction   277
    18.2   Inferring/estimating and combining (log) hazard ratios   278
    18.3   Calculation of the overall 'log-rank' odds ratio   278
    18.4   Calculation of pooled survival rates   279
    18.5   Method of Hunink and Wong   279
    18.6   Iterative generalized least squares for meta-analysis of
           survival data at multiple times   280
           18.6.1   Application of the model   281
    18.7   Identifying prognostic factors using a log (relative risk)
           measure   282

18.8   Combining quality of life adjusted survival data   282
18.9   Meta-analysis of survival data using individual patient
       data   283
       18.9.1 Pooling independent samples of survival data to
               form an estimator of the common survival
               function   283
       18.9.2 Is obtaining and using survival data necessary?   283
18.10 Summary/Discussion   284
References   284

19. **Cumulative Meta-analysis** **287**
     19.1   Introduction   287
     19.2   Example: Ordering by date of publication   288
     19.3   Using study characteristics other than date of
            publication   290
            19.3.1 Example: Ordering the cholesterol trials by
                    baseline risk in the control group   290
     19.4   Bayesian approaches   291
     19.5   Issues regarding uses of cumulative meta-analysis   291
     19.6   Summary/Discussion   292
     References   292

20. **Miscellaneous and Developing Areas of Application in
     Meta-analysis** **295**
     20.1   Introduction   295
     20.2   Alternatives to conventional meta-analysis   295
            20.2.1 Estimating and extrapolating a response surface 295
            20.2.2 Odd man out method   296
            20.2.3 Best evidence synthesis   296
     20.3   Developing areas   297
            20.3.1 Prospective meta-analysis   297
            20.3.2 Economic evaluation through meta-analysis   298
            20.3.3 Combining meta-analysis and decision analysis 299
            20.3.4 Net benefit model synthesizing disparate
                    sources of information   299
     References   299

Appendix I: Software Used for the Examples in this Book **301**

Index **309**

Preface

The move towards evidence-based health care and practice is to a large extent underpinned by the role of meta-analysis (the quantitative synthesis of a number of study results). The methods by which meta-analysis is undertaken have developed substantially in medical applications over the last 5 to 10 years.

The origins of this book are in a systematic review of the methodological literature relating to systematic review and meta-analysis methods recently carried out by the authors (Sutton *et al.*, 1998). The methodological literature concerning meta-analysis identified by this review was considerable – for example, over 1000 references were identified.

While compiling this survey of practice and problems in the area of meta-analysis, it became clear that no single source was available which described the methods available in this fast developing branch of science. Hence, it has been our aim to plug this gap, by producing a book which describes in detail the methods of meta-analysis, as applied to the medical and related literature. The original text of the review has been heavily revised, and updated. Many worked examples are now included which, we hope, add clarity, and de-mystify many of the statistical methods. Further, our aim was to produce a book which did not just present the theory of meta-analysis, but also helped researchers implement the methods themselves, without the need for heavy statistical programming. Hence, a description of how to carry out the analysis for each illustrative example is provided in an Appendix. In addition, a companion website has been developed from which much of the software used can be downloaded (see page 13 for details).

Alex J. Sutton
Keith R. Abrams
David R. Jones
Trevor A. Sheldon
Fujian Song
March 2000

Reference

Sutton, A.J., Abrams, K.R., Jones, D.R., Sheldon, T.A. and Song, F. (1998) Systematic reviews of trials and other studies. *Health Technology Assessment*, **2**(19).

Acknowledgements

We are indebted to many people and groups for the assistance we have received during the course of writing this book. In particular, we wish to thank: The NHS Health Technology Assessment (HTA) Programme, who funded the original work on which this book is based; Julie Glanville and Janette Boynton at the Centre for Reviews and Dissemination, York, for their advice on searching for the literature; The Cochrane Collaboration for supplying us with copies of their database of literature, and dealing with several queries; Teresa Prevost from the Biostatistics Unit at Cambridge for help and advice; Dr Martin Hellmich, a Visiting Fellow in the Department of Epidemiology and Public Health at Leicester, for shaping the chapter on meta-analysis of diagnostic test accuracy; Cindy Billingham, for her advice and support; Sage publications for supplying several books without charge; Julian Higgins for useful advice and allowing reproduction of one of his figures in Chapter 10; David Moher, for advice on assessing study quality, and reporting the results of a meta-analysis; Sue Duval and Richard Tweedie, for assisting with the method of 'Trim and Fill'; and Sharon Clutton and Robert Calver at Wiley for their support and patience.

Part A

Meta-Analysis Methodology:
The Basics

CHAPTER 1

Introduction – Meta-analysis: Its Development and Uses

1.1 Evidence-based health care

There has been a growing international interest in the development of measures to help ensure that public policy and practice decision making is better informed by the results of relevant and reliable research. This interest has been fuelled by evidence that some health and social interventions which have been commonly applied in the belief that they were doing good are actually harmful, that others are largely ineffective, and thus wasteful of public resources, and furthermore, that some interventions which reliable research shows to have significant benefits have been largely ignored. By bringing together the results of research in a systematic way, appraising its quality in the light of the question being asked, synthesizing the results in an explicit way and making this knowledge base more accessible, it is hoped to foster a greater sensitivity to the evidence by researchers, policy makers, practitioners and the public.

It is assumed by most of the public and patients that the activities of health care professionals is 'scientific', based firmly on the results of scientific research and subject to resource constraints of high quality. However, when practice patterns are studied, it is often found that health care interventions have not been evaluated, and even when they have, the evidence on effectiveness is not evenly applied; practice often reflecting local or national fashions or tradition.

The absence of an adequate knowledge base for much of care was highlighted in 1972 by the British epidemiologist Archie Cochrane, in his seminal monograph *Effectiveness and Efficiency* [1]. He highlighted the need for evidence from rigorous

3

evaluations to inform choices made by policy makers, professionals and consumers of services, and bemoaned the lack of ready access to the results of evaluations:

> It is surely a great criticism of our profession that we have not organised a critical summary, by speciality or subspeciality. Adapted periodically, of all relevant randomised controlled trials [2].

This criticism was one of the sparks which led to a 10 year long international effort to put together the results of all Randomized Controlled Trials (RCTs) relevant to care during pregnancy and childbirth [3]. This work set the stage for a general international initiative to summarize the results of experimental evaluation in all of health care – the Cochrane Collaboration [4].

The main thrust of this argument also provided the impetus for the rise of the Evidence Based Medicine (EBM) movement which stresses the 'conscientious, explicit and judicious use of current best evidence in making decisions about the care of individual patients' [5, 6]. The rise of EBM reflects in large part the increased scrutiny of the value of spending on health services, a response to the spiralling costs of health care systems over the past years. Policy makers have been seeking ways of containing these costs, and one approach adopted is to look for savings in practices that have been shown to be ineffective or relatively ineffective. This trend has been strengthened by research showing unjustified variations in treatment and inappropriate care. The practice of medicine in particular and healthcare in general has, therefore, come under a critical spotlight. Policy makers feel safer when rationing health care if they can invoke the use of research evidence when so doing [7].

1.2 Evidence-based Everything!

Over the last few years, the themes of evidence reviews, evidence-based health care, systematic reviews and meta-analysis which occur and re-occur in several countries, academic disciplines and policy areas have started to converge (for a good summary, see the special issue of *Public Money & Management*, 1999, vol. 19). An example of this is the development of an international collaboration – the Campbell Collaboration – which will prepare and maintain systematic reviews of research on the effects of interventions in areas such as education, criminal justice, social policy and social care. The development of this international collaboration alongside and in recognition of the contribution to knowledge and practice that the Cochrane Collaboration has made will help to meet the surge of interest in research synthesis and systematic reviews[1].

---

[1] The Campbell Collaboration will be chaired by Robert Boruch (Graduate School of Education, University of Pennsylvania: robertb@gse.upenn.edu).

1.3 Pulling Together the Evidence – Systematic Reviews

There has been a massive growth in the volume of scientific knowledge relevant to health care, the results being spread over many reports and journals. Studies use a variety of methods, are of variable quality, and may appear to have contradictory findings. Single evaluations and stand-alone studies add data to the knowledge base, but are rarely definitive in that they are often context specific and too small. Multi-site studies often show differential effects. Research information can seem to users like small jigsaw pieces in a box where there may be several pictures, several duplicates and several missing pieces. Under these circumstances, it can be difficult for the researcher, research funder or user to make sense of the research, or take stock of the knowledge base. This can have serious negative consequences, for example:

- Scientific progress can be impeded, and it becomes harder to make sense of divergent results. Research then tends to be less based on the accumulated knowledge of a field, rather than on limited and possibly biased views of it [8].
- Too often, proposals for research have not sufficiently explored what has already been done. There appears at times to be little collective memory, there is duplication, and some research funding decisions may not be clearly based on what research will contribute to the totality of knowledge. Since policy often moves in cycles, it is important to have access to the results of research from several decades previous. Much research is uncoordinated, lacking a cumulative character, and needs instead to build more effectively on earlier work. Researchers need to appraise more carefully what is known in a field before starting new research [9].
- Policy makers and practitioners can rarely make sense of the 'hotch potch' of findings, thus making it more likely that they will ignore research or use it in a biased and selective way. When research information has been poorly collected together and organized, the resulting confusion provides more fertile ground for 'experts' and gurus peddling the latest fashion, theory or ideology, and gives more influence to those lobbying on behalf of vested interests who can 'sandbag' – find bits of research which support an argument and then quote them.
- As a result, significant investments in a policy can be made in the mistaken belief that it will result in the desired effects. Similarly, practitioners may be left unaware of potentially harmful (beneficial) practices. Consider, for example, systematic reviews of four different educational interventions to improve maths performance [10] which revealed greater improvements with cross-age tutoring compared to changes in class size or use of computer assisted instruction.

Traditional mechanisms for professionals and policy makers to keep abreast of this growing knowledge base are now inadequate. An integral part of EBM is, therefore, the technology available to pull together the existing evidence in a form that can be used by practitioners – the systematic review.

The systematic review methodology is one of the first steps in the chain by which research evidence can inform policy and practice [11]. Until the late 1980s, there was little interest in the methods used to carry out reviews of the literature. A watershed was Mulrow's paper on the review article [12], which laid out the need for a more explicit approach to summarizing the evidence. Traditionally, reviews were carried out in a rather *ad hoc* fashion, depending on implicit and often idiosyncratic methods of collecting together the primary studies. Traditional reviews, although often heroic exercises offering handy updates, rarely provide synthetic accounts constructed in a scientifically explicit and defensible way. Too often, the authors have decided what they want to say and then seek the data to justify the conclusions. In contrast, systematic reviews use explicit and rigorous methods to identify, critically appraise, include and synthesize relevant research studies [13, 14].

Some of the characteristic features of a systematic review are shown in Table 1.1. Systematic reviews or research syntheses can be thought of as a technique for sorting out the bits of the jigsaw, weighing up where they might go, and putting bits together. They also help us see more clearly where there are gaps. Systematic reviews are more cumulative and more critically robust and, along with the related idea of evidence-based policy and practice, have been likened to ice breakers sailing through the pack ice of opinion and assertion. Thus, the conduct of a systematic review is a scientific process which needs to be approached with the same degree of care as a primary study. Several texts are available which discuss the methods of conducting a systematic review [15–18].

The importance of basing practice more firmly on the results of systematic reviews was starkly demonstrated by a paper in the early 1990s, which compared the results of meta-analysis of trials of treatments for people who have suffered a heart attack as the trials were published with the recommendations of experts published in review articles and textbooks over the same time period. This showed a significant divergence between recommendations and the summaries of the trials. Ineffective treatments were being recommended, and highly effective treatments were not. There were significant time delays between the publication of the studies and changes in the recommendations of the experts [19]. As a result, lives that could have been saved were lost, and resources were wasted.

Advantages of the systematic review or research synthesis approach include the following:

- Helps to democratize the research and the uses of research, for it makes the knowledge base more open, accessible and contestable by a range of stakeholders, including the public. This can reduce the power and

Table 1.1 Characteristics of a systematic review.

Aim	Characteristics
Clarity about aim of review	Protocol stating objectives, research questions, scope and methods prior to the conduct of the review.
Avoid omitting relevant research	Comprehensive, sensitive and documented search strategy using a range of bibliographical databases, index and text search terms, possibly hand searching, possibly attempting to identify unpublished research, not restricted by country or language.
Avoid biased selection/ inclusion of studies	Explicit and verifiable inclusion and exclusion criteria developed prior to examination of the research results.
Accurate abstraction of study data	Use of data extraction sheets/tables with some checking for accuracy.
Assess the validity of the research results	Develop and use explicit quality criteria to assess the validity of the research by examining the design, conduct and analysis of the research to assess bias, confounding and chance.
Estimate the size of associations and sources of study variation	Explore the reasons why study results may vary using appropriate quantitative models to examine the role of factors such as patient characteristics, dose, duration and nature of the intervention. Where possible, pool the results to produce an overall effect.
Assess the robustness of review results	Test the sensitivity of the results to choices and assumptions made in the review, such as the inclusion and validity criteria, the likely impact of publication bias and the modelling approach used to pool the results.
Aid the critical appraisal or the replication of the review	Document and report the key aspects of the review process, methods, analysis and results. This should include a summary of the protocol, the search strategy and a table of the key elements of each study included. This can be aided by graphical displays.
Help the reader to assess the implications of the review for policy, practice and research	Discuss the methodological limitations of both the primary research included and the review. Make explicit the research evidence being used to justify any stated implications or recommendations.

mystique of the 'expert' as the repository of knowledge which, in the past, has resulted in the proliferation of implicit biases.

- Provides an invaluable resource for the research community which helps to make sense of the past, contextualize new material and develop methodology, identify themes and key issues. This facilitates international replication of research and testing out of theories.
- Provides support for research funders who wish to identify key gaps in the research, place new proposals into the context of existing knowledge, and prevent unnecessary duplication.
- Provides a knowledge base for practitioners and policy makers who can use this to assess the likely effectiveness of different policies and practices.
- Aids the cumulative development of the sciences. Rarely does a study break completely new ground; we stand on the shoulders of previous research.
- Helps to identify more clearly what we do not know, and the degrees of uncertainty. This is generally useful, and particularly so for those wishing to 'comprehend' a field.

The use of systematic review methodology is not only likely to produce more reliable results, but the systematic nature of the reviews, the explicitness and grading of research quality has increased their credibility. This has helped to foster a 'sensitivity to the evidence' amongst policy makers and practitioners, and has 'sharpened the congressional appetite for impact evaluation' [20].

However, the fact that people have followed a set of techniques does not guarantee quality; following rules can still produce rubbish! In addition, as with analysis and interpretation of any research, there are judgement calls when conducting research syntheses. Systematic reviews do not produce uncontested monoliths of 'truth'. The scientific and open way they are carried out, however, makes them more susceptible to criticism than the expert opinion or the expert review paper which does not follow scientific methods. A key advantage of systematic reviews is that judgements and assumptions are more explicit, need to be justified and, if done properly, the sensitivity of the results to different assumptions and judgements (e.g. inclusion criteria or quality grading systems) are presented or can be assessed by others. The methods are explicit and are open to scrutiny, so that others can see what was done and question the results.

1.4　Why Meta-analysis?

If systematic reviews provide the research evidence input into the process of evidence-based decision making, then meta-analysis is the analytical or statistical part of systematic reviews. Figure 1.1 outlines a widely agreed approach to the whole systematic review process [15, 16, 21].

(1) Specification in a protocol of the objectives, hypotheses (in both biologic and health care terms), scope, and methods of the systematic review, before the study is undertaken.

(2) Compilation of as comprehensive a set of reports as possible of relevant primary studies, having searched for all potentially relevant data, clearly documenting all search methods and sources. Any selection of studies should be based on clearly stated *a priori* specifications.

(3) Assessment of the methodological quality of the set of studies (the method being based on the extent to which susceptibility to bias is minimized – and the specific system used reported). The reproducibility of the procedures in (2) and (3) should also be assessed.

(4) Identification of a common set of definitions of outcome, explanatory and confounding variables, which are, as far as possible, compatible with those in each of the primary studies.

(5) Extraction of estimates of outcome measures and of study and subject characteristics in a standardized way from primary study documentation, with due checks on extractor bias. Procedures should be explicit, unbiased and reproducible.

(6) Where warranted by the scope and characteristics of the data compiled, meta-analysis (quantitative synthesis of primary study results) using appropriate methods and models (clearly stated), in order to explore and allow for all important sources of variation (e.g. differences in study quality, participants, in the dose, duration, or nature of the intervention, or in the definitions and measurement of outcomes). Confidence intervals around pooled point estimates should be reported.

(7) Where data are too sparse, or of too low quality, or too heterogeneous to proceed with a statistical aggregation, a narrative or qualitative summary should be performed and the formal meta-analysis omitted. In such cases, the process of conduct and reporting should still be rigorous and explicit.

(8) Exploration of the robustness of the results of the systematic review to the choices and assumptions made in all of the above stages. In particular, the following should be explained or explored:
  (a) the impact of study quality/inclusion criteria,
  (b) the likelihood and possible impact of publication bias,
  (c) the implications of the effect of different model selection strategies, and exploration of a reasonable range of values for missing data from studies with uncertain results.

(9) Clear presentation of key aspects of all of the above stages in the study report, to enable critical appraisal and replication of the systematic review. These should include a table of key elements of each primary study. Graphical displays can also assist interpretation, and should be included where appropriate.

(10) Methodological limitations of both the primary studies and the systematic review should be appraised. Any clinical or policy recommendations should be practical and explicit, and make clear the research evidence on which they are based. Proposal of a future research agenda should include clinical and methodologic requirements as appropriate.

Figure 1.1 Guidelines for good meta-analytic practice.

Meta-analysis is the part of the review process, that which concerns itself with the analysis of the data extracted from the primary research included, and which uses quantitative methods to explore the heterogeneity of study results, estimate overall measures of association or effect and assess the sensitivity of the results to possible threats to validity such as publication bias and study quality. This corresponds principally to point (6) onward in Figure 1.1, and is the main subject of this text. Other elements of a systematic review have been documented in detail elsewhere [15, 16], and are not considered further in this text.

At about the same time that Archie Cochrane was urging a more evidence-based approach, social scientists were developing the tools of meta-analysis ('analysis of analyses') in the social sciences, particularly education and psychology [8, 22, 23]. Meta-analysis – here taken to mean the quantitative pooling of results of more than one study – dates back to the beginning of the century when Karl Pearson developed a statistical technique for making sense of the divergent results from small studies of the effectiveness of inoculation against typhoid fever [24]. Statistical techniques for combining study results were also used in agriculture in the 1930s [25], and in the medical area in 1955 [26].

Over the last 25 years there has been considerable development of the application of meta-analysis techniques in social sciences in the USA in particular, centred around the fields of education, psychology and criminology. The Russell Sage Foundation, for example, sponsored focused research into research synthesis in a number of areas, including education, economics and criminology, and this lead to a series of texts [8, 27, 28]. A number of other texts which help to set standards in and develop the field of research synthesis in the social sciences, including qualitative research, have been produced [29, 30]. One of the first meta-analyses in medicine in the 'modern' era was by Thomas Chalmers [31] in 1977. However, it was not until the mid-1980s that meta-analysis started to be used more frequently in the health care field, when Yusuf [32] published his meta-analysis on beta blockers in myocardial infarction, and it really took off in the early 1990s [33].

When several studies have examined the same question, such as the effectiveness of a treatment, there is likely to be chance fluctuations between their results, particularly if they are small studies, with low power. Thus, some may show results favouring the treatment, and others show no benefits. Very few traditional reviews consider the statistical power of any statistical test carried out in each study, and thus their likelihood of a type II error (failing to find a meaningful result because the study was too small). These reviews will often fail to take into account the fact that, by chance, a number of small studies are likely to produce conflicting findings. By pooling all the relevant studies together, one can reduce the effect of random error and so produce more reliable and precise estimates of effect.

In instances when the benefit of an intervention may at best be modest, the required sample size for an individual trial to detect a significant difference between groups may be extremely large. In the treatment of myocardial infarction, for example, it would be very difficult to detect a 10% reduction in the risk of death in a single trial, as at least 1000 *deaths* (N.B. not *patients*) would be required. However, if a widely practicable treatment did reduce deaths from myocardial infarction by 10%, because the condition is common, it would have the potential to save tens of thousands of lives worldwide. In situations such as this, it is the increased power gained from synthesizing results from different studies that makes the systematic review an important tool [34].

Meta-analyses include more patients than any single constituent study, and so may also reduce the random errors in the assessment of treatment [35]. Meta-analysis, by including studies carried out in different places, possibly with different entry criteria, may also produce more generalizable results which average over a range of settings and contexts.

Traditional reviews are also inefficient ways to extract useful information from the studies because they lack any systematic way of integrating relationships among the variables in different studies and understanding differences between the studies [29]. One of the most important contributions of the statistical approaches in meta-analysis, as well as estimating overall effects, is the capability of exploring variation between studies and assessing factors which might modify treatment effects [36].

Despite the obvious advantages of meta-analysis outlined above, this approach has been the subject of considerable debate [37, 38]. It has been described as 'a new *bete noir*' which should be 'stifled at birth' [39] and as a 'statistical alchemy for the 21st century [40]. Some of this debate has been fuelled by poor practice in the field: the indiscriminate pooling of diverse sets of studies with differing outcomes, the use of statistical technique without sufficient primary content expertise, and with little attention to context and theory [41].

To carry out systematic reviews in any area, it is important to develop a sensible framework of standards by which one can judge the intrinsic validity of individual studies, and their adequacy for answering the particular question being studied. Medical and related areas which can be subjected to a systematic review can range from examining the effectiveness of a particular intervention in improving prognosis, exploring risk-factors associated with a certain disease, methods of measuring organizational performance, to women's experiences of childbirth. These will encompass studies which have used a broad range of methods. Each method informs the question in different ways, and is more or less adequate (alone or with other designs) to answer different types of questions [42]. However, certain areas of research, in particular that are involving qualitative data, have less developed frameworks for assessing the quality or reliability of the study design, data collection and analysis, and work on this is developmental [43, 44]. This makes them less easy to critically appraise. In addition, qualitative data are less easy to synthesize using statistical techniques. Therefore, meta-analysis of qualitative data is poorly developed, and is not dealt with in this text.

The rise of evidence-based health care and the use of meta-analysis as a tool for summarizing the results of research to fuel EBHC runs the risk that only the sort of evidence which is quantitative and which is derived from formal research (especially experimental research) is valid in informing clinical and professional and policy decisions. It is important to remember that even though we do not give much coverage to more experiential and qualitative evidence in this text, that more informal observations from clinical experience and patient's experiences are also important.

1.5   Aim of this Book

This book has been written with the aim of enabling the health researcher to understand the basic principles of, and to apply, the methods of meta-analysis. Fortunately, the basic principles underlying the most commonly used methods are reasonably straightforward, and it is hoped that readers who are numerically/ statistically literate, but who are not experts in statistics, will be able to understand a large proportion of this book.

The text is divided into two parts. Part A provides comprehensive coverage of the basics, and many worked examples are provided to illustrate the various methods described. Where possible, intermediate stages of calculations are shown for these worked examples; however, the accuracy of figures reported for these intermediate stages has been reduced in the text from those actually used in the calculations, in order to increase the clarity. A small number of sections in Part A are marked with an asterisk; this indicates more technical material, which can be omitted on first reading. It is hoped that Part A will provide a solid foundation from which the reader will feel confident about critically appraising or undertaking a meta-analysis.

Part B covers more advanced topics. These include: (a) Bayesian methods, which can be seen as an appealing, but less familiar, approach to statistics than the often used Classical methods on which Part A is based; (b) methods for specific types of data not covered in Part A; and (c) advanced methods, such as the combination of results from studies with different designs. The coverage in Part B is less thorough than Part A due to its breadth, and aims to give an overview of the methods which have been developed. If the reader wishes to know further details concerning a particular method, then the reference lists provided at the end of each chapter enable easy identification of the original sources.

The software options for performing meta-analysis have improved considerably over the last few years. Appendix I provides details of the computer software and code used to carry out each example in the book. This enables all the methods in Part A, and a proportion in Part B, to be implemented by the reader without the need for any computer programming, mostly using freely available programs, or code routines written for popular commercial statistical packages. We believe this information will greatly facilitate the implementation of the methods described herein. A companion website to this book has been set up from which much of the code and the freely available programs can be accessed.[1]

1.6 Concluding Remarks

Meta-analysis is currently a very active area of research interest, with new additions published on a very regular basis. Although we have attempted to make this book as up to date as possible, new methods will undoubtedly have been described by the time it is published, and researchers should examine the published literature periodically, perhaps using search strategies described elsewhere [45], to keep informed of the new methodological developments.

References

1. Cochrane, A.L. (1972). *Effectiveness and Efficiency: Random reflections on the health service*. London: Nuffield Provincial Hospitals Trust.

2. Cochrane, A.L. 1931–1971: A critical review, with particular reference to the medical profession. In: *Medicines for the year 2000*. London: Office of Health Economics, (1979). 1–11.

3. Chalmers, I., Enkin, M. and Keirse, M. (editors). *Effective Care in Pregnancy and Childbirth*. Oxford: Oxford University Press (1989).

4. Chalmers, I., Sackett, D., Silagy, C., Maynard, A., Chalmers, I. (editors). *Non-Random Reflections on Health Services Research*. London: BMJ Publishing Group; 11, (The Cochrane Collaboration, pp. 231–49) (1997).

5. Sackett, D.L., Rosenberg, W.M., Gray, J.A., Haynes, R.B., Richardson, W.S. (1996). Evidence-based medicine: what it is and what it isn't. *Br. Med. J.* **312**: 71–2.

6. Gray, J.A.M. (1997). *Evidence-based Healthcare: How to make health policy and management decisions*. Edinburgh: Churchill Livingstone.

7. Grayson, L. (1997). *Evidence-based Medicine: An overview and guide to the literature*. London: The British Library.

8. Hunt, M. (1997). *How Science Takes Stock: The story of meta-analysis*. New York: Russell Sage Foundation.

9. Oakley, A., Roberts, H. (1996). *Evaluating Social Interventions*. Essex Barnardo's.

---

[1] *http://www.prw.le.ac.uk/epidemio/personal/ajs22/meta/book.html*

10. Levin, H.M., Glass, G.V., Meister, G.R. (1984). Cost-effectiveness of four educational interventions: Project report number 84–A11. Stanford Institute for Research on Eductional Finance and Government.

11. Cook, D.J., Mulrow, C.D., Haynes, R.B. (1997). Systematic reviews: Synthesis of best evidence for clinical decisions. *Annals. Intern. Med.* **126**: 376–80.

12. Mulrow, C.D. (1987). The medical review article: state of the science. *Ann. Intern. Med.* **106**: 485–8.

13. Altman, D., Chalmers, I. (editors). (1995). *Systematic Reviews.* London: BMJ Publishing Group.

14. Thacker, S.B. (1988). Meta-analysis. A quantitative approach to research integration. *J. Am. Medical Assoc.* **259**: 1685–9.

15. Deeks, J., Glanville, J., Sheldon, T. (1996). Undertaking systematic reviews of research on effectiveness: CRD guidelines for those carrying out or commissioning reviews. Centre for Reviews and Dissemination, York. York Publishing Services Ltd. Report #4.

16. Oxman, A.D. (editor). (1996). *The Cochrane Collaboration Handbook: Preparing and maintaining systematic reviews. 2nd ed.* Oxford: Cochrane Collaboration.

17. Cooper, H. (1998). *Synthesizing Research: A guide for literature reviews. 3rd ed.* Newbury Park, CA: Sage Publications.

18. Wolf, F.M. (1986). *Meta-Analysis: Quantitative methods for research synthesis.* Newbury Park, CA: Sage Publications.

19. Antman, E.M., Lau, J., Kupelnick, B., Mosteller, F., Chalmers, T.C. (1992). A comparison of results of meta-analyses of randomized control trials and recommendations of clinical experts: Treatments for myocardial infarction. *J. Am. Medical Assoc.* **268**: 240–248.

20. Chelimsky, E. Politics, policy and research synthesis. Keynote address before National Conference on Research Synthesis, sponsored by the Russell Sage Foundation, Washington DC, 21 June 1994.

21. Cook, D.J., Sackett, D.L., Spitzer, W.O. (1995). Methodologic guidelines for systematic reviews of randomized control trials in health care from the Potsdam Consultation on Meta-Analysis. *J. Clin. Epidemiol.* **48**: 167–71.

22. Glass, G.V. (1976). Primary, secondary and meta-analysis of research. *Educ. Res.* **5**: 3–8.

23. Glass, G.V., McGraw, B., Smith, M.L. (1981). *Meta-analysis in Social Research.* Newbury Park, CA: Sage.

24. Pearson, K. (1904). Report on certain enteric fever inoculation statistics. *Br. Med. J.* **3**: 1243–6.

25. Yates, F., Cochran, W.G. (1938). The analysis of groups of experiments. *J. Agricultural Sci.* **28**: 556–80.

26. Beecher, H.K. (1995). The powerful placebo. *J. Am. Medical Assoc.* **159**: 1602–6.

27. Cook, T.D., Cooper, H., Cordray, D.S. *et al.* (1992). *Meta-analysis for Explanation: A Casebook.* New York: Russel Sage Foundation.

28. Cooper, H., Hedges, L.V. (editors). (1994). *The Handbook of Research Synthesis.* New York: Russel Sage Foundation.

29. Light, R.J., Pillemer, D.B. (1984). *Summing Up: The Science of Reviewing Research.* Cambridge, MA: Harvard University Press.

30. Noblit, G.W., Hare, R.D. (1988). *Meta-ethnography: Synthesising qualitative studies.* Beverly Hills, CA: Sage.

31. Chalmers, T.C., Matta, R.J., Smith, H. Jr., Kunzler, A.M. (1977). Evidence favoring the use of anticoagulants in the hospital phase of acute myocardial infarction. *New Engl. J. Med.* **297**: 1091–6.

32. Yusuf, S., Peto, R., Lewis, J., Collins, R., Sleight, P. *et al.* (1985). Beta blockade during and after myocardial infarction: an overview of the randomised trials. *Progress in Cardiovascular Diseases* **27**: 335–71.

33. Bausell, R.B., Li, Y., Gau, M., Soeken, K.L. (1995). The growth of meta-analytic literature from 1980 to 1993. *Evaluation & the Health Professions* **18**: 238–51.

34. Peto, R. (1987). Why do we need systematic overviews of randomised trials? *Stat. Med.* **6**: 233–40.

35. Collins, R., Peto, R., Gray, R., Maynard, A., Chalmers, I. (editors). (1997). *Non-random Reflections on Health Services Research.* London: BMJ Publishing Group; 10, Large-scale randomised evidence: trials and overviews; 197–230.

36. Thompson, S.G. (1994). Why sources of heterogeneity in meta-analysis should be investigated. *Br. Med. J.* **309**: 1351–5.

37. Eysenck, H.J. Problems with meta-analysis. In: Altman, D. and Chalmers, I. (editors). (1995). *Systematic Reviews.* London: BMJ Publishing Group; ch 6, 64–74.

38. Bangert-Drowns, R.L. (1995). Misunderstanding meta-analysis. *Evaluation & the Health Professions* **18**: 304–14.

39. Oakes, M. (1986). *Statistical Inference: A commentary for the social and behavioural sciences.* Chichester: John Wiley & Sons.

40. Feinstein, A.R. (1995). Metanalysis, statistical alchemy for the 21st century. *J. Clin. Epidemiol.* **48**: 71–9

41. Naylor, D.C. (1988). Two cheers for meta-analysis: problems and opportunities in aggregating results of clinical trials. *Canadian Med. Assoc. J.* **138**: 891–5.

42. Davies, P. (1999). What is evidence-based education? *Br. J. Educational Studies* **47**: 108–21.

43. Popay, J., Rogers, A., Williams, G. (1998). Rationale and standards for the review of qualitative literature in health services research. *Qualitative Health Res.* **8**: 341–51.

44. Popay, J., Williams, G. (1998). Qualitative research and evidence-based health care. *J. Roy. Soc. Med.* **91**(Suppl. 35):32–7.

45. Sutton, A.J., Abrams, K.R., Jones, D.R., Sheldon, T.A., Song, F. (1998). Systematic reviews of trials and other studies. *Health Technology Assessment* 2(19).

CHAPTER 2

Defining Outcome Measures used for Combining via Meta-analysis

2.1 Introduction

Measures of outcome need to be calculated for each of the studies in a meta-analysis before they can be quantitatively combined. In a meta-analysis of RCTs, a comparative estimate of treatment effect such as the odds ratio is often chosen, while in cohort (observational) studies it is common to employ a comparative measure of the risk between groups such as the relative risk. This section provides a brief summary of these and other measures which are commonly used in meta-analyses of health care data. Details of how to calculate each of them from summary data, which should be supplied in study reports, is also provided. Formulae for the standard measure of variation, or uncertainty (the variance), of each measure are also given.

Outcomes have been categorized into one of three groups, depending on the type of data from which they are derived. Those based on binary data, such as whether patients are alive or dead, diseased/non-diseased, are discussed first. Next, outcomes derived from continuous data (e.g. blood pressure) are described. Finally, outcomes based on ordered categorical (ordinal) data, for example, from a disease severity scale, are discussed. The vast majority of outcomes reported in medicine are comparative, with two groups having been compared. However, simpler, non-comparative binary outcomes are also discussed. Consideration of which particular scale to use within these data types for a given meta-analysis is discussed. For a statistically rigorous derivation of most of the outcome measures considered here, see Fleiss [1].

Unfortunately, due to inconsistent and poor reporting of studies, it may not be obvious how to derive the required estimate of size of the effect for certain studies.

Section 14.3 considers this problem further, and Section 16.2 considers methods specifically for extracting the results of observational studies, which are notoriously problematic. Other, more specific, outcomes are covered in Part B of this book; for example, outcomes used for diagnostic test data are outlined in Section 14.4, and outcomes derived from survival data in Chapter 18.

The reader may be aware that more complex studies, such as factorial designed RCTs, which include more than two groups, are sometimes carried out. Few meta-analysis methods have been developed to deal with such studies (and those which have are complex). It is common practice to merge similar arms of studies, or exclude arms, in an effort to reduce the study to two groups which can be compared.

2.2   Non-comparative binary outcomes

For meta-analysis, the majority of outcomes used are a comparative measure, that is one where the outcome in two groups is being compared. This is natural for combining most study types; RCTs are inherently comparative, as are many observational studies, including case-control, where people with disease are compared with non-diseased, and cohort studies where the exposed are compared with the non-exposed. For binary outcomes these are described in Section 2.3. However, in certain situations, descriptive, as opposed to comparative, estimates may be the primary interest. These are generally simpler than comparative measures, and are hence described first.

2.2.1   Odds

The odds outcome is sometimes used in non-controlled trials, where patients on a new treatment are compared with historical rather than randomized controls. In meta-analysis the historical control estimate may be discarded; a method for combining such studies with standard RCTs is given in Section 17.2. Here the odds of an event in a single group is the outcome of interest, rather than the comparative estimator between two groups – the odds ratio (discussed in Section 2.3.1). Synthesis of odds is usually carried out on the log scale. Calculation of log odds is easy, and is simply defined as:

$$\ln{(ODDS)} = \ln{\left(\frac{No.\ of\ patients\ having\ event}{No.\ of\ patients\ not\ having\ event}\right)}. \tag{2.1}$$

The variance of the ln $(ODDS)$ is:

$$\text{var}\left(\ln{(ODDS)}\right) = \frac{1}{No.\ of\ patients\ having\ event} + \frac{1}{No.\ of\ patients\ not\ having\ event}. \tag{2.2}$$

Example calculation of odds

Consider a study where 204 patients were enrolled, of whom 176 survived, and 28 died (this is in fact the numbers for the *treatment* arm of the first study in the cholesterol lowering dataset described in section 2.3.1). The ln(*ODDS*) is calculated (from equation (2.12)) by:

$$\ln (ODDS) = \ln (28/176)$$
$$= -1.838$$

The variance for ln (*ODDS*) can be calculated using equation (2.13):

$$\text{var} (\ln (ODDS)) = 1/28 + 1/176$$
$$= 0.041$$

A 95% confidence interval for the odds can be calculated using:

$$\ln (ODDS) \pm 1.96 \times \sqrt{\text{var} (\ln (ODDS))}, \tag{2.3}$$

since $\sqrt{\text{var}(\ln(ODDS))}$ gives the standard error (s.e.) of ln(*ODDS*). For the study considered above, the 95% confidence interval of the ln(*ODDS*) is given by:

$$-1.838 \pm 1.96 \times \sqrt{0.041} = (-2.234 \text{ to } - 1.441).$$

This confidence interval can be converted back to the odds ratio scale by taking the exponential of the interval on the log odds ratio scale:

$$95\%\text{CI for OR} = (e^{-2.234} \text{ to } e^{-1.441})$$
$$= (0.11 \text{ to } 0.24).$$

This indicates the range of most likely values for the odds of an event.

2.2.2 Incidence rates

If the rate at which new cases of a disease, for example cases of AIDS, occur in a population, then incidence rates from observational studies may be combined. Again, the log scale is used; the log incidence is given by:

$$\ln (incidence\ rate) = \ln (d/q), \tag{2.4}$$

where d events are observed in q person-years. The approximate variance of this outcome is:

$$\text{var} (\ln (incidence\ rate)) = 1/d. \tag{2.5}$$

Confidence intervals can be calculated using these values in parallel to equation (2.3).

	Failure /Dead	Success /Alive
New Treatment	a	b
Control	c	d

(a)

	Diseased (cases)	Non-diseased (controls)
Exposed	a	b
Not Exposed	c	d

(b)

Figure 2.1   Outcome data from a single (a) RCT, (b) Case-control study.

2.3   Comparative binary outcomes

If a comparative binary outcome is being considered, generally it will be possible to construct, or find in the report, a 2 × 2 table, for each study, which displays all the information required to work out the commonly used outcome measures. Typical 2 × 2 tables for an RCT and a case-control study are presented in Figures 2.1(a) and 2.1(b), respectively.

Commonly used outcome measures which can be derived from these types of tables are outlined below.

2.3.1   The odds ratio

The Odds Ratio (OR) can be calculated by the formula:

$$OR = \frac{ad}{bc},$$ (2.6)

where a, b, c and d relate to the cells of Figures 2.1(a) and 2.1(b). This outcome gives a relative measure of chance of the event of interest in the form of the ratio of the odds (defined in Section 2.2.1) of an event (a:b and c:d) in the two groups [2]. In an RCT setting, for an outcome considered undesirable, an odds ratio estimate of less than one, when comparing a new treatment to the control, would indicate an improvement on the new treatment; while a ratio greater than one would imply the new treatment was less effective than any treatment received by the control group (the converse is true for desirable outcomes).

For the purposes of combining, it is common (and recommended) to transform the data by taking the natural logarithm of the odds ratio and work with log odds ratios, as this should provide a measure which is approximately normally distributed [3]. The large sample variance of log odds ratio is [1]:

$$\mathrm{var}_{Ln(OR)} = \frac{1}{a} + \frac{1}{b} + \frac{1}{c} + \frac{1}{d}. \tag{2.7}$$

An important problem that needs addressing is that equation (2.7) is undefined if there are no events/all events in either of the treatment arms (i.e. one or more of $a, b, c, d = 0$). One way to get round this problem is to add 0.5 to each cell of Figure 2.1 before calculating the odds/log odds ratio. It has been suggested that this also reduces bias caused by one or more small cells; it can be seen as a continuity correction factor for converting discrete data into a continuous scale [4]. This has been done in the calculations, yielding Table 2.2.

Example calculation of an odds ratio

Consider the cholesterol lowering intervention meta-analysis dataset described in Table 2.1. This dataset, consisting of 34 RCTs, was originally analysed by Davey Smith et al. [5]

We are going to calculate an odds ratio for total mortality in the first RCT in the dataset. Here, 204 patients were enrolled into the treatment arm, and 202 into the control arm. There were a total of 28 deaths in the treatment arm and 51 deaths in the control arm, implying 176 and 151 patients survived in the treatment and control arms, respectively. In Figure 2.2 this data is arranged in a 2×2 table of the form of Figure 2.1a.

The odds ratio for this study is calculated, using equation (2.6), by:

$$OR = (28 \times 151)/(176 \times 51)$$
$$= 0.47,$$

(suggesting patients were approximately half as likely to die if they were in the new treatment group compared to the control group), and log transforming the data gives:

$$\ln(OR) = \ln(0.47)$$
$$= -0.753$$

The variance for $\ln(OR)$, can be calculated using equation (2.2):

$$\mathrm{var}(\ln(OR)) = 1/28 + 1/176 + 1/51 + 1/151$$
$$= 0.068.$$

Using these figures, it is possible to construct a confidence interval for the treatment estimate of this individual trial. This interval indicates a range of the

	Dead	Alive	Total
New Treatment	28	176	204
Control	51	151	202

Figure 2.2 2×2 table for the first RCT in the cholesterol lowering dataset.

Table 2.1   Total mortality data from 34 RCTs investigating the effect of cholesterol lowering interventions.

Trial	No. patients in treatment arm	No. patients in control arm	Total deaths in treatment arm	Total deaths in control arm	Coronary heart disease deaths in treatment arm[a]	Coronary heart disease deaths in control arm[a]
1	204	202	28	51	25	45
2	285	147	70	38	62	35
3	156	119	37	40	34	39
4	88	30	2	3	2	2
5	30	33	0	3	0	2
6	279	276	61	82	47	73
7	206	206	41	55	37	50
8	123	129	20	24	17	20
9	1018	1015	111	113	97	97
10	427	143	81	27	71	23
11	244	253	31	51	25	44
12	50	50	17	12	13	10
13	47	48	23	20	13	5
14	30	60	0	4	0	4
15	5552	2789	1025	723	826	632
16	424	422	174	178	41	50
17	199	194	28	31	25	25
18	350	367	42	48	34	35
19	79	78	4	5	2	4
20	1149	1129	37	48	19	31
21	221	237	39	28	35	26
22	54	26	8	1	8	1
23	71	72	5	7	5	6
24	4541	4516	269	248	61	54
25	421	417	49	62	32	44
26	94	94	0	1	0	1
27	311	317	19	12	17	8
28	1906	1900	68	71	32	44
29	2051	2030	44	43	14	19
30	6582	1663	33	3	28	3
31	5331	5296	236	181	91	77
32	48	49	0	1	0	0
33	94	52	1	0	1	0
34	23	29	1	2	1	0

[a] Coronary heart disease mortality data is used in Chapter 11.

most plausible values for the true odds ratio, taking into account the uncertaintyassociated with the trial. If we assume the $\ln(OR)$ outcome is normally distributed, then a 95% confidence interval (the interval expected to include 95 out of 100 estimated odds ratios if the trial were replicated 100 times) is given by:

$$\ln(OR) \pm 1.96 \times \sqrt{\text{var}(\ln(OR))}. \tag{2.8}$$

Since $\sqrt{\text{var}(\ln(OR))}$ gives the standard error (s.e.) of $\ln(OR)$. For the trial considered above, the 95% confidence interval of the $\ln(OR)$ is given by:

$$-0.753 \pm 1.96 \times \sqrt{0.068} = (-1.263 \text{ to} - 0.243).$$

This confidence interval can be converted back to the odds ratio scale by taking the exponential of the interval on the log odds ratio scale:

$$95\%\text{CI for OR} = (e^{-1.263} \text{ to } e^{-0.243})$$
$$= (0.28 \text{ to } 0.78).$$

This indicates the range of likely values for the odds of dying in the treatment group compared to the control group. As the confidence interval only contains values less than one, it could be concluded that the treatment effect in this trial statistically significant at the 5% level. Confidence intervals will be discussed further in Chapter 4, where they are constructed for pooled outcomes derived from meta-analyses. Table 2.2 provides the odds ratio, variance and 95% confidence interval for each trial in the cholesterol lowering trials dataset derived from the raw data in Table 2.1, although in the calculations, allowance has been made for zero cells by adding 0.5 to each entry before calculation. Thus for the first study, the calculation of the odds ratio becomes:

$$OR = (28.5 \times 151.5)/(176.5 \times 51.5)$$
$$= 0.48,$$

which is slightly larger than the uncorrected value (0.47) calculated above. Calculation of the variance of the log odds ratio, and hence confidence intervals, also uses these adjusted cell figures.

2.3.2 Relative risk (or rate ratio/relative rate)

The Relative Risk (RR) is simply defined as the probability of an event in the treatment group $(a/(a+b))$ divided by the probability of an event in the control group $(c/(c+d))$. Thus:

$$RR = (a/(a+b))/(c/(c+d)). \tag{2.8}$$

This can be seen as a ratio of the risks in the two groups being compared. As for odds ratios, it is usual to use the (natural) log scale when combining studies; the variance of the log relative risk is given by [1]

Table 2.2 Odds ratio calculations for total mortality in the cholesterol lowering RCTs.[a]

Study	var(ln(OR))	ln(OR)(95%CI)	OR(95% CI)
1	0.07	−0.74 (−1.25 to −0.24)	0.48 (0.29 to 0.79)
2	0.05	−0.07 (−0.53 to 0.38)	0.93 (0.59 to 1.47)
3	0.07	−0.48 (−1.01 to 0.04)	0.62 (0.36 to 1.04)
4	0.73	−1.48 (−3.16 to 0.20)	0.23 (0.04 to 1.22)
5	2.35	−1.95 (−4.95 to 1.06)	0.14 (0.01 to 2.89)
6	0.04	−0.41 (−0.79 to −0.03)	0.66 (0.45 to 0.97)
7	0.05	−0.38 (−0.84 to 0.08)	0.68 (0.43 to 1.08)
8	0.11	−0.16 (−0.81 to 0.49)	0.85 (0.45 to 1.63)
9	0.02	−0.02 (−0.30 to 0.25)	0.98 (0.74 to 1.29)
10	0.06	0.00 (−0.48 to 0.48)	1.00 (0.62 to 1.61)
11	0.06	−0.54 (−1.03 to −0.06)	0.58 (0.36 to 0.94)
12	0.19	0.48 (−0.39 to 1.34)	1.61 (0.68 to 3.81)
13	0.17	0.29 (−0.51 to 1.09)	1.33 (0.60 to 2.97)
14	2.27	−1.58 (−4.54 to 1.37)	0.21 (0.01 to 3.95)
15	0.00	−0.44 (−0.54 to −0.33)	0.65 (0.58 to 0.72)
16	0.02	−0.05 (−0.32 to 0.23)	0.95 (0.73 to 1.25)
17	0.08	−0.15 (−0.70 to 0.40)	0.86 (0.50 to 1.50)
18	0.05	−0.10 (−0.54 to 0.34)	0.91 (0.58 to 1.41)
19	0.43	−0.23 (−1.51 to 1.06)	0.80 (0.22 to 2.88)
20	0.05	−0.29 (−0.72 to 0.15)	0.75 (0.49 to 1.16)
21	0.07	0.46 (−0.06 to 0.99)	1.59 (0.94 to 2.68)
22	0.85	1.13 (−0.67 to 2.94)	3.11 (0.51 to 18.83)
23	0.35	−0.33 (−1.48 to 0.83)	0.72 (0.23 to 2.29)
24	0.01	0.08 (−0.10 to 0.26)	1.08 (0.91 to 1.29)
25	0.04	−0.28 (−0.68 to 0.12)	0.76 (0.51 to 1.13)
26	2.69	−1.11 (−4.32 to 2.10)	0.33 (0.01 to 8.20)
27	0.14	0.49 (−0.24 to 1.22)	1.63 (0.79 to 3.37)
28	0.03	−0.05 (−0.39 to 0.29)	0.95 (0.68 to 1.34)
29	0.05	0.01 (−0.41 to 0.44)	1.01 (0.66 to 1.55)
30	0.32	0.89 (−0.22 to 1.99)	2.43 (0.81 to 7.31)
31	0.01	0.27 to 0.07 to 0.47)	1.31 (1.07 to 1.59)
32	2.71	−1.10 (−4.32 to 2.13)	0.33 (0.01 to 8.39)
33	2.70	0.52 (−2.70 to 3.74)	1.68 (0.07 to 42.10)
34	1.15	−0.31 (−2.41 to 1.79)	0.73 (0.09 to 5.99)

[a] Note: 0.5 has been added to every cell of every 2 × 2 table to allow for the arms of trials with 0 events.

$$v_{(Ln(RR))} = \frac{1}{a} - \frac{1}{a+b} + \frac{1}{c} - \frac{1}{c+d}. \tag{2.9}$$

Example calculation of a relative risk

Again focusing on study 1 from the cholesterol dataset, using Table 2.1 and equation (2.8), the relative risk is calculated by:

$$RR = (28/(28 + 176))/(51/(51 + 151))$$
$$= 0.54.$$

The log relative risk is hence $\ln(0.54) = -0.609$. The variance for $\ln(RR)$ is calculated using equation (2.9):

$$\text{var}\,(\ln(RR)) = \frac{1}{28} - \frac{1}{(28 + 176)} + \frac{1}{51} - \frac{1}{(51 + 151)}$$
$$= 0.045.$$

If one assumes the $\ln(RR)$ outcome to be normally distributed, then in parallel to equation (2.8) a 95% confidence interval for this estimate is given by:

$$\ln(RR) \pm 1.96 \times \sqrt{\text{var}(\ln(RR))}$$
$$-0.609 \pm 1.96 \times \sqrt{0.045}$$
$$= (-1.027 \text{ to } -0.192).$$

Converting back to the relative risk scale:

$$95\%\text{CI for } RR = (e^{-1.027} \text{ to } e^{-0.192})$$
$$= (0.36 \text{ to } 0.83).$$

The relative risk has a similar interpretation to the odds ratio, and for rare diseases they are approximately equal to each other. It is the ratio of the probabilities (rather than odds) of dying in the two groups.

2.3.3 Risk differences between proportions (or the absolute risk reduction)

The Risk Difference (RD) is defined by:

$$RD = (a/(a+b)) - (c/(c+d)). \tag{2.10}$$

It can be interpreted as the difference between the probabilities of an event occurring in the two groups. The variance of this measure is calculated by [1]:

$$v_{RD} = \frac{p_1(1-p_1)}{n_1} + \frac{p_2(1-p_2)}{n_2} \tag{2.11}$$

where p_1 and p_2 are the observed rates of occurrence of the given event in the treatment and control groups, respectively. Thus, in terms of Figure 2.1,

$p_1 = a/a + b$ and $p_2 = c/c + d$, $n_1 = a + b$ and $n_2 = c + d$ and the estimate of the $RR = p_1/p_2$. It is worth noting that the absolute risk difference or reduction is a measure of the impact of the treatment or exposure on the number of events, since it takes into account the prevalence of events, i.e. how common it is. This is in contrast to the odds or risk ratios, which are measures of association between treatment and outcome, but do not give an indication of the *impact* of the intervention. For example, a RR of 0.5 for a common event would result in a lot greater impact on health of the population than a RR of 0.5 on a rare disease or a disease with rare events.

Example calculation of the risk difference

Calculating for study 1 from the cholesterol dataset, using Table 2.1, and equation (2.10), the risk difference is calculated by:

$$RD = (28/(28 + 176)) - (51/(51 + 151))$$
$$= -0.115.$$

The variance for the risk difference is calculated using equation (2.11), first calculating p_1, p_2, n_1, n_2:

$$p_1 = 28/(28 + 176)$$
$$= 0.137,$$
$$p_2 = 51/(51 + 151)$$
$$= 0.252.$$
$$n_1 = (28 + 176)$$
$$= 204,$$
$$n_2 = (51 + 151)$$
$$= 202,$$
$$\text{var}(RD) = 0.137(1 - 0.137)/204 + 0.252(1 - 0.252)/202$$
$$= 0.0015.$$

A risk difference can be positive or negative; hence, converting to the log scale is not done. If one assumes the RD outcome to be normally distributed, then in parallel to equation (2.8), a 95% confidence interval can be calculated for this estimate:

$$RD \pm 1.96 \times \sqrt{\text{var } RD}$$
$$-0.115 \pm 1.96 \times \sqrt{0.0015}$$
$$= (-0.19 \text{ to } -0.04).$$

The corresponding odds ratio and relative risk confidence interval only included numbers less than one; here the risk difference 95% confidence interval only contains negative numbers, suggesting that across the most plausible range of values for the risk of death is always greater in the control group.

2.3.4 The number needed to treat

Another way of expressing the impact of the treatment is to use the Number Needed to Treat (NNT), [6] which is simply the reciprocal of the risk difference:

$$NNT = \frac{1}{Risk\ Difference}$$
$$= \frac{1}{((a/a+b)) - (c/c+d))}.$$

(2.12)

In a clinical trials setting, which is where it is usually used, it can be interpreted as the number of patients that need to be treated using the experimental/new treatment, rather than the placebo/old treatment in order to prevent 1 additional adverse outcome. It is uncommon for meta-analyses to combine studies using the NNT, but an NNT can be calculated for the pooled result after the data has been combined using some other scale [7]. For example, since the NNT is the reciprocal of the risk difference, a meta-analysis could be carried out using risk differences, and the pooled estimate along with its corresponding confidence interval could be transformed to the NNT scale by taking their reciprocal [6]. Similarly, it is possible to calculate NNTs from the odds ratio or relative risk scales [8].

However, it is important to note that NNT derived from meta-analyses directly can be misleading due to differences in study characteristics [9]. A preferable approach to deriving NNT is to apply the relative risk estimate from a meta-analysis to estimates of prognosis from cohort studies representative of the groups for whom treatment decisions are to be made (rather than those from the meta-analysis) [10].

Example of calculating NNTs from risk differences

In Section 2.3.3 the risk difference and confidence interval for the first cholesterol lowering trial were calculated to be $-0.115(-0.19$ to $-0.04)$. The NNT for this trial is hence

$$NNT = 1/-0.115$$
$$= (-)8.7.$$

This suggests that approximately nine patients would need treating on the new treatment to prevent one of them from dying who would of died had they not been given the new treatment (the negative sign indicating the treatment is beneficial). The 95% confidence for this NNT is produced by taking the reciprocal of the confidence interval values for the RD, and is (5.2 to 25.7). Hence, the results of this trial suggest that between 5 and 26 patients would need treating on the new treatment to prevent one of them from dying who would have died had they not been given the new treatment. However, as stated above, it is important to note that deriving NNTs in this way from a meta-analysis can be misleading, and is not recommended [9].

2.3.5   Comparisons of rates

Section 2.2.2 discussed incidence rates; the measure of interest may not be one, but a ratio of two of these rates. The ratio of two incidence rates (sometimes called the *rate ratio*) can be calculated as the log of the division of two rates [11]. An approximate estimate of the variance is given by

$$\text{Var}[\log(\overline{T}.)] \approx \frac{1}{A + 1/2} + \frac{1}{B + 1/2}, \tag{2.13}$$

where A and B are the number of cases in each group, and the addition of the two halves can be viewed as a continuity correction (a more accurate variance estimate is given elsewhere [11]).

2.3.6   Other scales of measurement used in summarizing binary data

The above measures are those used most commonly in health research, although the relative risk reduction (defined as 1 minus the relative risk) is also used [12]. The Phi coefficient can be used when a measure of the correlation (rather than difference) is required between two dichotomous variables; see Fleiss [1, 3] for the formulae.

2.3.7   Which scale to use?

When deciding on the choice of measure to summarize studies, there are two considerations: (1) whether it is statistically appropriate and convenient to work with; and (2) whether it conveys the necessary clinically useful information [13]. For example, the number needed to treat is a measure that is being used increasingly when reporting the results of clinical trials. The motivation for its use is that it is more useful than the odds ratio and the relative risk for clinical decision making [6]. However, its role as a measure of comparative treatment effects is limited [14]. Detailed discussions of the relative merits of using the other different scales are given elsewhore [3, 13]. Historically, the odds ratio is the most common measure to report for RCTs and case control studies, while relative risks are often used for reporting cohort studies. The reader should be aware that while analyses using different scales will often come to similar conclusions, in some circumstances, considerably different results will be obtained [15].

2.4   Continuous data

There are many different continuous scales used to measure outcome in the health care literature (e.g. lung function, pulse rate, weight and blood pressure). They are all measured on a positive scale, and so it is common practice to transform the data logarithmically if they are skewed (as they often are – with a long tail of extreme positive values). In a clinical trials setting, the parameter of interest is usually the difference in average effect between the treatment and control groups. If it can be assumed that all the studies estimated the same parameter, then studies can be combined directly using the original scale (or the log transformed scale if deemed more appropriate) used to measure the outcome variable. If different studies measure their outcomes on different scales, then synthesis is still possible, but the data first needs transforming on to a single standardized scale. It should be noted that the resulting estimate may be difficult to interpret clinically.

These measures could easily be applied to an observational study setting. Additionally, although only differences between continuous measures, and hence comparative outcomes are described below, non-comparative continuous outcomes could also be combined. For example, if non-comparative observational studies measured blood pressure, then the mean blood pressure from each study (and corresponding variance) could be used as the outcome measure.

2.4.1 Outcomes defined on their original metric (mean difference)

The measure of treatment effect (T) is given by

$$T = \mu_t - \mu_c, \tag{2.14}$$

where μ_t and μ_c are the mean responses in the treatment and control groups, respectively. The variance of this treatment difference is

$$\text{var}(T) = \sigma^2(1/n_t + 1/n_c), \tag{2.15}$$

where n_t is the within-study sample size in the treatment group, n_c is the within-study sample size for the control group, and σ^2 is the variance, assumed common to both groups. σ^2 is normally estimated from the data, and is denoted $s^{2*} \cdot s^*$ (i.e. $\sqrt{s^{2*}}$), can be defined in different ways, each of which will yield a different estimate. Common and intuitive choices for s^* are s^{t*} and s^{c*}, which are the standard deviations of the response in the treatment and control groups, respectively.

Alternatively, a pooled standard deviation combining both s^{t*} and s^{c*} could be used. Hedges and Olkin [16, p. 78] suggest using the pooled estimate for the standard deviation, if it is reasonable to assume equal population variances, which has both smaller bias and variance than using s^{c*}, suggested by Glass [17]. A thorough treatment of the alternative measures is given elsewhere [16, 18]. The formula for Hedges and Olkin's pooled sample standard deviation is

$$s = \sqrt{\frac{(n_t - 1)(s^t)^2 + (n_c - 1)(s^c)^2}{n_t + n_c - 2}}, \tag{2.16}$$

where n_t and n_c are the treatment and control group sample sizes, respectively.

Example calculation of a mean difference measured on the original metric

Data from a meta-analysis comparing sodium fluoride (NaF) with sodium mono-flouorophosphate (SMFP) dentifrices for the purpose of reducing dental decay, originally meta-analysed by Johnson [19] is used for illustration. The outcome used is difference from baseline in terms of a decayed, missing or filled teeth score (DMFS). Fortunately, all the RCTs used the same scale to report their results. Data for the nine RCTs included in the meta-analysis is given in Table 2.3.

The outcome of interest, the difference between the two groups, is calculated using equation (2.14). Calculating for study 1:

$$\text{mean difference} = 6.82 - 5.96$$
$$= 0.86.$$

The variance of this difference can be calculated using equation (2.15); however, first it is necessary to calculate the estimate of the standard error of the two groups. The weighted average formulae (2.16) is used here for study 1:

$$s = \sqrt{\frac{(134 - 1)(4.24)^2 + (113 - 1)(4.72)^2}{(134 + 113 - 2)}}$$
$$= 4.47.$$

Hence, the variance of this difference, T, is:

$$\text{var(T)} = 4.47(1/134 + 1/113)$$
$$= 0.33.$$

Table 2.3   RCTs comparing NaF with SMFP in terms of differences from baseline in DMFS dental index.

		NaF			SMFP	
Study	N	Mean	SD	N	Mean	SD
1	134	5.96	4.24	113	6.82	4.72
2	175	4.74	4.64	151	5.07	5.38
3	137	2.04	2.59	140	2.51	3.22
4	184	2.70	2.32	179	3.20	2.46
5	174	6.09	4.86	169	5.81	5.14
6	754	4.72	5.33	736	4.76	5.29
7	209	10.10	8.10	209	10.90	7.90
8	1151	2.82	3.05	1122	3.01	3.32
9	679	3.88	4.85	673	4.37	5.37

Table 2.4 Outcome, and related values for nine studies comparing NaF with SMFP in terms of differences from baseline in DMFS dental index.

Study	Difference (SMFP -NaF)	s	var (Difference)	95% CI for difference
1	+0.86	4.47	0.33	(−0.26 to + 1.98)
2	+0.33	5.00	0.31	(−0.76 to + 1.42)
3	+0.47	2.93	0.12	(−0.22 to + 1.16)
4	+0.50	2.39	0.06	(+0.01 to + 0.99)
5	−0.28	5.00	0.29	(−1.34 to + 0.78)
6	+0.04	5.31	0.08	(−0.50 to + 0.58)
7	+0.80	8.00	0.61	(−0.73 to + 2.33)
8	+0.19	3.19	0.02	(−0.07 to + 0.45)
9	+0.49	5.12	0.08	(−0.06 to + 1.04)

An approximate 95% confidence interval can be calculated as before using these values. Table 2.4 provides the difference, s, the variance of the difference, and a 95% confidence interval for the difference for all nine studies.

2.4.2 Outcomes defined using standardized mean differences

The effect size of an experiment, d, is defined as

$$d = (\mu_t - \mu_c)/s^*, \tag{2.17}$$

where μ_t and μ_c are the sample means of the treated and control arms, respectively, and s^* is the estimate of the standard deviation of the study (discussed in the previous section). The estimate d has small sample bias (systematically gives the wrong answer), and a correction equation has been derived (see Hedges and Olkin [16, p81]). The variance of d is difficult to compute exactly. However, if the underlying data can be assumed to be normal, the conditional variance of d can be estimated as

$$\text{var}(d) = \frac{n_t + n_c}{n_t n_c} + \frac{d^2}{2(n_t + n_c)}, \tag{2.18}$$

where n_t and n_c are the numbers in the treatment and control groups, respectively, and d is the observed standardized mean difference. More exact methods are possible using computer intensive methods [16]. Fleiss [1] states that if the sample sizes in the two treatment groups (n_t and n_c) are both large, and the population variances are equal, then the simpler variance approximation

$$\text{var}(d) = \frac{(n_t + n_c)}{n_t n_c} \tag{2.19}$$

can be used. If the data appear to be non-normally distributed, a non-parametric estimate has been developed [16, 20, 21] which may be used instead.

The use of the standardized effect measure has been criticized, it being suggested that studies with identical results may spuriously appear to yield different results. Such a transformation can even make a study whose original estimate was smaller in magnitude than another studies appear greater, and vice versa [22]. Other continuous outcome measures, including those of correlation, do exist, though they are rarely used in medicine. A large selection of these are discussed by Rosenthal [18].

Example calculation of a standardized mean difference

A meta-analysis comparing clozapine to 'typical' drugs for schizophrenia [23] is used to illustrate the calculation of standardized mean differences and related quantities. The outcome of interest is mean change in mental state. However, different rating scales had been used to calculate the scores in the different studies, so it is necessary to calculate a standardized effect size. The data for the 11 studies included are given in Table 2.5.

Before calculating the standardized difference, the standard error of the mean for the two groups (s) needs estimating. This is done here using (2.16), as for the example in section 2.4.1 (calculations omitted here). The value of s so obtained for the first study is 10.59, hence using equation (2.17), the outcome for the first study is

$$d = (28.00 - 25.00)/10.59$$
$$= 0.28.$$

Table 2.5   RCTs comparing clozapine to 'typical' drugs for schizophrenia: change in mental state measured on different rating scales.

Study	Clozapine			'typical'		
	N	Mean	SD	N	Mean	SD
1	17	28.00	12.30	20	25.00	8.90
2	39	47.87	13.87	40	60.95	18.15
3	15	36.00	6.00	15	44.00	11.00
4	37	38.40	17.80	45	38.40	16.30
5	47	37.00	10.50	41	40.00	13.50
6	19	45.00	12.00	19	52.00	10.00
7	14	40.07	13.72	12	54.67	17.45
8	62	35.52	14.64	63	40.30	14.24
9	126	45.10	13.00	139	55.67	12.00
10	10	52.50	12.60	11	64.70	18.10
11	38	35.60	10.60	37	37.00	9.40

Table 2.6 Outcome and related values for 11 studies comparing clozapine to 'typical' drugs in terms of differences from baseline on the standardized mean different scale.

Study	d	s	var(d)	95% CI for d
1	0.28	10.59	0.11	(−0.37 to 0.93)
2	−0.81	16.18	0.05	(−1.27 to −0.35)
3	−0.90	8.86	0.15	(−1.65 to −0.15)
4	0.00	16.99	0.05	(−0.43 to 0.43)
5	−0.25	11.99	0.05	(−0.67 to 0.17)
6	−0.63	11.05	0.11	(−1.29 to 0.02)
7	−0.94	15.54	0.17	(−1.75 to −0.13)
8	−0.33	14.44	0.03	(−0.68 to 0.02)
9	−0.85	12.49	0.02	(−1.10 to −0.59)
10	−0.78	15.74	0.21	(−1.66 to 0.11)
11	−0.14	10.03	0.05	(−0.59 to 0.31)

The variance of d is now calculated for study 1 using equation (2.18):

$$\text{var}(d) = \frac{17 + 20}{(17 \times 20)} + \frac{0.28^2}{2(17 + 20)}$$
$$= 0.11.$$

Using the simpler equation (2.19) gives almost identical answers for all 11 studies. Confidence intervals can be calculated for the standardized mean difference, in a parallel manner to those previously. Table 2.6 provides d, s, var(d) and a 95% confidence interval for d for all 11 studies.

2.4   Ordinal outcomes

If the outcome of interest is measured on a categorical scale and the categories are ordered in terms of desirability, then one can consider the data as ordinal (e.g. gastrointestinal damage is often measured by the number of lesions). [24] Special methods are needed to combine ordinal data: these are described in Section 14.2.

2.5   Summary/Discussion

This chapter has considered the different types of outcome data (binary, continuous and ordinal) and the different scales outcomes are measured on. Both comparative and descriptive outcomes have been considered. The vast majority of RCT and observational studies report their results on one or more of these scales. This chapter forms the foundation for future chapters which describe the different methods of combining the outcomes discussed here.

References

1. Fleiss, J.L. (1993). The statistical basis of meta-analysis. *Stat. Methods Med. Res.* **2**: 121–45.

2. Meinert, C.L. (1996). *Clinical Trials Dictionary: Terminology and usage recommendations.* Baltimore, MD: The Johns Hopkins Center for Clinical Trials.

3. Fleiss, J.L. (1994). Measures of effect size for categorical data. In: *The Handbook of Research Synthesis.* Cooper, H., Hedges, L.V., editors. New York: Russell Sage Foundation; 245–60.

4. Gart, J.J., Zweifel, J.R. (1967). On the bias of various estimators of the logit and its variance, with application to quantal bioassay. *Biometrica* **54**: 471–5.

5. Smith, G.D., Song, F., Sheldon, T.A., Song, F.J. (1993). Cholesterol lowering and mortality: The importance of considering initial level of risk. *Br. Med. J.* **306**: 1367–73.

6. Cook, R.J., Sackett, D.L. (1995). The number needed to treat: a clinically useful measure of treatment effect. *Br. Med. J.* **310**: 452–4.

7. Chatellier, G., Zapletal, E., Lemaitre, D., Menard, J., Degoulet, P. (1996). The number needed to treat: a clinically useful nomogram in its proper context. *Br. Med. J.* **312**: 426–9.

8. McQuay, H.J., Moore, R.A. (1997). Using numerical results from systematic reviews in clinical practice. *Annals Internal Medicine* **126**: 712–20.

9. Smeeth, L., Haines, A., Ebrahim, S. (1999). Numbers needed to treat derived from meta-analyses – sometimes informative, usuall misleading. *Br. Med. J.* **318**: 1548–51.

10. Jackson, R.T., Sackett, D.L. (1996). Guidelines for managing raised blood pressure. *Br. Med. J.* **313**: 64–5.

11. Hasselblad, V.I.C., McCrory, D.C. (1995). Meta-analytic tools for medical decision making: A practical guide. *Med. Decis Making* **15**: 81–96.

12. Oxman, A.D., (editor). (1996). The *Cochrane Collaboration Handbook: Preparing and maintaining systematic reviews.* 2nd ed. Oxford: Cochrane Collaboration.

13. Sinclair, J.C., Bracken, M.B. (1994). Clinically useful measures of effect in binary analyses of randomized trials. *J. Clin. Epidemiol* **47**: 881–9.

14. Dowie, J. (1997). The number needed to treat and the 'adjusted NNT' in health care decision making. Presented at Workshop *From Research Evidence to Recommendations Exploring the Methods*, Harrogate 17–18 February.

15. Deeks, J.J., Altman, D.G., Dooley, G., Sackett, D.L.S. (1997). Choosing an appropriate dichotomous effect measure for meta-analysis: empirical evidence of the appropriateness of the odds ratio and relative risk. *Controlled Clin. Trials* **18**: 84s–5s.

16. Hedges, L.V., Olkin, I. (1985). *Statistical Methods for Meta-Analysis*. London: Academic Press.

17. Glass, G.V. (1976). Primary, secondary and meta-analysis of research. *Educ. Res.* **5**: 3–8.

18. Rosenthal, R. (1994). In: Parametric measures of effect size. Cooper, H., Hedges, L.V., editors. *The Handbook of Research Synthesis*. New York: Russell Sage Foundation; p. 231–44.

19. Johnson, M.F. (1993). Comparative efficacy of NaF and SMFP dentifrices in caries prevention; a meta-analytic overview. *Caries Res.* **27**: 328–36.

20. Kraemer, H.C., Andrews, G. (1982). A non-parametric technique for meta-analysis effect size calculation. *Psychol. Bull* **91**: 404–12.

21. Hedges, L.V., Olkin, I. (1984). Nonparametric estimators of effect size in meta-analysis. *Psychol. Bull* **96**: 573–80.

22. Greenland, S. (1987). Quantitative methods in the review of epidemiological literature. *Epidemiol Rev.* **9**: 1–30.

23. Wahlbeck, K., Cheine, M., Essali, M. (1999). Clozapine versus typical neuroleptic medication for schizophrenia (Cochrane Review). The Cochrane Library Issue 1: Oxford: Update Software.

24. Whitehead, A., Jones, N.M.B. (1994). A meta-analysis of clinical trials involving different classifications of response into ordered categories. *Stat. Med.* **13**: 2503–15.

CHAPTER 3

Assessing Between Study Heterogeneity

3.1   Introduction

In any meta-analysis the point estimates of the effect size from the different studies being considered will almost always differ, to some degree. This is to be expected, and is at least partly due to sampling error which is present in every estimate. When effect sizes differ, but *only due to sampling error* (i.e. the true effect is the same in each study), the effect estimates are considered to be homogeneous; in other words, differences between estimates are random variation, and not due to systematic differences between studies. This source of variation can be accommodated in meta-analysis by using a fixed effects model which is discussed in Chapter 4. However, often the variability in effect size estimates exceeds that expected from sampling error alone, i.e. there is not just the same true underlying effect for each study, but 'real' differences exist between studies. Variation between individual study estimates can be examined graphically using plots described in Section 3.3, such as a forest plot (Figure 3.2). If a formal synthesis is to be undertaken, this extra variability requires further consideration. When it is present, the effect size estimates are considered to be heterogeneous. Possible reasons for this heterogeneity are discussed in Section 3.4. The issue of heterogeneity of study results is fundamental in meta-analysis, and how to handle it is the source of much debate. Heterogeneity may exist when all or most studies indicate the same direction of treatment effect (i.e. either harmful or beneficial), but the size of this effect differs, as well as when the results of trials are in different directions.

This section considers heterogeneity exclusively at the study-level. This is appropriate if only summary results from each trial are to be pooled. If, however, individual patient data are available for synthesis (see Chapter 12), within study variation (as well as between study variation) may be investigated.

In the remainder of this chapter, statistical and graphical methods for detection of the presence of heterogeneity are outlined, followed by a discussion of their shortcomings. This is followed by a discussion of the various causes of heterogeneity, and then a section discussing how heterogeneity can be explored and managed. Consideration of how heterogeneity affects the results and interpretation of a meta-analysis concludes this chapter.

This chapter should be read in conjunction with at least Chapters 4 and 5, and preferably Chapter 6, before any assessment of heterogeneity is made when conducting a meta-analysis.

3.2   Hypothesis tests for presence of heterogeneity

Probably the most commonly used method to assess the presence of heterogeneity is to carry out a simple statistical test. The standard test, together with several alternatives and extensions, are described below.

3.2.1   Standard χ^2 test

Several slightly different formulae for a general test are, for the most part, essentially equivalent, being based on χ^2 or F statistics [1]. The one devised by Cochran [2], which is widely used, is given below. It tests the hypothesis that the true treatment effects are the same in all the primary studies ($H_0: \theta_1 = \theta_2 = \ldots = \theta_k$, where the θ_i's are the underlying true treatment effects of the corresponding, $i = 1$ to k, studies in the meta-analysis), versus the alternative that at least one of the effect sizes (θ_i) differs from the remainder. Essentially, this is testing whether it is reasonable to assume that all the studies to be combined are estimating a single underlying population parameter, and whether variation in study estimates is likely to be wholly random. The test statistic is

$$Q = \sum_{i=1}^{k} w_i \left(T_i - \overline{T}. \right)^2, \tag{3.1}$$

where k is the number of studies being combined, T_i is the treatment effect estimate in the ith study,

$$\overline{T}. = \frac{\sum_i w_i T_i}{\sum_i w_i}$$

is the weighted estimator of treatment effect, and w_i is the weight attached to that study (usually, the reciprocal of the variance (v_i) of the outcome estimate from the

ith study – this is discussed in Chapter 4) in the meta-analysis. The formulae required to calculate the T_i and v_i for common outcome measures was given in Chapter 2. For example, if the (log) odds ratio scale were being used to combine the k studies in a meta-analysis, then each T_i could be calculated taking the logarithm of equation (2.6), and equation (2.7) could be used to calculate each corresponding v_i (these figures were calculated in Table 2.2).

A computationally convenient form of (3.1) is

$$Q = \sum_{i=1}^{k} w_i T_i^2 - \frac{\left(\sum_{i=1}^{k} w_i T_i\right)^2}{\sum_{i=1}^{k} w_i}. \tag{3.2}$$

Q is approximately distributed as a χ^2 distribution on $k-1$ degrees of freedom under H_0, hence statistical tables can be used to obtain the corresponding p-value [3]. Unfortunately, the interpretation of this test is often difficult for the following reasons:

1. The statistical power of tests for heterogeneity are, in most cases, very low due to the small number of combined trials [4], implying that heterogeneity may be present even if the Q statistic is not statistically significant at 'conventional' levels of significance. As a response to this, Fleiss [5] recommended using a cut-off significance level of 0.10, rather than the usual 0.05. This has become a customary practice in meta-analysis.
2. When the sample sizes in each study are very large, H_0 may be rejected even when the individual effect size estimates do not really differ much [6].
3. The likelihood of design flaws in primary studies (see Chapter 9) and of publication biases (Chapter 7) makes the interpretation of heterogeneity tests complex. If all the studies being considered share the same flaw, or if the studies with zero or negative effects are less likely to be published, then a consistent bias results across studies can make the effect sizes appear more consistent than they really are. Conversely, if all the studies have different design flaws, estimates of effects could be heterogeneous even though they actually reflect the same underlying population effect [7].

Caution is essential when interpreting the Q statistic. Some authors [6] have gone as far as to suggest it should not be used as a test at all, or only as a diagnostic tool to help researchers know whether they have accounted for all the variation between studies through modelling. This has led to the suggestion that estimating the magnitude of the between-study variation is a more sensible approach [8]. This is discussed further in Section 3.5.5.

*3.2.2 Extensions/alternative tests

1. Although the test based on the Q statistic is used most frequently, others do exist. An alternative approach to estimating between study variation is

available using one-way analysis of variance (ANOVA) to investigate hetero-geneity between and within groups of studies, where the groups are defined by study characteristics [9].

2. The contribution of each trial to the overall Q test statistic (q_i) can be investigated, and an approximate comparison of each q_i^2 to a χ_1^2 distribution can be made, provided the number of trials k is not too small [9, p. 256]. This is useful for the identification of potential outlying studies.

3. Gail and Simon [10] propose a test for qualitative interaction (testing for differences in *direction* of effect) based on a likelihood ratio statistic (outlined by Schmid *et al.* [11, p 109]). The implications of a qualitative interaction (in a trial setting) are that it suggests that the effect of a treatment may differ in direction in certain subsets of patients.

4. Recently, Biggerstaff and Tweedie [12] derived a likelihood ratio test for the general comparison of meta-analytic models. Hardy and Thompson [13] show how the maximum likelihood estimates required for such a test can be calculated, either via a relatively straightforward iterative procedure or by direct maximization of the likelihood.

5. Zelen [14] has devised an exact test for heterogeneity. This is a relatively new development, and one which Emerson [15] recommends. However, empirical evidence of its benefit over the Q statistic for meta-analysis is currently lacking.

6. When linear or logistic regression are used to combine studies (see Chapter 6), lack of homogeneity of treatment effect can be tested by computing tests of goodness of fit of the model with only main effects for study and treatment, or by testing the interaction of study and treatment [16].

7. A test to correct for variable reliability in the measuring of outcomes in different studies has been described [9, p 136].

8. Other tests exist which to the authors' knowledge have not been applied in meta-analysis, but which could be usefully applied. These include further use of likelihood method [9, 17] and modifications for sparse data [5, 18].

3.2.3   Example: Testing for heterogeneity in the cholesterol lowering trial dataset

The basic statistical (χ^2) test to detect heterogeneity is applied to the total mortality cholesterol log odds ratios for the 34 RCTs, as presented in Table 2.2. The test statistic is calculated using equation (3.2), where the T_i are the given in the $\ln(OR)$ column and w_i are calculated as $1/\mathrm{var}(\ln(OR))$. As for the majority of the calculations presented in this book, the use of a computer is recommended, but Q could be calculated by hand using the expression outlined below:

$$Q = \left(14.98 \times (-0.74)^2\right) + \ldots + \left(0.87 \times (-0.31)^2\right)$$
$$- \frac{\left((14.98 \times -0.74) + \ldots + (0.87 \times -0.31)\right)}{(14.98 + \ldots + 0.87)}$$

$$= 88.23.$$

This test statistic of 88.23 is compared to a chi-squared distribution on 33 $(k-1)$ degrees of freedom, which gives a p-value which is highly significant ($p < 0.0001$). Hence, we conclude that between study heterogeneity is almost certainly present, implying that the variability between study estimates is too great to assume that they are estimating the same underlying treatment effect.

3.3   Graphical informal tests/explorations of heterogeneity

Since the formal Q statistic has low power, the following exploratory methods should be considered to aid decisions on how to proceed with the synthesis, even when this statistic is non-significant [19]. Visual inspections of graphical representation of the data can be very useful, and complimentary to the statistical methods. In particular, they can give an indication of which studies are the source of the greatest heterogeneity, and also identify possible outliers.

3.3.1   Plot of normalized (z) scores

The z-scores or standardized residuals for each study can be calculated by

$$z_i = \left(T_i - \overline{T}.\right)/se(T_i), \tag{3.3}$$

where the T_i are the outcomes from the k studies, $\overline{T}.$ is the weighted average of those outcomes (weights given by $1/\mathrm{var}(T_i)$), and $se(T_i)$ is the standard error of each T_i (calculated by taking the square root of $\mathrm{var}(T_i)$). (Note: the calculation of $\overline{T}.$ is described in detail in Chapter 4.) Under the null hypothesis of only random differences among the studies, the histogram of these z-scores should have an approximately normal distribution, with mean zero and a variance of one. Large absolute z-scores can signal important departures of individual studies from the average result [19].

Calculation and histogram of normalized (z) scores for the cholesterol lowering RCTs

$\overline{T}.$, a weighted average of all 34 estimates of treatment effect, is calculated (using the method presented in Section 4.2). Hence, using this value (-0.17), and data contained within Table 2.2, the z-score for study 1 is

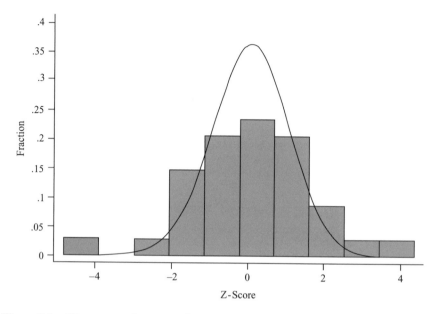

Figure 3.1   Histogram of z-scores for the 34 cholesterol trials.

$$z_i = (-0.744 - (-0.169))/\sqrt{0.067}$$
$$= -2.23.$$

The z-scores or standardized residuals for all 34 RCTs are given in Table 3.1 These are then used to create the histogram in Figure 3.1. In this figure, a normal distribution with mean equal to zero and variance equal to one is also plotted. This is the distribution from which the residuals should be drawn under the null hypothesis that all studies are estimating the same underlying effect. The spread of the data appears greater than that of the standardized normal distribution in Figure 3.1. This indicates that more variation than can be expected by sampling variation is present in the cholesterol lowering trials. Hence, this histogram supports the result of the statistical test of Section 3.2.3.

3.3.2   Forest plot

These plots are more commonly used as a way of presenting the results of a meta-analysis (see Chapter 10); however, they are also useful at the exploratory stage of a meta-analysis. Each study's effect estimate and respective confidence interval are plotted on one set of axes. For the purposes of reporting results, the pooled estimate, together with its confidence interval, are also plotted, but this is not

Table 3.1 Calculations used in the cholesterol lowering trial examples throughout this chapter.

Study	ln(OR)	var(ln (OR))	$w = 1/\text{var}$ (ln(OR))	Standardized residuals	z-scores	Mortality rate treatment	Mortality rate control
1	−0.74	0.07	14.98	−2.23	−2.23	0.14	0.25
2	−0.07	0.05	18.54	0.42	0.42	0.25	0.26
3	−0.48	0.07	13.83	−1.17	−1.17	0.24	0.34
4	−1.48	0.73	1.36	−1.53	−1.53	0.02	0.10
5	−1.95	2.35	0.43	−1.16	−1.16	0.00	0.09
6	−0.41	0.04	26.25	−1.24	−1.24	0.22	0.30
7	−0.38	0.05	18.26	−0.90	−0.90	0.20	0.27
8	−0.16	0.11	9.20	0.03	0.03	0.16	0.19
9	−0.02	0.02	50.03	1.03	1.03	0.11	0.11
10	0.00	0.06	16.64	0.67	0.67	0.19	0.19
11	−0.54	0.06	16.45	−1.52	−1.52	0.13	0.20
12	0.48	0.19	5.18	1.47	1.47	0.34	0.24
13	0.29	0.17	5.98	1.12	1.12	0.49	0.42
14	−1.58	2.27	0.44	−0.94	−0.94	0.00	0.07
15	−0.44	0.00	326.57	−4.82	−4.82	0.18	0.26
16	−0.05	0.02	51.51	0.88	0.88	0.41	0.42
17	−0.15	0.08	12.69	0.08	0.08	0.14	0.16
18	−0.10	0.05	19.79	0.32	0.32	0.12	0.13
19	−0.23	0.43	2.32	−0.09	−0.09	0.05	0.06
20	−0.29	0.05	20.36	−0.53	−0.53	0.03	0.04
21	0.46	0.07	14.15	2.38	2.38	0.18	0.12
22	1.13	0.85	1.18	1.42	1.42	0.15	0.04
23	−0.33	0.35	2.89	−0.27	−0.27	0.07	0.10

Continues

Table 3.1 Cont.

Study	ln(*OR*)	var(ln (*OR*))	$w = 1/\mathrm{var}$ (ln(*OR*))	Standardized residuals	z-scores	Mortality rate treatment	Mortality rate control
24	0.08	0.01	121.91	2.75	2.75	0.06	0.05
25	-0.28	0.04	23.98	-0.54	-0.54	0.12	0.15
26	-1.11	2.69	0.37	-0.57	-0.57	0.00	0.01
27	0.49	0.14	7.25	1.77	1.77	0.06	0.04
28	-0.05	0.03	33.70	0.70	0.70	0.04	0.04
29	0.01	0.05	21.52	0.84	0.84	0.02	0.02
30	0.89	0.32	3.16	1.88	1.88	0.01	0.00
31	0.27	0.01	98.72	4.35	4.35	0.04	0.03
32	-1.10	2.71	0.37	-0.56	-0.56	0.00	0.02
33	0.52	2.70	0.37	0.42	0.42	0.01	0.00
34	-0.31	1.15	0.87	-0.13	-0.13	0.04	0.07

required at the exploratory stage. Often the size of the plotting symbol used to mark the point estimate from each study is made proportional to the reciprocal of the variance of the estimate. Hence, the more precise estimates, which are generally most influential in a meta-analysis, are given the largest plotting symbols. From this plot, an idea of the variability between the estimates from the individual studies can be gained.

Example: Forest plot for the cholesterol lowering trials

The log odds ratios and confidence intervals for the 34 cholesterol lowering RCTs, as reported in Table 2.2, are plotted in Figure 3.2 This plot highlights the difference in precision (which is related to the size of the trial) between estimates from different studies. It should be remembered that the estimates with the widest confidence intervals, which thus have most visual impact, are the least precise and generally least influential. The vertical line on the plot corresponds to an odds ratio of one, where treatment and control are equally effective. Any confidence interval which includes this value implies that no statistically significant effect was found in the study. In this example, the majority of confidence intervals do include 1, and hence were individually inconclusive. Statistically significant beneficial effects of treatment were observed in some trials (e.g. study 1 and the large study 15), but statistically significantly harmful effects were also reported (study 31). The variability between estimates on the plot again highlights the heterogeneity between trials.

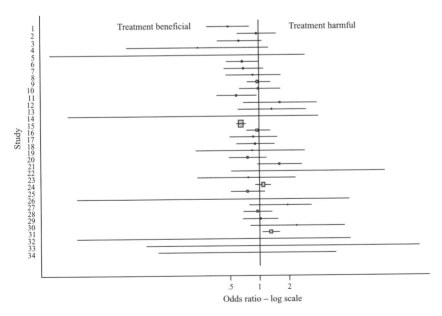

Figure 3.2   Forest plot for 34 cholesterol lowering trials.

3.3.3 Radial plot (Galbraith diagram)

Radial plots, first described by Galbraith [20], plot the outcome from each study, divided by the square root of its variance ($T_i/\sqrt{v_i}$) (known as the z statistic), against the reciprocal of the standard error ($1/\sqrt{v_i}$). The interpretation of this plot may not be instantly obvious. Consider the format of Figure 3.3, an example plot. The position of each study on the horizontal axis gives an indication of the weight allocated to it in the meta-analysis (the larger the standard error, the smaller the weight, and hence the closer it is to the y-axis). The gradient obtained from drawing a line from the origin to a particular study corresponds to the estimated outcome for that study. Additionally, if one were to fit an unweighted regression line constrained through the origin (the central line of Figure 3.3), then the gradient of this line would correspond directly with the pooled estimate from a fixed effect meta-analysis (see Chapter 4). Points which form a homogeneous set of trials will scatter homoscedastically (with constant variance, i.e. the 'cloud' of trials does not widen or narrow along the length of the line), with unit standard deviation, about this line. Points a long way from the line of best fit indicate outliers that will contribute considerably to the between study heterogeneity; in fact, the distance between each study point and the regression line is q_i^2, defined in Section 3.2.2, as the contribution of each study to the heterogeneity statistic. Plots may differentiate subsets of studies

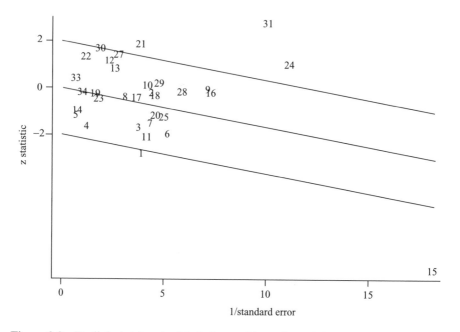

Figure 3.3 Radial plot for the 34 cholesterol lowering trials.

through the use of different symbols/colours, which may aid in identifying possible sources of heterogeneity.

Example radial plot for the cholesterol trials

The z-statistics for the first cholesterol trial are calculated as follows:

$$z\text{-statistic}_1 = -0.74/\sqrt{0.07}$$
$$= -2.88.$$

Table 3.1 provides z-statistics for all 34 trials. Figure 3.3 presents a radial plot generated using these figures.

This plot is more informative than the plot of normalized (z) scores of Section 3.3.1. In addition to the pooled estimate line, lines two standard errors (forming an approximate 95% confidence region) above and below it have been included. Studies from points outside these lines contribute to the heterogeneity considerably. Each study has been plotted using its id number, so individual outliers can be identified. Five studies have estimates outside these bounds (studies 1, 15, 21, 24 and 31).

3.3.4   L'Abbé plot

This plot was first described by L'Abbé *et al.* [21], and is usually used for meta-analysis of RCTs where the outcome is a binary variable. The event risk (number of events in an arm/total number of patients in arm) in the treatment groups are plotted against the risk for the controls for each trial. If the trials are fairly homogeneous the points would form a 'cloud' close to a line, the gradient of which would correspond to the pooled treatment effect. Large deviations, or scatter, would indicate possible heterogeneity [22]. In a similar manner to the radial plot, different subgroups of studies can be plotted using different symbols, and the size of the plotting symbol can be made proportional to the precision of the study estimate.

Example L'Abbé plot for the cholesterol lowering trials

Table 2.1 provides the total number of patients and total number of deaths in each arm of each of the 34 RCTs. From these figures, the event rates are calculated for the treatment and control groups, as given in columns 7 and 8, respectively, of Table 3.1. These are plotted in Figure 3.4 to form the L'Abbé plot. The line of equal event rates in each group (in this example mortality), and hence of no treatment effect has been included on the graph. Points below the line indicate trials for which the event rate in the treatment group is higher than in the control group, and hence the new treatment appeared less effective than the standard/placebo. Conversely, points above the line indicate trials for which the new treatment appeared effective, due to them having a lower risk in the treatment compared to the control group. There is a suggestion that the new treatment is more effective in the trials in which risks in both groups were high, while at low levels of risk (near the origin) the treatment may actually be harmful. This issue of an intervention's effectiveness being related to the

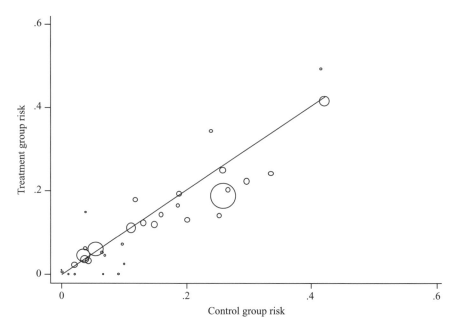

Figure 3.4   L'Abbé plot for the cholesterol lowering trials.

underlying risk of patients treated is explored further via regression methods in
Section 6.3.5.

3.4   Possible causes of heterogeneity

As well as investigating for the presence of heterogeneity, it is also important to
consider its underlying cause. It may then be possible to adjust the analysis accord-
ingly (options are discussed in Section 3.5). Bailey [23] suggests heterogeneity can be
categorized as: (i) due to chance; (ii) spurious, due to the scale used to measure the
treatment effect; (iii) due to treatment characteristics which can be investigated; (iv)
due to patient-level covariates which can only be investigated if the researcher has
individual patient data; (v) unexplainable, if none of the above account for it; or (vi)
characteristics of the design and conduct of the studies being considered. Ways of
exploring (ii) and (iii) are given in Section 3.5, while (iv) is considered in Chapter 12.

   Specific factors, which come under (vi), when combining RCTs, are described
below (heterogeneity between observational studies are considered in Chapter 16).
For further information on making an assessment of the general quality of studies
being meta-analysed (as opposed to specific factors) see Chapter 8. It should be

stressed that if studies with different designs are included, then because they are subject to different biases, this may be a cause of heterogeneity. The topic of combining studies with different designs is the focus of Chapter 17.

3.4.1   Specific factors that may cause heterogeneity in RCTs

Differences in design and conduct of the studies may explain some of the heterogeneity between study results. Figure 3.5 indicates possible ways in which apparently similar trials in a meta-analysis may differ:

All these can be considered when carrying out a meta-analysis. It is worth noting that points 2 and 4 of Figure 3.5 may reflect differences in when and where the trials were conducted. In the sections below, we discuss how certain factors from the above list may affect heterogeneity; in some cases, specific methodology exists for dealing with them.

- *Impact of underlying risk on heterogeneity*: Thompson *et al.* [25], Brand [26] and Davey Smith and Egger [27] have all emphasized the importance of ascertaining whether the treatment benefit varies according to the underlying risk of the patients in different RCTs. In addition to the L'Abbé plot, several formal methods have been proposed to investigate this; these are described in Section 6.3.5.
- *Impact of size of intervention dose on heterogeneity*: studies may have used different dose levels of the intervention under investigation. While this can be explored in the same manner is other trial characteristics, dose-response modelling is possible in meta-analysis, and is covered in section 16.7.2 (as it is usually performed on observational studies).
- *Compliance rate*: Gelber and Goldhirsch [28] highlight the problem of compliance in the primary studies. They give a mathematical justification

1. Differences in inclusion and exclusion criteria.
2. Other pertinent differences in baseline states of available patients despite identical selection criteria.
3. Variability in control or treatment interventions (e.g. doses, timing and brand).
4. Broader variability in management (e.g. pharmacological co-interventions, responses to intermediate outcomes including cross-overs, different settings for patient care).
5. Differences in outcome measures, such as variability in follow-up times, or (subtle) differences in outcome definition.
6. Variation in analysis, especially in handling withdrawals, drop-outs, and cross-overs.
7. Variation in quality of design and execution, with bias or imprecision in individual estimates of treatment effect.

Figure 3.5   Ways in which apparently similar trials may differ (modified from Naylor [24]).

of how reduced compliance could change the effect estimate, and hence increase heterogeneity.

- *Length of follow-up*: the length of follow-up of a trial may have an influence on the estimate of treatment effect [28]. The following issues need to be considered when investigating the impact of this factor: (a) treatment effects may vary over time; (b) trials with the longest follow-up may be selective because they may have been designed and conducted earlier; and (c) a summary measurement based on an overall risk reduction that assumes constant annual risk ratios may differ from actuarial estimates based on yearly assessment. Thompson observes that when trial duration is modelled, a longer trial would include information on events both soon after and a long time after randomisation, so any true effect of duration would be diluted in such an analysis [29]. A possible solution to this problem would be to use a survival type analysis (see Chapter 18).

- *Impact of early stopping rules on heterogeneity*: RCTs are sometimes stopped early if it is clear from interim analyses that one of the treatment arms is clearly superior to the other(s). Hughes *et al.* [30] investigated the effect of stopping a trial early would have on heterogeneity, and concluded that if the true treatment effect being studied is small, then artificial heterogeneity is introduced, which increases the Type I error rate in the heterogeneity test. Conversely, if the true mean effect is large, then between-trial heterogeneity may be underestimated. Hence, reviewers should ascertain whether stopping rules have been used in the primary studies, and repeat heterogeneity tests excluding studies stopped early, in order to ascertain their influence [30]. Green *et al.* [31] also investigate the effect of early stopping rules on overviews of clinical trials. As well as investigating three different stopping rule methods, they also considered the effect of the inappropriate use of early stopping rules, specifically when no significance level adjustments have been made, to take into account multiple testing. The authors report that the bias induced by this latter mechanism would inflate the effect of the new treatment in overview results in a similar way to that induced by publication bias. However generally, the bias induced should be of limited concern.

3.5   Methods for investigating and dealing with sources of heterogeneity

When conducting a meta-analysis, it is important to investigate the possible sources of heterogeneity. Too often, analysts are content to pool the results to obtain an overall average estimate of effect

In a meta-analysis, documenting heterogeneity of effect (by identifying sources of variability in results across studies) can be as important as reporting averages. Heterogeneity may point to situations in which an intervention works and those in which it does not. Finding systematic variation in results and identifying factors that may account for such variation, in this way, aids in the interpretation of existing data and the planning and execution of future studies [32].

Anello and Fleiss [33] seek to distinguish two types of meta-analysis: analytic and exploratory. In 'analytic' meta-analysis there is little or no heterogeneity, and the aim of the analysis it to improve an estimate of effect or test a hypothesis. When the goal is to resolve controversy, or pose and answer new questions, the main concern of the meta-analysis is to explain the variation in the effect sizes; the authors call this an 'exploratory' meta-analysis, where the characteristics of the different studies become the focus of the analysis. They further suggest that protocols for a meta-analysis should reflect its goals and how the results are to be used.

However, it may not always be easy or possible to explore potential sources of heterogeneity in a meta-analyses, and such an investigation can be unrewarding unless the number of trials is large or individual patient data are available [25]. Analysis can still proceed when heterogeneity has not been explained, but efforts to do so should be made first. In some cases, it is more appropriate to conclude that the results of the studies are too heterogeneous to combine and interpret meaningfully (see Section 3.6). Different methods proposed to deal with heterogeneity are discussed below.

3.5.1 Change scale of outcome variable

It may be sufficient to simply change the scale on which the study outcomes are measured, to remove heterogeneity [34]. For binary outcomes, changing the outcome from a absolute measure (such as the risk difference) to a relative measure (such as the odds ratio) can reduce the degree of heterogeneity. For continuous outcomes, a transformation such as taking logarithms is common practice, though there may be a trade-off between statistical homogeneity and clinical interpretability.

3.5.2 Include covariates in a regression model *(meta-regression)*

A regression analysis can be performed to examine whether the heterogeneity between studies can be explained by one or more factors across all studies. These factors may describe patient or study characteristics; several potential study characteristics were discussed in Section 3.4. Full details of meta-regression models are given in Chapter 6. As epidemiological studies tend to vary more in their design and conduct than RCTs, the scope of the adjustment methods is greater in meta-analysis of epidemiological studies [1].

If the overall effectiveness of a new treatment is related to the severity of the disease, then studies may appear heterogeneous because of differences in the baseline risk of the patients included. Unfortunately, including baseline risk as a covariate in a meta-regression model can produce misleading answers, and hence specialist methods have been developed. This topic is described fully in Section 6.3.5.

3.5.3   Exclude studies

As noted above, the contribution of each study to the heterogeneity statistic can be assessed by comparing its contribution to the Q statistic to the chi-square distribution on 1 df (an approximate test) [9]. The studies which contribute most variation can then be excluded. This corresponds to removing outliers or extreme results at an early stage of any analysis of data. However, this could introduce bias, perhaps substantial, into the estimates. As Fleiss states:

> It is invalid to delete from the set of studies to be meta-analyzed those whose results are in the 'wrong direction', for the opportunity for bias in identifying the 'deviant' studies is too great [32].

The effect of removing extreme studies should be explored in a sensitivity analysis (see Chapter 9).

3.5.4   Analyse groups of studies separately

If the studies are too heterogeneous to be combined sensibly, there may be groups of studies that seem similar, and thus a decision to combine just these may be justified. However, as with subgroup analyses of individual studies, it is important to define possible subgroups prior to carrying out the meta-analysis (e.g. based on theoretically sensible considerations such as drug type, dosage or clinical context), to avoid the temptation of 'data dredging'. Subgroup analyses defined *post hoc* amounts to no more than exploratory data analysis (see Chapter 6).

3.5.5   Use of random effects models

Rather than seeking to explain or adjust explicitly for variation between studies, one can pool studies using a random effects model, which allows for variation in the underlying effect size between studies to be taken into account. This approach is often used when the source of variation cannot be identified. Chapter 5 is devoted to this type of meta-analysis model. It should be stressed, however, that by using a random effects model, no investigation of the *causes* of heterogeneity is made.

3.5.6   Use of mixed-effect models

If one or more variables appear to account for a proportion of the variation, but evidence that some level of heterogeneity (above the level of random variation) remains, then a random effects term can be included in the model to account for this 'residual' heterogeneity. This model is called a *mixed effect-model*, as it can be viewed as a combination of a meta-regression and a random effects model. A description of the various mixed models proposed is given in Chapter 6.

3.6   The validity of pooling studies with heterogeneous outcomes

So far this section has outlined ways to detect, and up to a point, deal with heterogeneity in study estimates. It would be wrong, however, to give the impression that heterogeneity between studies can always be dealt with satisfactorily and without controversy. This issue of whether the results of separate trials/studies are homogeneous enough to be meaningfully combined is problematic [35]. It has been argued that producing an overall combined estimate for heterogeneous studies is wrong, and leads to a result which is misleading, and impossible to interpret. A much used quote is that it is equivalent to 'combining apples and oranges and the occasional lemon' [36]. No clear guidelines exist (tests of heterogeneity apart) for how variable study results have to be before it is deemed invalid to combine them. The decision as to whether estimated differences are large enough to preclude combination or averaging across studies should not just rest on statistical considera-tion, but also on understanding of the science and context of the questions being considered [37].

3.7   Summary/Discussion

This chapter has considered the issue of between study variability, or heterogeneity, and its implications for meta-analysis. There is no one best strategy for dealing with heterogeneity; however, the authors consider it is essential to look for it, test for it and explore possible reasons for its presence using the techniques described above when carrying out any meta-analysis. When a sizeable amount of unex-plained heterogeneity is still present after this, a judgement has to be made on whether it is still appropriate to combine the results to produce an overall estimate. If so, which of the models described in the remainder of this book should be used, and what conclusions can be draw from it need to be decided. These decisions inevitably involve a large degree of subjectivity on the part of the meta-analyser.

References

1. Dickersin, K., Berlin, J.A. (1992). Meta-analysis: state-of-the-science. *Epidemiol. Rev.* **14**: 154–76.

2. Cochran, W.G. (1954). The combination of estimates from different experiments. *Biometrics* **10**: 101–29.

3. White, J., Yeats, A., Skipworth, G. (1979). *Tables for Statisticians. 3rd ed.* Cheltenham: Stanley Thornes.

4. Boissel, J.P., Blanchard, J., Panak, E., Peyrieux, J.C., Sacks, H. (1989). Considerations for the meta-analysis of randomized clinical trials: summary of a panel discussion. *Controlled Clin. Trials* **10**: 254–81.

5. Fleiss, J.L. (1986). Analysis of data from multiclinic trials. *Controlled Clin. Trials* **7**: 267–75.

6. Shadish, W.R., Haddock, C.K. (1994). Combining estimates of effect size. In: Cooper, H., Hedges, L.V., editors. *The Handbook of Research Synthesis*. New York: Russell Sage Foundation; 261–84.

7. Matt, G.E., Cook, T.D. (1994). Threats to the validity of research synthosis. In: Cooper, H., Hedges, L.V., editors. *The Handbook of Research Synthesis*. New York: Russell Sage Foundation; 503–20.

8. National Research Council. *Combining Information: Statistical Issues and Opportunities for Research.* Washington D.C. National Academy Press, 1992.

9. Hedges, L.V., Olkin, I. (1985). *Statistical Methods for Meta-Analysis*. London: Academic Press.

10. Gail, M., Simon, R. (1985). Testing for qualitative interaction between treatment effects and patient subsets. *Biometrics* **41**: 361–72.

11. Schmid, J.E., Koch, G.G., LaVange, L.M. (1991). An overview of statistical issues and methods of meta-analysis. *J. Biopharm. Stat.* **1**: 103–20.

12. Biggerstaff, B.J., Tweedie, R.L. Incorporating variability in estimates of heterogeneity in the random effects model in meta-analysis. *Stat. Med.* 1997; **16**: 753–68.

13. Hardy, R.J., Thompson, S.G. (1996). A likelihood approach to meta-analysis with random effects. *Stat. Med.* **15**: 619–29.

14. Zelen, M. (1971). The analysis of several 2×2 contingency tables. *Biometrika* **58**: 129–37.

15. Emerson, J.D. (1994). Combining estimates of the odds ratio: the state of the art. *Stat. Methods Med. Res.* **3**: 157–78.

16. Rosenthal, R. Parametric measures of effect size. In: Cooper, H., Hedges, L.V. (editors). (1994). *The Handbook of Research Synthesis*. New York: Russell Sage Foundation; p. 231–44.

17. Matuzzi, M., Hills, M. (1995). Estimating the degree of heterogeneity between event rates using likelihood. *Am. J. Epidemiol.* **141**: 369–74.

18. Jones, M.P., O'Gorman, T.W., Lemke, J.H., Woolson, R.F. (1989). A Monte Carlo investigation of homogeneity tests of the odds ratio under various sample size configurations. *Biometrics* **45**: 171–81.

19. Greenland, S. (1987). Quantitative methods in the review of epidemiological literature. *Epidemiol. Rev.* **9**: 1–30.

20. Galbraith, R.F. (1988). A note on graphical presentation of estimated odds ratios from several clinical trials. *Stat. Med.* **7**: 889–94.

21. L'Abbé, K.A., Detsky, A.S., O'Rourke, K. (1987). Meta-analysis in clinical research. *Annals of Internal Medicine* **107**: 224–33.

22. Sharp, S.J., Thompson, S.G., Altman, D.G. (1996). The relation between treatment benefit and underlying risk in meta-analysis. *Br. Med. J.* **313**: 735–8.

23. Bailey, K.R. (1987). Inter-study differences – how should they influence the interpretation and analysis of results. *Stat. Med.* **6**: 351–60.

24. Naylor, C.D. (1989). Meta-analysis of controlled clinical trials. *J. Rheumatology* **16**: 424–6.

25. Thompson, S.G., Smith, T.C., Sharp, S.J. (1997). Investigation underlying risk as a source of heterogeneity in meta-analysis. *Stat. Med.* **16**: 2741–58.

26. Brand, R., Kragt, H. (1992). Importance of trends in the interpretation of an overall odds ratio in the meta-analysis of clinical trials. *Stat. Med.* **11**: 2077–82.

27. Davey Smith, G., Egger, M. (1994). Commentary on the cholesterol papers: statistical problems. *Br. Med. J.* **308**: 1025–7.

28. Gelber, R.D., Goldhirsch, A. (1987). The evaluation of subsets in meta-analysis. *Stat. Med.* **6**: 371–88.

29. Thompson, S.G. (1993). Controversies in meta-analysis: the case of the trials of serum cholesterol reduction. *Stat Methods Med. Res.* **2**: 173–92.

30. Hughes, M.D., Freedman, L.S., Pocock, S.J. (1992). The impact of stopping rules on heterogeneity of results in overviews of clinical trials. *Biometrics* **48**: 41–53.

31. Green, S.J., Fleming, T.R., Emerson, S. (1987). Effects on overviews of early stopping rules for clinical-trials. *Stat. Med.* **6**: 361.

32. Colditz, G.A., Burdick, E., Mosteller, F. (1995). Heterogeneity in meta-analysis of data from epidemiologic studies: Commentary. *Am. J. Epidemiol.* **142**: 371–82.

33. Anello, C., Fleiss, J.L. (1995). Exploratory or analytic meta-analysis: Should we distinguish between them? *J. Clin. Epidemiol.* **48**: 109–16.

34. Fleiss, J.L. (1993). The statistical basis of meta-analysis. Stat *Methods Med. Res.* **2**: 121–45.

35. Morris, M.C., Sacks, F., Rosner, B. (1993). Does fish oil lower blood pressure? A meta-analysis of controlled trials. *Circulation* **88**: 523–33.

36. Furberg, C.T., Morgan, T.M. (1987). Lessons from overviews of cardiovascular trials. *Stat. Med.* **6**: 295–303.

37. Blair, A., Burg, J., Foran, J., Gibb, H., Greenland, S., Morris, R., Raabe, G., Savitz, D., Teta, J., Wartenberg, D., *et al.* (1995). Guidelines for application of meta-analysis in environmental epidemiology. ISLI Risk Science Institute. *Regul. Toxicol. Pharmacol.* **22**: 189–97.

CHAPTER 4

Fixed Effects Methods for Combining Study Estimates

4.1 Introduction

Using a fixed effect model to combine treatment estimates assumes no heterogeneity between the study results; the studies are assumed all to be estimating a single true underlying effect size. Clearly, in many instances this may not be realistic, and methods outlined in Chapter 3, including formal tests of heterogeneity, can be used to help the analyst decide whether the use of a fixed effect model is appropriate.

The general approach to fixed effect analyses, which can be adapted to combine most data types/effect estimates, including all the binary and continuous outcomes described in Chapter 2, is presented first. Methods *specific* to the pooling of odds ratios have been developed, and these are also described in this chapter. Specialist methods for combining other outcome measures are dealt with later in Chapters 14 and 18.

An important point concerning the combination of data on binary outcomes is that it is erroneous to merge 2×2 tables (such as Figures 2.1(a) and 2.1(b)) from individual studies in order to produce one overall table from which the desired outcome measure (OR, RR, etc.) can be calculated. Doing so can produce misleading results. This issue has been well documented [1], and is often referred to as Simpson's or Yule's paradox.

4.2   General fixed effect model – the inverse variance-weighted method

The inverse variance-weighted method was first described by Birge [2] and Cochran [3] in the 1930s, and is conceptually simple. Each study estimate is given a weight directly proportional to its precision (that is inversely proportional to its variance). For $i = 1, \ldots, k$ independent studies to be combined, let T_i be the observed effect size, θ_i the underlying population effect size, with variance v_i, for the ith study. For a fixed effect model, all population effect sizes are assumed equal, i.e. $\theta_1 = \ldots = \theta_k = \theta$, where θ is the true common underlying effect size. (This was the null hypothesis of the heterogeneity test of Section 3.2.) Thus, a pooled estimate of the treatment effect is given by

$$\overline{T}. = \frac{\sum_{i=1}^{k} w_i T_i}{\sum_{i=1}^{k} w_i}. \tag{4.1}$$

The weights that minimize the variance of $\overline{T}.$, and hence are routinely used, are inversely proportional to the conditional variance in each study [4]:

$$w_i = \frac{1}{v_i}. \tag{4.2}$$

The detail of the variance formulae depends upon the effect measure being combined; the commonly used ones were given in Chapter 2, others are reported elsewhere [5, 6]. An estimate of the variance of the pooled estimate $\overline{T}.$ is given by the reciprocal of the sum of the weights, i.e.

$$\operatorname{var}(\overline{T}.) = 1 / \sum_{i=1}^{k} w_i. \tag{4.3}$$

If $\overline{T}.$ is assumed to be normally distributed, an approximate $100(1 - \alpha)\%$ confidence interval for the population effect, θ, is given by

$$\overline{T}. - z_{\alpha/2} \sqrt{1 / \sum_{i=1}^{k} w_i} \leq \theta \leq \overline{T}. + z_{\alpha/2} \sqrt{1 / \sum_{i=1}^{k} w_i}, \tag{4.4}$$

where $z_{\alpha/2}$ is the $\alpha/2$ percentage point of a Standardized Normal Distribution. For example, for a 95% confidence interval ($\alpha = 0.05$), $z (\alpha = 0.05/2) = 1.96$ (as in Section 2.3, etc.).

This is essentially all there is to the inverse variance method. It should, however, be noted that the variance estimation formula for the standard inverse variance-weighted method can sometimes be biased and too sensitive to the minimum of the estimates of the variances in the K studies; an adjusted variance formula is available [7]. For further coverage of the inverse variance-weighted method, see Shadish and Haddock [4].

4.2.1 Example: Combining odds ratios using the inverse variance-weighted method

To illustrate how binary data can be combined using the inverse variance weighted method, a meta-analysis of antibiotics versus placebo for the common cold is used [8]. The outcome considered here is cure or general improvement in the first seven days after presenting symptoms. Data from the five trials included in the meta-analysis are given in Table 4.1

The odds ratio for each study is calculated using equation (2.1). As mentioned previously, it is advisable to log transform the data before combining, so each of these are transformed onto the log scale. The variance of each log odds ratio is calculated using equation (2.2). Confidence intervals for each study's log odds ratio can then be constructed using equation (2.3), and these can simply be transformed back to the odds ratio scale (e.g. study 1: $[e^{(-0.49)}$ to $e^{(0.41)}] = [0.61$ to $1.50]$).

All five of the studies' confidence intervals are wide, and all include the value one, indicating no statistically significant increased benefit from being given antibiotics over placebo was observed in any trial. (Note: unlike the cholesterol example, the event of interest – cure or general improvement – is desirable, and hence odds ratios greater than 1 indicate the intervention is superior to the control.) By combining these studies, a more precise estimate of the comparative treatment effect can be obtained (through reducing the effect of sampling variation in estimates based on smaller numbers).

Before deciding whether to combine the studies, a full assessment of between study heterogeneity should be done. Applying the test of Section 3.2.1 to these five trials produces a test static, $Q = 1.039$, which when compared to a χ_4^2 distribution gives a p-value of 0.90. Hence, the null hypothesis that all studies are estimating the same underlying treatment effect is plausible. Informal inspection of the data via the various graphical plots of Section 3.3 would appear to agree with this conclusion. Figure 4.1 displays the radial plot, which indicates a small degree of between study variation (the forest plot of Figure 4.2 can also be inspected). Hence, it would appear that a fixed effects model is appropriate for combining these studies.

To do this, the weighting given to each study in the analysis (the w_i's) needs to be calculated. When the inverse variance-weighted method is used, this is simply equal to the inverse of the variance of each estimate as denoted in equation (4.2). These weights are displayed in the final column of Table 4.1 The log odds ratios can now be combined using equation (4.1), the outline of the calculation required is given below:

$$\ln(OR_{Pooled}) = \frac{[(19.03 \times (-0.04)) + \ldots + (2.26 \times (-0.65))]}{(19.03 + \ldots + 2.26)} = -0.047.$$

The variance of this combined estimate is given by evaluating equation (4.3), so the standard error (square root of the variance) is given by

Table 4.1 Data from five trials investigating the effectiveness of antibiotics versus placebo for the common cold.

Study	No. patients in treatment arm	No. patients in control arm	No. cured/ improved in treatment arm (= a)	No. cured/ improved in control arm (= c)	No. not cured/ improved in treatment arm (= b)	No. not cured/ improved in control arm (= d)	Estimate of odds ratio (=(a × d)/(b × c))	Log odds ratio (ln(OR))	Variance of ln(OR) (var(ln(OR)) =1/a + 1/b +1/c + 1/d	95% Confidence Interval for ln(OR)	95% Confidence Interval for OR	Weight (= 1/var(ln(OR))
1	154	155	67	69	87	86	0.96	−0.04	0.05	−0.49 to 0.41	0.61 to 1.50	19.03
2	146	142	49	48	97	94	0.99	−0.01	0.06	−0.50 to 0.48	0.61 to 1.61	16.08
3	174	87	166	83	8	4	1.00	0.00	0.39	−1.23 to 1.23	0.29 to 3.42	2.54
4	13	16	9	10	4	6	1.35	0.30	0.63	−1.25 to 1.85	0.29 to 6.38	1.59
5	129	59	117	56	12	3	0.52	−0.65	0.44	−1.95 to 0.66	0.14 to 1.93	2.26

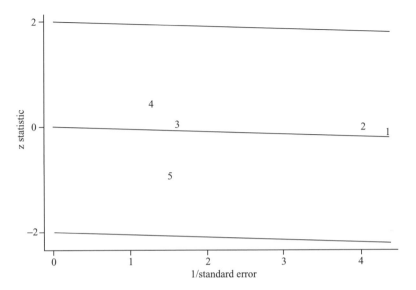

Figure 4.1 Radial plot for the antibiotics versus placebo RCTs.

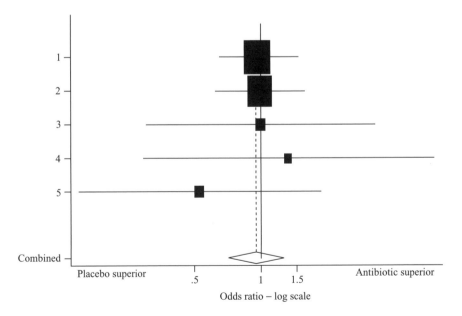

Figure 4.2 Forest plot for meta-analysis of antibiotics versus placebo for the common cold.

$$SE(\ln(OR_{Pooled})) = \sqrt{1/(19.03 + \ldots + 2.26)} = 0.16.$$

Hence, using equation (4.4), a 95% confidence interval for this pooled log odds ratio is given by

$$0.04 \pm (1.96 \times 0.16) = (-0.35 \text{ to } 0.26).$$

Taking the exponential of the combined log odds ratio gives the pooled result on the odds ratio scale which is more easily interpretable. Hence

$$OR = \exp^{(-0.047)}$$
$$= 0.95,$$

and the 95% confidence interval for this estimate is given by

$$(\exp(-0.35) \text{ to } \exp(0.26)) = (0.70 \text{ to } 1.29).$$

The combined odds ratio is very close to 1, indicating that patients' recovery is very comparable in the antibiotic and placebo groups. The 95% confidence interval around this estimate is reasonably wide, indicating considerable uncertainty in the pooled result. This interval reflects the fact that only a small number of relatively small trials were pooled. The conclusion that there is no evidence to suggest that antibiotics are effective at treating the common cold appears reasonable.

The results of this meta-analysis are displayed graphically in Figure 4.2 as a forest plot. Forest plots, which were introduced in Section 3.3.2, are used routinely for displaying the results of meta-analyses, and are discussed further in Section 10.3.1. The pooled estimate and its 95% confidence interval are plotted below the individual study results in this figure.

4.2.2   Example: Combining standardized mean differences using a continuous outcome scale

For completeness, an example combining continuous outcomes using the inverse weighted-method is also provided here. This uses the same principles as those for pooling the odds ratios in Section 4.2.1. The meta-analysis comparing sodium fluoride (NaF) with sodium monoflouorophosphate (SMFP) dentifrices, described in Section 2.4.1, is used for illustration. Table 4.2 includes the differences and corresponding standard errors for each study (calculated previously), together with the weighting given to each study derived using equation (4.2). Note the large weighting given to trial 8 (almost half of the total weight) due to its relatively large size.

The test for heterogeneity for these nine trials produces a test statistic, $Q = 5.40$, which when compared to a χ^2_8 distribution gives a p-value of 0.71 (see Figure 4.2 for the forest plot). Hence, there is no evidence to reject the null hypothesis, that all studies are estimating the same underlying treatment effect and it would seem valid

Table 4.2 Quantities required for meta-analysis comparing sodium fluoride (NaF) with sodium monoflouorophosphate (SMFP).

Study	Mean difference	se(mean difference)	weight (w_i)
1	0.86	0.57	3.07
2	0.33	0.55	3.25
3	0.47	0.35	8.09
4	0.5	0.25	15.88
5	−0.28	0.54	3.43
6	0.04	0.28	13.21
7	0.8	0.78	1.63
8	0.19	0.13	55.97
9	0.49	0.28	12.92

to combine the studies using a fixed effect model. The mean differences are combined using equation (4.1), as outlined below:

$$weighted\ mean\ difference = \frac{[(3.07 \times (0.86)) + \ldots + (12.92 \times (0.49))]}{(3.07 + \ldots + 12.92)} = 0.28.$$

The standard error of this pooled estimate is calculated by taking the square-root equation of (4.3):

$$SE(weighted\ mean\ difference) = \sqrt{1/(3.07 + \ldots + 12.92)} = 0.09.$$

Evaluating (4.4) gives a 95% confidence interval for this pooled estimate of the mean difference:

$$0.28 \pm (1.96 \times 0.09) = (0.10\ to\ 0.46).$$

This confidence interval does not include 0, the treatment difference is hence statistically significant ($p = 0.02$). The results of this analysis are plotted in Figure 4.3. The meta-analysis suggests that NaF provides an advantage over SMFP. Our best estimate is that NaF protects, on average, 0.28 more dental surfaces than SMFP (over 2–3 years of daily use). Note that a direct clinical interpretation of this meta-analysis' results is possible because it was possible to pool using an original scale (DMFS). If it had been necessary for the effects to be standardized, such an interpretation would be less easy.

4.3 Specific methods for combining odds ratios

Other fixed effect methods specific to combining odds ratios have been developed. Under most conditions the estimates obtained from each method should be very

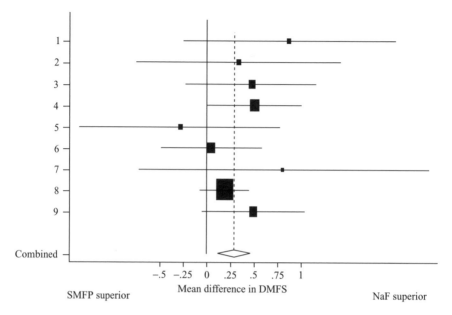

Figure 4.3   Forest plot of meta-analysis comparing NaF with SMFP.

similar to one another. However, when the data are sparse, results may differ, and some traditional methods may break down altogether. For this reason, new computer intensive methods have been developed to be used in situations where the traditional methods have questionable validity [9]. Both the standard methods and the newer ones are outlined below.

4.3.1 Mantel–Haenszel method for combining odds ratios

This method was first described by Mantel and Haenszel [10] for the use in combining odds ratios for stratified case-control studies. Later, Mantel [11] reported the method could be used for a wider class of problems, including prospective studies [12]. In meta-analysis each study forms an individual stratum [13]. The pooled estimate is calculated by

$$\overline{T}_{MH\,(OR)} = \frac{\sum_{i=1}^{k} a_i d_i \big/ n_i}{\sum_{i=1}^{k} b_i c_i \big/ n_i}, \tag{4.5}$$

where a_i, b_i, c_i, and d_i are the four cells of the 2×2 table for the $i = 1 \ldots k$ studies, as given in Figures 2.1(a) and (b), and n_i is the total number of people in the ith study.

A variance estimate for the estimated summary odds ratio, $\overline{T}_{MH(OR)}$, is required to calculate a confidence interval around this point estimate. The formula commonly used was derived by Robins *et al.* [14, 15] This formula computes a variance estimate for the log of $\overline{T}_{MH(OR)}$, as:

$$v_{MH(\ln(OR))} = \frac{\sum_{i=1}^{k} P_i R_i}{2\left(\sum_{i=1}^{k} R_i\right)^2} + \frac{\sum_{i=1}^{k} (P_i S_i + Q_i R_i)}{2\left(\sum_{i=1}^{k} R_i\right)\left(\sum_{i=1}^{k} S_i\right)} + \frac{\sum_{i=1}^{k} Q_i S_i}{2\left(\sum_{i=1}^{k} S_i\right)^2}, \qquad (4.6)$$

where $Pi = (a_i + d_i)/n_i$, $Q_i = (b_i + c_i)/n_i$, $R_i = a_i d_i/n_i$, and $S_i = b_i c_i/n_i$.

A $100(1 - \alpha)\%$ confidence interval for the summary odds ratio θ, is thus given by

$$\exp\left[\ln(\overline{T}_{MH(OR)}) - z_{\alpha/2}(v_{MH(OR)})^{1/2}\right] \leq \theta$$
$$\leq \exp\left[\ln(\overline{T}_{MH(OR)}) + z_{\alpha/2}(v_{MH(OR)})^{1/2}\right], \quad (4.7)$$

where $z_{\alpha/2}$ is the $\alpha/2$ percentage point of a Standardized Normal Distribution. Several other variance estimates have been put forward; these are further explored elsewhere [9, 14–16]. Sato [17] has developed a method that works directly on the odds ratio scale (as opposed to ln *OR*). Simulations have shown [17] that this works as well as the method of Robins *et al.* given above.

Example: Combining odds ratios using the Mantel–Haenszel method

The five trials comparing antibiotics with placebo (Table 4.1) are pooled using the Mantel–Haenszel estimate. Using equation (4.5), the calculation required to compute the pooled odds ratio is outlined below:

$$\overline{T}_{MH(OR)} = \frac{\left[\left(\frac{67 \times 86}{154 + 155}\right) + \ldots + \left(\frac{117 \times 3}{129 + 59}\right)\right]}{\left[\left(\frac{87 \times 69}{154 + 155}\right) + \ldots + \left(\frac{12 \times 56}{129 + 59}\right)\right]}$$
$$= 0.95.$$

The variance of the log of this estimate is calculated using equation (4.6), and the required calculations are outlined below:

$$v_{MH\,\ln(OR)} = \frac{[(0.50 \times 18.65) + \ldots + (0.64 \times 1.87)]}{2(18.65 + \ldots + 1.87)^2} +$$

$$\frac{[(0.50 \times 19.43 + 0.50 \times 18.65) + \ldots + (0.64 \times 3.57 + 0.36 \times 1.87)]}{2(18.65 + \ldots + 1.87)(19.43 + \ldots + 3.57)} +$$

$$\frac{[(0.50 \times 19.43) + \ldots + (0.36 \times 3.57)]}{2(19.43 + \ldots + 3.57)^2}$$

$$= \frac{20.99}{2(40.91)^2} + \frac{42.23}{2(40.91)(43.09)} + \frac{20.79}{2(43.09)^2}$$

$$= 0.024.$$

Using equation (4.7), a 95% confidence interval for $\overline{T}_{MH(OR)}$, is given by

$$\exp\left[-0.05 - 1.96(0.024)^{1/2}\right] \leq \theta \leq \exp\left[-0.95 + 1.96(0.024)^{1/2}\right] = [0.70, 1.29].$$

Comparing this estimate and confidence interval to that obtained using the inverse variance-weighted method in Section 4.2.1, it can be seen that in this case the two methods give identical answers to two decimal places.

4.3.2   Peto's method for combining odds ratios

This method was first described by Peto [18], and more thoroughly by Yusuf *et al.* [19]. It can be regarded as a modification of the Mantel–Haenszel method. An advantage over the Mantel–Haenszel method is that it can still be used when cells in individual studies 2 × 2 tables are zero; it is also easy to calculate. Unfortunately, this method may produce serious under estimates [6], when the odds ratio is far from unity (large treatment or exposure effects). This is unlikely to be a problem in clinical trials, but could be so in the meta-analysis of epidemiological studies [20].

    Defining n_i as the number of patients in the *i*th trial and n_{ti} as the number in the new treatment group of the *i*th trial, let d_i equal the total number of events from both treatment and control groups, and O_i the number of events in the treatment group. Then E_i, the 'expected' number of events in the treatment group (in the *i*th trial), can be calculated as $E_i = (n_{ti}/n_i)d_i$. For each study, two statistics are calculated: (1) O-E, the difference between the observed and the number expected to have done so under the hypothesis that the treatment is no different from the control, E; and (2) v, the variance of the difference O-E. For k studies the pooled estimate of the odds ratio is given by [21]:

$$\overline{T}_{PETO(OR)} = \exp\left[\sum_{i=1}^{k}(O_i - E_i) \middle/ \sum_{i=1}^{k} v_i\right], \tag{4.8}$$

where $v_i = E_i[(n_i - n_{ti})/n_i][(n_i - d_i)/(n_i - 1)]$.

    An estimate of the approximate variance of the natural log of the estimated pooled odds ratio is provided by

Table 4.3 2 × 2 Table including marginal values for study 1.

	Cured/improved	No improvement	Total
Antibiotic	67	87	154
Placebo	69	86	155
Total	136	173	309

$$\text{var}\left(\ln \overline{T}_{\text{PETO(OR)}}\right) = \left(\frac{1}{\sum_{i=1}^{k} v_i}\right), \tag{4.9}$$

A $100(1 - \alpha)\%$ (non symmetric) confidence interval is thus given by

$$\exp\left(\frac{\sum_{i=1}^{k}(O_i - E_i) \pm z_{\alpha/2}\sqrt{\sum_{i=1}^{k} v_i}}{\sum_{i=1}^{k} v_i}\right), \tag{4.10}$$

where $z_{\alpha/2}$ is the $\alpha/2$ percentage point of a Standardized Normal Distribution.

Example: Combining odds ratios using the Peto method

Returning again to the antibiotics for common cold trials, for this method it is necessary to calculate the marginal values for each studies 2 × 2 table. Table 4.3 illustrates this for the first study.

From tables like Table 4.3, the values needed to calculate Peto's method can be calculated. For study 1:

$$O_i = 67$$
$$E_i = \left(\frac{154}{309}\right) \times 136 = 67.78$$
$$v_i = 67.78\left[\frac{(309 - 154)}{309}\right]\left[\frac{(309 - 136)}{(309 - 1)}\right] = 51.45$$

Table 4.4 presents these values for all five studies.

Entering the values from Table 4.4 into equation 4.8 gives the combined estimate

$$\overline{T}_{PETO(OR)} = \exp\left[\frac{-0.78 + \ldots + (-1.71)}{19.10 + \ldots + 2.99}\right]$$
$$= 0.95.$$

Table 4.4 Intermediate values needed to calculate the Peto estimate.

Study	O_i	E_i	v_i	$O_i - E_i$
1	67	67.78	19.10	−0.78
2	49	49.17	16.14	−0.17
3	166	166.00	2.55	0.00
4	9	8.52	1.68	0.48
5	117	118.71	2.99	−1.71

The variance of this estimate is given by equation (4.9):

$$\mathrm{var}\left(\ln \overline{T}_{PETO(OR)}\right) = (19.10 + \ldots + 2.99)$$
$$= 42.45.$$

A 95% confidence interval is calculated using equation (4.10):

$$\exp\left(\frac{-2.18 \pm 1.96\sqrt{42.45}}{42.45}\right)$$

95% CI [0.70 to 1.28].

This result is almost identical to that given by the Mantel–Haenszel estimate in the previous section. Hence, in this example, all three methods lead to the same conclusions. This is not always the case; Section 4.3.5 discusses conditions under which the results of these methods may differ, and examines which methods are then superior.

*4.3.3   Combining odds ratios via maximum-likelihood techniques

Maximum likelihood estimates are difficult to compute exactly, but they are the most efficient for large sample sizes. Unfortunately, there is no way of knowing how large the sample sizes must be for this property to hold [12]. The likelihood of the k studies is given by [12]:

$$L \propto \prod_{i=1}^{k} \theta_{ci}^{b_i}(1 - \theta_{ci})^{d_i}\theta_{ti}^{a_i}1 - \theta_{ti})^{c_i}, \tag{4.11}$$

from which the unconditional estimate

$$OR_{Umle} = \theta_{ti}(1 - \theta_{ci})/\theta_{ci}(1 - \theta_{ti}) \tag{4.12}$$

is obtained. Emerson [9] reports that Breslow found that unconditional maximum likelihood estimation, which had earlier been investigated by Gart, not to be consistent for estimating the odds ratio when the number of counts remained bounded.

Conditional maximum likelihood estimates also exist. They use the conditional distribution of the data in each table, given the fixed values for the total counts in the margins. This leads to an estimator that is consistent and asymptotically normal [9, 22]. It is superior to the unconditional maximum likelihood estimator, and equal or superior to the Mantel–Haenszel estimator in both bias and precision [22].

This is a very brief outline of these methods; a recommended starting point for further investigation is the excellent review by Emerson [9].

*4.3.4 Exact methods of interval estimation

The above methods for interval estimation are all asymptotic; their justification assumes either that the counts are large, or that the number of strata is large. Exact methods do exist that are not restrained in this way, and are based on exact distribution theory. Although these methods have long been available in principle, modern computer power (using network algorithms) now makes them routinely available. A detailed description of these methods are beyond the scope of this book; the interested reader is referred to Emerson [9] for a review of this topic.

4.3.5 Discussion of the relative merits of each method

With a number of different approaches to combine odds ratios available, it would be desirable to have guidelines indicating which a particular method is most appropriate.

As mentioned previously, the Peto method has come under strong criticism. It has been demonstrated that this method may produce seriously biased odds ratios and corresponding standard errors when there is severe imbalance in the numbers in the two groups being compared [23]. Bias is also possible when the estimated odds ratio is far from unity [24]. Fleiss [25] describes conditions under which the inverse-weighted and the Mantel–Haenszel method are to be preferred: if the number of studies to be combined is small, but the within-study sample sizes per study are large, the inverse-weighted method should be used. If there are many studies to combine, but the within-study sample size in each study is small, the Mantel–Haenszel method is preferred.

A comparison between the Mantel–Haenszel and (conditional and unconditional) maximum likelihood techniques has been carried out. Generally, if the sample sizes of the studies are large (all cells $>= 5$) the methods will give almost identical results, otherwise differences between the methods will be small. As there seems to be no clear benefit to be reaped from the difficult computation of the maximum likelihood method, using the inverse-weighted and Mantel–Haenzel methods when indicated would seem the best strategy in most cases. If, however, samples sizes are small for individual studies exact methods may be preferred [23].

Another consideration when deciding on which method to use is whether any of the trials arms have zero observed events. Using the Mantel–Haenszel estimate, a study with zero total events is completely excluded from the analysis if no continuity correction is used. This is unappealing as a trial with zero events from 200 subjects would then be equally as non-informative as a trial with only 20 subjects. A recent investigation into this problem recommended that a continuity correction (adding 0.5 to each cell) should be used for sparse data in meta-analysis, except in the situation when there is strong evidence suggesting that very little heterogeneity exists among component studies [26]. This correction has already been applied to the cholesterol lowering trials examples of Chapter 3. At the time of writing, simulation studies appeared to show that the Peto method outperformed other simple methods, including the Mantel–Haenszel method and the inverse variance-weighted method when there were small numbers of events in one or more cells of studies 2×2 tables [27].

A further factor has been identified that may be important when carrying out a fixed effects meta-analysis. Mengersen *et al.* [28] compared the ways in which confidence intervals for odds ratios were calculated for individual studies. They compared the calculation of the odds ratios in epidemiological studies investigating the effect of exposure to environmental smoke on lung cancer. An exact test (Fisher's) was compared to the Mantel–Haenszel method, and the logit variance approximation (used in this instance to calculate OR's from each individual stratified study as opposed to across studies to combine estimates). The investigation concluded that exact methods might increase estimated confidence interval widths by 5–20% over standard approximate (logit and Mantel–Haenszel) methods, and these methods themselves may differ by this order of magnitude [28].

Emerson [9] gives detailed formal guidelines on procedures to follow when combining odds ratios in a general context. There appears to be no reason why they should not be applicable for meta-analysis. Finally, further methods for combining odds ratios are discussed in later chapters; Section 6.3.5 describes how logistic regression can be used, and Section 11.4 describes the use of Bayesian methods. Both of these combine the raw data directly, and hence circumvent some of the problems encountered with the simpler methods discussed above.

4.4   Summary/Discussion

This chapter has considered the so-called fixed effect approach to meta-analysis. This assumes that all the studies in a meta-analysis are estimating the same underlying unknown true intervention effect. For combining odds ratios, a variety of estimation methods have been proposed; whilst in many situations they give qualitatively similar results, in some circumstances differences can be serious. In terms of binary data, problems with a number of methods occur if there are zero events in any treatment arms in any study. The use of continuity correction factors and of exact methods have been discussed as means for overcoming this problem. When reporting a meta-analysis, it is important to report precisely what methods have been used.

References

1. Simpson, E.H. (1951). The interpretation of interaction in contingency tables. *J. Roy. Statist. Soc.* Series B **13**: 238–41.

2. Birge, R.T. (1932). The calculation of errors by the method of least squares. *Phys. Rev.* **16**: 1–32.

3. Cochran, W.G. (1937). Problems arising in the analysis of a series of similar experiments. *J. Roy. Statist. Soc.* **4**:(Supplement) 102–18.

4. Shadish, W.R., Haddock, C.K. (1994). Combining estimates of effect size. In: Cooper, H., Hedges, L.V., (editors). *The Handbook of Research Synthesis*. New York: Russell Sage Foundation; 261–84.

5. Rosenthal, R. (1994). Parametric measures of effect size. In: Cooper, H., Hedges, L.V., editors. *The Handbook of Research Synthesis*. New York: Russell Sage Foundation; 231–44.

6. Fleiss, J.L. Measures of effect size for categorical data. In: Cooper, H., Hedges, L.V., (editors). (1994). *The Handbook of Research Synthesis*. New York: Russell Sage Foundation; 245–60.

7. Li, Y.Z., Shi, L., Roth, H.D. (1994). The bias of the commonly-used estimate of variance in metaanalysis. *Comm. Statistics–Theory and Methods* **23**: 1063–85.

8. Arroll, B., Kenealy, T. (1999). Antibiotics versus placebo for the common cold (Cochrane review). In the Cochrane Library Issue 2 Oxford: Update Software.

9. Emerson, J.D. (1994). Combining estimates of the odds ratio: the state of the art. *Stat. Methods Med. Res.* **3**: 157–78.

10. Mantel, N., Haenszel, W. (1959). Statistical aspects of the analysis of data from retro-spective studies of disease. *J. Nat. Cancer Inst.* **22**: 719–48.

11. Mantel, N. (1963). Chi-square tests with one degree of freedom: extensions of the Mantel–Haenszel procedure. *J. Am. Statist. Assoc.* **58**: 690–700.

12. Hasselblad, V., McCrory, D.C. (1995). Meta-analytic tools for medical decision making: A practical guide. *Med. Decis. Making* **15**: 81–96.

13. Dickersin, K., Berlin, J.A. (1992). Meta-analysis: state-of-the-science. *Epidemiol. Rev.* **14**: 154–76.

14. Robins, J., Breslow, N., Greenland, S. (1986). Estimators of the Mantel–Haenszel variance consistent in both sparse data and large-strata limiting models. *Biometrics* **42**: 311–23.

15. Robins, J., Greenland, S., Breslow, N.E. (1986). A general estimator for the variance of the Mantel–Haenszel odds ratio. *Am. J. Epidemiol.* **124**: 719–23.

16. Phillips, A., Holland, P.W. (1987). Estimators of the variance of the Mantel–Haenszel log-odds-ratio estimate. *Biometrics* **43**: 425–31.

17. Sato, T. (1990). Confidence limits for the common odds ratio based on the asymptotic distribution of the Mantel–Haenszel estimator. *Biometrics* **46**: 71–80.

18. Peto, R., Pike, M.C., Armitage, P., Breslow, N.E., Cox, D.R., Howard, S.V., Mantel, N., McPherson, K., Peto, J., Smith, P.G. (1977). Design and analysis of randomized clinical trials requiring prolonged observation of each patient. II: Analysis and examples. *Br. J. Cancer* **35**: 1–39.

19. Yusuf, S., Peto, R., Lewis, J., Collins, R., Sleight, P. *et al.* (1985). Beta blockade during and after myocardial infarction: an overview of the randomised trials. *Progress in Cardiovascular Diseases* **27**: 335–71.

20. Spector, T.D., Thompson, S.G. (1991). Research methods in epidemiology 5. The potential and limitations of meta-analysis. *J. Epidemiol. Comm. Hlth.* **45**: 89–92.

21. Berlin, J.A., Laird, N.M., Sacks, H.S., Chalmers, T.C. (1989). A comparison of statistical methods for combining event rates from clinical trials. *Stat. Med.* **8**: 141–51.

22. Hauck, W.W. (1984). A comparative study of conditional maximum likelihood estimation of a common odds ratio. *Biometrics* **40**: 1117–23.

23. Greenland, S., Salvan, A. (1990). Bias in the one-step method for pooling study results. *Stat. Med.* **9**: 247–52.

24. Fleiss, J.L. (1993). The statistical basis of meta-analysis. *Stat Methods Med Res* **2**: 121–45.

25. Fleiss, J.L. (1981). *Statistical Methods for Rates and Proportions. 2nd ed.* New York: Wiley.

26. Sankey, S.S., Weissfeld, L.A., Fine, M.J., Kapoor, W. (1996). An assessment of the use of the continuity correction for sparse data in metaanalysis. *Comm. Statistics – Simulation and Computation* **25**: 1031–56.

27. Deeks, J., Bradburn, M., Localio, R., Berlin, J. (1999). Much ado about nothing: statistical methods for meta-analysis with rare events. Presented at *Systematic Reviews: Beyond the basics,* January Oxford. (Abstract available at: http://www. ihs.ox.ac.uk/csm/talks.html#p23.) (Paper in preparation at time of press.)

28. Mengersen, K.L., Tweedie, R.L., Biggerstaff, B.J. (1995). The impact of method choice in meta-analysis. *Aust. J. Stats.* **37**: 19–44.

CHAPTER 5

Random Effects Methods for Combining Study Estimates

5.1 Introduction

The fixed effect model described in Chapter 4 assumes that all the studies are estimating the same underlying effect size $\theta = \theta_1 = \theta_2 = \cdots = \theta_k$ The test for heterogeneity (Section 3.2.1) provides a way of testing this assumption. Using a fixed effect model under conditions of heterogeneity, the confidence interval for the overall treatment effect reflects the random variation within each trial, but not potential heterogeneity *between* trials, so that the confidence interval is artificially narrow [1]. Random effects models have been advocated as a more conservative alternative. This approach assumes the studies are estimating different (underlying) effect sizes, and takes into account the extra variation implied in making this assumption. More specifically, these underlying effects are assumed to vary at random, and typically, the distribution of such effects is assumed to be normally distributed. Hence, this model includes two sources of variation, the between and within study variance.

The standard random effects model used in meta-analysis was described by DerSimonian and Laird [2]. The model assumes that the study specific effect sizes come from a random distribution of effect sizes with a fixed mean and variance. This assumption has caused much dispute, as some researchers have difficulty with the notion of a population of studies from which the observed sample are drawn. It should be noted, however, that random effects models have a long history in other fields of application. Expressed algebraically, where T_i is an estimate of effect size and θ_i is the true effect size in the ith study:

$$T_i = \theta_i + e_i, \tag{5.1}$$

where e_i is the error with which T_i estimates θ_i, and

$$\text{var}(T_i) = \tau_\theta^2 + v_i, \qquad (5.2)$$

where τ_θ^2 is the random effects variance and v_i is the variance due to sampling error in the ith study. If the random effects variance was zero, the above model would reduce exactly to the fixed effects model of Chapter 4.

Formulae can be derived using both a weighted and an unweighted approach; these can be solved using four different methods; Weighted and Unweighted Least Squares (WLS, UWLS), and Maximum and Restricted Maximum Likelihood (ML & REML); the latter two assume normality of the underlying effect parameters. The likelihood to be maximised is slightly modified using REML (from that of ML), to adjust for the fact that the underlying mean and variance are being estimated from the same data. The REML are the iterative equivalent to the weighted estimators [2]. The relative merits of each of the above methods have not been widely investigated [3].

5.2   Algebraic derivation for random effects models by the weighted method

Formulae can be derived using two different approaches: the weighted and un-weighted methods. Only the weighted is outlined here (following an account elsewhere [3]). Also, see Shadish and Haddock [3] for a derivation of the unweighted approach.

Starting with the general inverse weighted variance model (equations (4.1)–(4.3)), under the assumption that the studies are a random sample from a larger population of studies, there is a mean population effect size about which the underlying study-specific effect sizes vary [4]. This is the parameter we primarily wish to estimate. Let $\hat{\tau}^2$ denote the between study variance of the studies effect sizes (an estimate for τ_θ^2 a quantity yet to be determined). Further, define $\bar{w}$ and s_w^2 to be the mean and variance of the weights from the k studies:

$$\bar{w} = \sum_{i=1}^{k} w_i/k, \qquad (5.3)$$

and

$$s_w^2 = \frac{1}{k-1}\left(\sum_{i=1}^{k} w_i^2 - k\bar{w}^2\right). \qquad (5.4)$$

Further, define

$$U = (k-1)\left(\bar{w} - \frac{s_w^2}{k\bar{w}}\right). \qquad (5.5)$$

The estimated component of variance due to inter-study variation in effect size, $\hat{\tau}^2$, is calculated as

$$\hat{\tau}^2 = 0 \text{ if } Q \leq k - 1$$

and

$$\hat{\tau}^2 = (Q - (k - 1))/U \text{ if } Q > k - 1, \tag{5.6}$$

where Q is the heterogeneity test statistic defined in equations (3.1 and 3.2). Adjusted weights w_i^* for each of the studies may now be calculated as

$$w_i^* = \frac{1}{[(1/w_i) + \hat{\tau}^2]}. \tag{5.7}$$

Thus the random effects study weighting is given by the reciprocal of the sum of the between and within study variances, confirming that the random effect model reduces to the fixed effect model when the between study variance τ^2 is estimated as 0. The treatment point estimate for the mean treatment effect of all studies $\bar{\theta}$ can then be computed by

$$\overline{T}_{.RND} = \sum_{i=1}^{k} w_i^* T_i / \sum_{i=1}^{k} w_i^*. \tag{5.8}$$

The variance of this estimate is simply

$$\text{var}(\overline{T}_{.RND}) = 1 / \sum_{i=1}^{k} w_i^*, \tag{5.9}$$

and, if normality is assumed, a $100(1 - \alpha)\%$ confidence interval can be calculated by

$$\overline{T}_{.RND} - z_{\alpha/2} / \sqrt{\sum_{i=1}^{k} w_i^*} \leq \bar{\theta} \leq \overline{T}_{.RND} + z_{\alpha/2} / \sqrt{\sum_{i=1}^{k} w_i^*}. \tag{5.10}$$

It is worth noting that the random effects confidence interval for the treatment effect will always be larger than for a fixed effects analysis, as long as the between study variance $\hat{\tau}^2$ is estimated to be greater than zero.

*5.3 Maximum likelihood and restricted maximum likelihood estimate solutions

If it is assumed that each of the underlying effect parameters, the θ_is, come from a normal distribution, with mean μ and variance τ^2 (and T_i is N (θ_i, s_i^2)) [2], then the likelihood is proportional to [5]

$$L \propto \exp\left[-\sum_{i=1}^{k} \left[\left(\hat{\theta}_i - \mu\right)^2 / \left(\tau^2 + \sigma_i^2\right) + \ln\left(\tau^2 + \sigma_i^2\right) \right] / 2 \right]. \qquad (5.11)$$

Approximate solutions to this model have been given by DerSimonian and Laird [2] and Hedges [6] using the EM algorithm [7] to calculate MLE [8] and REML [9] solutions. In REML estimation, the likelihood to be maximized is slightly modified to adjust for μ and τ being estimated from the same data [2]. Additionally, it is possible to calculate maximum-likelihood estimates directly, or Bayesian estimates can be calculated with the specification of a prior (see Chapter 11).

*5.4   Comparison of estimation methods

DerSimonian and Laird compared the estimates produced by the four methods [2], and concluded that the WLS and the REML estimation procedures consistently yield slightly higher values of $\hat{\tau}^2$ (the random effects variance) than the maximum likelihood procedure. This is because both these procedures adjust for $\overline{T}_{.RND}$ and $\hat{\tau}^2$ being estimated from the same data, whereas the MLE procedure does not. The estimates of $\hat{\tau}^2, \overline{T}_{.RND}$ and its standard error from the unweighted method (not developed above) differ from the estimates from the other three methods [2]. This lead the authors to suggest that the WLS method is an attractive procedure because of the comparability of its estimates with those of the maximum likelihood methods, and because of its relative simplicity. It has become the commonly used method, and is often simply referred to as the DerSimonian and Laird method.

5.5   Example: Combining the cholesterol lowering trials using a random effects model

In Chapter 3 heterogeneity was assessed in 34 cholesterol lowering trials using a formal test and several graphical tools. The test for heterogeneity led to a test statistic of $Q = 88.23$, which has a corresponding p-value of < 0.0001. The overall conclusion of this exploration was that there appeared to be considerable heterogeneity between the trials. If one still wished to pool the studies without exploring sources of this heterogeneity, it would appear that a random effects, as opposed to a fixed effects, analysis, would be more appropriate. The weighted non-iterative (DerSimonian and Laird) approach is calculated below.

The first step is to calculate the mean and variance of the within study weights, using the weights given to each individual study presented in Table 3.1 and equation (5.3):

$$\bar{w} = \frac{(14.98 + \cdots + 0.87)}{34} = 28.27,$$

and equation (5.4):

$$s_w^2 = \frac{1}{34 - 1}\left((224.29 + \cdots + 0.76) - 34 \times 28.27^2\right) = 3488.95.$$

Calculating U defined in equation (5.5):

$$U = (34 - 1)\left(28.27 - \frac{3488.95}{34 \times 28.27}\right) = 813.22.$$

In calculating the estimated component of variance due to between-study variation in the values of the odds ratios ($\hat{\tau}^2$) from equation (5.6), we note that $Q = 88.23$ (calculated in Section 3.2.3), and $k = 34$. Thus, $Q > k - 1$, so

$$\hat{\tau}^2 = (88.23 - (34 - 1))/813.22 = 0.068.$$

Now the weights $w_i^*, \cdots, w_{34}^*$ used in the random effects model can be calculated using equation (5.7). So, for the first study,

$$w_1^* = \frac{1}{[(1/14.98) + 0.068]} = 7.43.$$

Table 5.1 Displays these weights, in addition to the original inverse variance weights, used in a fixed effect analysis, for all 34 studies.

It is instructive to examine how the relative weighting has changed between these two weighting schemes corresponding to those used by fixed and the random effects models. It can be seen, by examining the percentage of the total weight allocated to each study, that in using the random effects model, the relative weighting of the larger studies has been reduced, while the relative weighting of the smaller studies is increased. For example, study 15, the most influential study, is given a weight equivalent to almost 34% of the total in a fixed effect model, but this reduces to under 7% in a random effects model. Conversely, study 17, one of the smaller and least influential studies, given 1.3% of the total weight under a fixed effect model, is given 3.3% under a random effects one. This trend generally holds true for all meta-analyses.

Using the log odds ratio estimates of each individual study shown in Table 3.1, and the weights calculated above, the pooled point estimate of the odds ratio can be calculated using equation (5.8):

$$\bar{T}_{RND}(\ln(OR)) = [(7.43 \times (-0.74)) + \cdots + (0.82 \times (-0.31))]/(7.43 + \cdots + 0.82)$$
$$= -0.117.$$

The variance of this estimate is calculated using equation (5.9):

$$\mathrm{var}(\bar{T}_{RND(\ln(OR))}) = 1/(7.43 + \cdots + 0.82)$$
$$= 0.005.$$

Table 5.1   Weighting of individual studies used in the fixed effect and weighted non-iterative random effects model.

Study	w_i-fixed effect analysis (% of total)	w_i^*-random effect analysis (% of total)
1	14.98 (1.56)	7.43 (3.56)
2	18.54 (1.93)	8.21 (3.93)
3	13.83 (1.44)	7.14 (3.42)
4	1.36 (0.14)	1.25 (0.60)
5	0.43 (0.04)	0.41 (0.20)
6	26.25 (2.73)	9.44 (4.52)
7	18.26 (1.90)	8.16 (3.91)
8	9.20 (0.96)	5.66 (2.71)
9	50.03 (5.20)	11.39 (5.45)
10	16.64 (1.73)	7.82 (3.74)
11	16.45 (1.71)	7.78 (3.72)
12	5.18 (0.54)	3.83 (1.84)
13	5.98 (0.62)	4.25 (2.04)
14	0.44 (0.05)	0.43 (0.20)
15	326.57 (33.97)	14.11 (6.76)
16	51.51 (5.36)	11.46 (5.49)
17	12.69 (1.32)	6.82 (3.27)
18	19.79 (2.06)	8.45 (4.05)
19	2.32 (0.24)	2.01 (0.96)
20	20.36 (2.12)	8.55 (4.10)
21	14.15 (1.47)	7.22 (3.46)
22	1.18 (0.12)	1.10 (0.52)
23	2.89 (0.30)	2.42 (1.16)
24	121.91 (12.68)	13.15 (6.30)
25	23.98 (2.49)	9.13 (4.37)
26	0.37 (0.04)	0.36 (0.17)
27	7.25 (0.75)	4.86 (2.33)
28	33.70 (3.51)	10.26 (4.91)
29	21.52 (2.24)	8.75 (4.19)
30	3.16 (0.33)	2.60 (1.25)
31	98.72 (10.27)	12.83 (6.14)
32	0.37 (0.04)	0.36 (0.17)
33	0.37 (0.04)	0.36 (0.17)
34	0.87 (0.09)	0.82 (0.39)

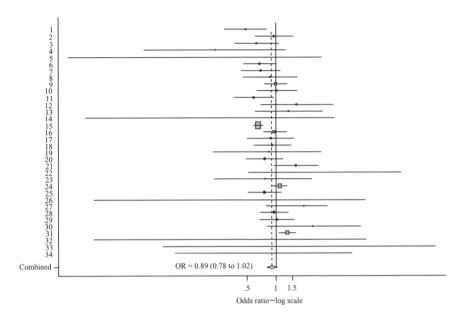

Figure 5.1 Plot of cholesterol trials combined using a random effects model.

An approximate 95% confidence interval is now calculated using equation (5.10):

$$-0.117 \pm 1.96 * \sqrt{0.005} = (-0.253 \text{ to } 0.018).$$

Converting back to odds ratio scale gives $\overline{T}_{RND(OR)} = 0.89$, with approximate 95% confidence interval [0.78, 1.02]. This compares to 0.85 [0.79, 0.90] using the fixed effect model (inverse variance-weighted method). A Forest plot of these results is given in Figure 5.1.

The confidence interval for this fixed effect estimate is markedly narrower than that for the random effects. The extra width is due to the between study variation being taken into account in the random effects analysis. As mentioned previously, this is typical. However, in this particular instance the choice of model is critical, because the fixed effect model gives a statistically significant treatment effect, while the random effect interval contains 1, and thus the treatment effect is not statistically significant (at the 5% level). Due to the large between study heterogeneity, the random effect model would appear more appropriate if pooling to find an overall estimate was to be carried out.

5.6   Extensions to the random effects model

*5.6.1   Including uncertainty induced by estimating the between study variance

Although the random effects model gives wider confidence intervals than that of a corresponding fixed effect analysis, they are still too narrow, because the method assumes the between study variance is known, when in fact it is estimated from the data [10–12]. Two recent advances have attempted to address this problem. Hardy and Thompson [10] developed an approach using profile likelihood methods, which assumes normality of the data, to calculate appropriate confidence regions. A sensitivity plot of $\hat{\tau}^2$ against $T_{.RND}$ can be used to investigate the robustness of $T_{.RND}$ to changes in the value of $\hat{\tau}^2$; this can be used to provide insight into whether this method is required, or whether the simpler standard random effects analysis using a moment estimator of the between-study variance is adequate. A full explanation of this plot is provided in Section 9.3.1. This method still assumes that the individual study variances are known, when in practice they too must be estimated. If a full likelihood method were pursued, the confidence intervals for the overall treatment effect would be expected to be even wider. Except when all the trials are small, the additional uncertainty would not be expected to have a great impact on the results, and so pursuing a full likelihood approach is unnecessarily sophisticated for most practical purposes [10]. However, this approach in the case of binomial data (which includes the conditional distribution of each 2×2 frequency table given its margins), is discussed below [13].

Biggerstaff and Tweedie [11] address the same problem by developing a variance estimator for Q, that leads to an interval estimation of $\hat{\tau}^2$, utilizing an approximating distribution for Q. They also develop asymptotic likelihood methods for the same estimate. This information is then used to give a new method of calculating the weight given to the individual studies, which takes into account variation in these point estimates of $\hat{\tau}^2$. These new weights are between the standard fixed and random effects, in terms of down-weighting the results of large studies and up-weighting those of small. (A past concern has been that when $\hat{\tau}^2$ is large, the standard random effects model gives too much weight to the relatively small studies.) These new weights will differ most from those of the standard random effects model when the number of studies to be combined is small; the results being similar to the standard random effects model when 20 or more studies are to be combined. A comparison of interval estimates for $\hat{\tau}^2$, using these methods and several alternative ones, through simulation studies has been carried out [12].

It should be noted that a fully Bayesian approach would automatically take into account this source of extra uncertainty (see Chapter 11) [14].

Example: Meta-analysis including the uncertainty induced by estimating the between study variance

Recall the dataset comparing clozapine to 'typical' drugs for schizophrenia first described in Section 2.3.2. The heterogeneity statistic ($Q = 27.1$ (10 df)) is highly significant ($p = 0.003$) for this dataset. A random effects analysis, using the standardized mean difference outcome measure, produces a pooled estimate of -0.46 (95% CI -0.70 to -0.21). The between study variance parameter estimate ($\hat{\tau}^2$) is 0.098. The fixed effect estimate is $-0.50(-0.64$ to $-0.36)$. Hence, not unexpectedly, the confidence interval for the random effect analysis is wider, taking into account the estimated between study heterogeneity. Because the between study heterogeneity is considerable, and there are relatively few trials in the meta-analysis [11], it is appropriate to explore the influence of incorporating the extra uncertainty associated with estimating $\hat{\tau}^2$ from the data.

Both the profile likelihood method of Hardy and Thompson [10] and the moment based estimator of Biggerstaff and Tweedie [11] are applied to the dataset. As with many of the examples from this point on in the book, the calculations required are too involved to outline in the text, but details of software used for each example, and where it can be obtained are provided in Appendix I. The profile likelihood approach produces a pooled estimate of -0.46 (-0.70 to -0.22) for the treatment effect. The estimate of $\hat{\tau}^2$ is 0.09 (0.03 to 0.30); notice that a confidence interval is now given because the value is no longer assumed known, but is estimated. The moment-based estimator produced similar, but not identical results: the pooled standardized mean difference is -0.47 (-0.72 to -0.22), and the estimate for $\hat{\tau}^2$ is 0.10 (0.01 to 0.51). Hence, the point estimate for the treatment effect from both approaches is around the same value as for the standard random effects model, but the respective confidence interval is fractionally wider. If fewer studies were combined, the impact of accounting for the uncertainty associated with the between study variance could be considerably greater.

Finally, it is instructive to examine the weightings assigned to each study using different analyses. Table 5.2 displays these for the fixed, random and 'random incorporating the uncertainty in estimating the between study variance' approaches. We have already seen that using a random effect model gives more equal weighting to each study compared to the fixed effect model. When the random effect model which accounts for the uncertainty in estimating the between study variance is used, the weighting is somewhere between the fixed and the random effect approach – here it is closer to the random effect weights.

5.6.2 Exact approach to random effects meta-analysis of binary data

Van Houwelingen developed a likelihood based approach to random effects, for binary data [13], which avoids use of approximating normal distributions,

Table 5.2 Absolute and percentage weightings given to each study using a fixed effect, standard random effect and moment based method of incorporating uncertainty in estimating the between study variance.

Fixed		Random		Random including uncertainty in est. $\hat{\tau}^2$	
Absolute	(%)	Absolute	(%)	Absolute	(%)
9.10	(4.4)	4.81	(7.5)	5.39	(6.5)
18.26	(8.9)	6.55	(10.2)	8.26	(9.9)
6.81	(3.3)	4.08	(6.4)	4.43	(5.3)
20.30	(9.8)	6.80	(10.6)	8.77	(10.5)
21.73	(10.5)	6.95	(10.9)	9.11	(10.9)
9.05	(4.4)	4.80	(7.5)	5.37	(6.4)
5.82	(2.8)	3.71	(5.8)	3.96	(4.7)
30.83	(15.0)	7.67	(12.0)	10.97	(13.1)
60.67	(29.4)	8.74	(13.7)	15.38	(18.4)
4.87	(2.4)	3.30	(5.2)	3.47	(4.2)
18.70	(9.1)	6.61	(10.3)	8.38	(10.0)

an assumption which is rarely checked in meta-analyses. Solutions are obtained via the EM algorithm [7]. An extension is given to a bivariate random effects model, in which the effects in both groups are supposed random. In this way, inference can be made about the relationship between improvement and baseline effect. This is a non-parametric procedure, and has been recommended [10] when the normality assumption is violated.

*5.6.3 Miscellaneous extensions to the random effects model

Several other extensions and alternatives for the random effects model have been advocated:

1. The use of sample survey methods have been suggested [15]; this approach differs from the random-effects model by not involving an explicit estimate of the subject or study variance.
2. The idea of using trimmed versions of meta-analytic estimators for the risk difference has been investigated [16].
3. When the number of studies being combined is small, it is difficult to assess whether the normality assumption has been violated. In this situation, a robust estimation procedure has been developed which allows the analyst to assume that random effects are t rather than normally distributed [17].

4. The standard random effects model inversely weights studies by the sum of the between-study variance and the conditional within study variance. It has been argued that because these weights are not independent of the risk differences, the procedure sometimes exhibits bias and unnatural behaviour [18] (though it should be stressed that this is only a problem in a limited number of cases – see multiple effect sizes in Chapter 15). To avoid this, a modified weighting scheme uses unconditional within-study variance to avoid this source of bias by removing the correlation between the risk differences and their weights [18].

Other extensions of the methodology presented in this section are given in other sections of this book, where they fit more naturally. For example, random effect dose response models are discussed in Chapter 16 under observational studies.

Finally, it may be noted that a unified parametric approach for meta-analysis methodology has been developed [19], which allows random effect models and fixed effects models to be conceived as variants of a single model.

5.7   Comparison of random with fixed effect models

The argument over which model is theoretically and/or practically superior has been running for many years, with many comments scattered through the literature. Random effects models have been criticized on grounds that unrealistic/unjustified distributional assumptions have to be made [20]. However, it has also been argued that they are consistent with the standard specific aims of generalization [21]. A further consideration is that random effects models are more sensitive to publication bias, because of the greater relative weight given to smaller studies [22]. Perhaps it is wise to conclude that neither fixed nor random effect analyses can be considered ideal [23]. Investigations into the differences between results obtained from the two methods have been made [24, 25] For example, Berlin *et al.* [24] compared the results of 22 meta-analyses; in three, different conclusions would have been drawn about the treatment effect [26], the Peto fixed effect method suggesting a beneficial treatment effect while the random effect method did not.

It is worth noting that Peto [20], one of the strongest opponents of random effect models, advocates a stringent criterion of significance (i.e. he suggests a critical value of $p < 0.01$, instead of the usual $p < 0.05$, to decide whether a treatment effect is statistically significant) for a fixed-effect estimate. This more conservative approach has the effect of reducing the differences between fixed and random effect models.

The question 'When should random effects models (opposed to fixed effect models) be used?' has no simple answer. However, many believe that if there is evidence of heterogeneity, which cannot be explained, this extra variation needs to be accounted for when estimating the pooled estimate and confidence interval [2], and that in such situations the fixed effect methods will be insufficiently conservative. It is recognized that the heterogeneity test lacks power, so that studies may be regarded as homogeneous when in fact there is a degree of heterogeneity. This implies a random effects model may still be worth considering as it cannot be assumed that true homogeneity exists [1].

Other models for meta-analysis do exist, which may be more appropriate than fixed or random effect models. Indeed, the random and fixed effects models can be considered special cases within a hierarchical model framework, of which other models (such as mixed models and cross-design synthesis (see Chapters 6 and 17, respectively)) are simply extensions. Additionally, the option of using an alternative Bayesian formulation is available; here one does not necessarily have to choose between the two models (fixed and random), but rather we can average across models using Bayes factors (see Chapter 11).

5.8 Summary/Discussion

This section has described a more sophisticated alternative to the fixed effect model for combining studies which allows for between study heterogeneity. Not allowing for this heterogeneity when it exists can result in over confident estimates, and hence the random effects model is recommended for routine use. Extensions to this method which take into account further uncertainty related to the fact that the estimate of between study variation is estimated from the data have been described. These methods are recommended when small heterogeneous sets of studies are being combined. Although several different estimation methods exist, a Weighted Least Squares (WLS) approach has become the commonly used method, and is often referred to as the method of DerSimonian and Laird. Further extensions to random effect modelling are described in the following chapters.

References

1. Thompson, S.G., Pocock, S.J. (1991). Can meta-analyses be trusted? *Lancet* **338**: 1127–30.

2. DerSimonian, R., Laird N. (1986). Meta-analysis in clinical trials. *Controlled Clin Trials* **7**: 177–88.

3. Shadish, W.R., Haddock, C.K. (1994). Combining estimates of effect size. In: Cooper, H., Hedges, L.V., editors. *The Handbook of Research Synthesis*. New York: Russell Sage Foundation; 261–84.

4. Fleiss, J.L. (1993). The statistical basis of meta-analysis. *Stat. Methods Med. Res.* **2**: 121–45.

5. Hasselblad, V.I.C., McCrory, D.C. (1995). Meta-analytic tools for medical decision making: A practical guide. *Med. Decis Making* **15**: 81–96.

6. Hedges, L. (1981). Distribution theory for Glass's estimator of effect size and related estimators. *J. Educ. Stat.* **6**: 107–28.

7. Dempster, A.P., Laird, N.M., Rubin, D.B. (1977). Maximum likelihood from incomplete data via the EM algorithm. *J. R. Stat. Soc. B* **39**: 1–38.

8. Rao, P.S., Kaplan, J., Cochran, W.G. (1981). Estimators for the one-way random effects model with unequal variances. *J. Am. Stat. Assoc.* **76**: 89–97.

9. Harville, D.A. (1977). Maximum likelihood approaches to variance component estimation and to related problems. *J. Am. Stat. Assoc.* **72**: 320–38.

10. Hardy, R.J., Thompson, S.G. (1996). A likelihood approach to meta-analysis with random effects. *Stat. Med.* **15**: 619–29.

11. Biggerstaff, B.J., Tweedie, R.L. (1997). Incorporating variability in estimates of heterogeneity in the random effects model in meta-analysis. *Stat. Med.* **16**: 753–68.

12. Biggerstaff, B.J. (1997). Confidence intervals in the one-way random effects model for meta-analytic applications. Technical Report, Department of Statistics, Colorado State University.

13. Van Houwelingen, H.C., Zwinderman, K.H., Stijnen, T. (1993). A bivariate approach to meta-analysis. *Stat. Med.* **12**: 2273–84.

14. Louis, T.A., Zelterman, D. (1994). Bayesian approaches to research synthesis. In: Cooper, H., Hedges, L.V., editors. *The Handbook of Research Synthesis*. New York: Russell Sage Foundation; 411–22.

15. Schmid, J.E., Koch, G.G., LaVange, L.M. (1991). An overview of statistical issues and methods of meta-analysis. *J. Biopharm Stat.* **1**: 103–20.

16. Emerson, J.D., Hoaglin, D.C., Mosteller, F. (1996). Simple robust procedures for combining risk differences in sets of 2×2 tables. *Stat. Med.* **15**: 1465–88.

17. Seltzer, M. (1991). The use of data augmentation in fitting hierarchical models to education data. Unpublished doctorial dissertation, University of Chicago.

18. Emerson, J.D., Hoaglin, D.C., Mosteller, F. (1993). A modified random-effect procedure for combining risk difference in sets of 2×2 tables from clinical trials. *J. Italian Stat. Soc.* **2**: 269–90.

19. Whitehead, A., Whitehead, J. (1991). A general parametric approach to the meta-analysis of randomised clinical trials. *Stat. Med.* **10**: 1665–77.

20. Peto R. (1987). Why do we need systematic overviews of randomised trials? *Stat. Med.* **6**: 233–40.

21. Raudenbush, S.W. (1994). Random effects models. In Cooper, H., Hedges, L.V., (editors). *The Handbook of Research Synthesis*. New York: Russell Sage Foundation. 301–22.

22. Greenland, S. (1994). Invited commentary: a critical look at some popular meta-analytic methods. *Am. J. Epidemiol.* **140**: 290–6.

23. Thompson, S.G. (1993). Controversies in meta-analysis: the case of the trials of serum cholesterol reduction. *Stat. Methods Med. Res.* **2**: 173–92.

24. Berlin, J.A., Laird, N.M., Sacks, H.S., Chalmers, T.C. (1989). A comparison of statistical methods for combining event rates from clinical trials. *Stat. Med.* **8**: 141–51.

25. Mengersen, K.L., Tweedie, R.L., Biggerstaff, B.J. (1995). The impact of method choice in meta-analysis. *Aust. J. Stats.* **37**: 19–44.

26. Dickersin, K., Berlin, J.A. (1992). Meta-analysis: state-of-the-science. *Epidemiol. Rev.* **14**: 154–76.

CHAPTER 6

Exploring Between Study Heterogeneity

6.1 Introduction

Random effect models account for heterogeneity between studies, but they do not provide a method of exploring and potentially of explaining the reasons study results vary. This chapter describes methods which can be used for this purpose. Investigating why study results vary systematically may lead to the identification of associations between study or patient characteristics and the outcome measure, which would not have been possible in single studies. For meta-analyses of RCTs, this in turn may lead to clinically important findings, and may eventually assist in individualizing treatment regimes [1]. Two related approaches are described; first, subgroup analyses are considered; this is followed by a description of the use of regression methods in meta-analysis.

Both study and patient characteristics can be explored using these methods. It should be stressed, however, that this type of analysis should be treated as exploratory as associations between study characteristics, and their outcomes can occur purely by chance, or due to the presence of confounding factors. Further, regression analysis of this type is also susceptible to aggregation bias, which occurs if the relation between patient characteristic study means and outcomes do not directly reflect the relation between individuals' values and individuals' outcomes [2]. A further restriction is that the data available for analysis from original study reports may be limited.

Very importantly, subgroup and regression methods can be employed in a sensitivity analysis; the sensitivity of inferences to variations in or violations of certain assumptions can be investigated (see Chapter 10) [3, 4].

6.2   Subgroup analyses

Two different types of subgroup analysis are theoretically possible in a meta-analysis. One can investigate subsets defined by study or patient characteristics. Studies being pooled may differ with respect to many factors including; treatments applied, control groups, patient eligibility criteria, quality control, study conduct, and follow-up maturity (see Section 3.4) [1].

Alternatively, one can consider subsets of patients within the studies being pooled. In an RCT setting, this type of subgroup analysis can be described as the investigation of the influence of factors other than treatment factors on the response variables or treatment effects – a search for evidence of effect modifications [5]. This procedure is commonly carried out on single RCTs, but it is also sometimes possible in a meta-analysis setting. Due to limitations on the data reported in study reports, further data (possibly individual patient data) will often be required from the original study investigators. It is important to have clear definitions of the subgroups [6], and be aware of the possible presence of misclassification, dilution, and other biases [1]. This type of analysis is not considered further here, but it is relatively straightforward to carry out, data availability permitting.

Due to increased amounts of data being included in a meta-analysis compared to individual RCTs, meta-analysis may have the statistical power to detect the likely existence of differential subgroup effects not detectable by individual trials. Alternatively, pooled data from several trials may refute the claim of a subgroup effect from a single trial [6].

When conducting subgroup analyses within a meta-analysis it has been stressed [6] that clear definitions of the subgroups are essential. Counsell *et al.* [7] carried out an experiment to determine whether inappropriate subgroup analysis, together with chance, could change the conclusion of a systematic review of several randomized trials of an ineffective treatment. Trials were simulated by throwing fair dice and recording outcomes as either death or survival. Publication bias (Chapter 7) was also simulated. The results showed that analysis of subsets can be misleading, with chance influencing the outcomes of clinical trials and systematic reviews of trials much more than many investigators realise. However, the larger the number of studies, the less this is a problem, but it is difficult to ascertain how large. Oxman *et al.* [8] state that the extent to which a clinician should believe and act on the results of subgroup analyses of data from randomized trials or meta-analyses is controversial. Their paper provides guidelines for making these decisions.

It may be helpful to note that combining results from subsets of studies defined by study/patient characteristics is analogous to including an indicator covariate in a fixed effect regression model (see later) which defines the subgroups. Regression models are necessary, however, if the effect of a factor measured on a continuous scale, such as average age of patients in the study, is to be investigated; or on practical grounds, if the effect of two or more characteristics are to be investigated simultaneously.

6.2.1 Example: Stratification by study characteristics

We return to the meta-analysis of cholesterol lowering interventions and their effect on total mortality, first described in Section 2.2. These trials differed considerably with respect to the actual intervention administered to the treatment groups. Treatments can be classified into three broad groups: drugs, diets and surgery. (Since different drugs and different types of diets were administered, these subgroups could be divided further, but this is not pursued in the analysis which follows.) Column two of Table 6.1 indicates which intervention was given in each of the 34 trials. Since only two trials administered surgical interventions, only pooling of the drug and diet trials is reported. As there is still considerable heterogeneity between studies within subgroups, random effect models were used. Forrest plots for the dietary and drug intervention subgroups are presented in Figures 6.1 and 6.2, respectively. The pooled odds ratios for the drug and diet trials, respectively, are 0.89 (0.73 to 1.07), and 0.93 (0.76 to 1.13). These can be compared with the odds ratio estimate produced by combining all the studies (see Section 5.5), which was 0.89 (0.78 to 1.02). Both subgroup estimates are very close to the overall pooled result, and there appears to be no major differences in effectiveness between diet and drug treatments for cholesterol lowering. Additionally, since the heterogeneity statistic was still statistically significant in both subgroups, stratifying by a broad definition of intervention type used does not appear to explain the majority of the heterogeneity between trials.

6.2.2 Example: Stratification by patient characteristics

Still focusing on the cholesterol dataset, another way in which trials differed was in the type of patient enrolled into the studies. In some studies, the majority of patients were healthy when they were randomized and hence cholesterol lowering was administered as a primary prevention measure. In other studies, the majority of patients had already had a myocardial infarction, and hence cholesterol lowering was administered as a secondary prevention measure. A subgroup analysis is carried out to check if the effectiveness of cholesterol lowering differs depending on whether it is used as a primary or secondary prevention measure. (Note: Trials 5 and 32 were carried out using very specific groups of patients, and have been excluded from this analysis.) Column 3 of Table 6.1 indicates which trials implemented cholesterol lowering as a primary and secondary intervention. Forrest plots for the primary and secondary intervention subgroups are presented in Figures 6.3 and 6.4, respectively. The meta-analyses have been carried out using a random effects model. The pooled odds ratios for the primary and secondary intervention trials, respectively, are 1.06 (0.92 to 1.23) and 0.82 (0.71 to 0.95). Hence, it would appear that cholesterol lowering is effective as a secondary prevention measure, but not so clearly as a

Table 6.1 Characteristics of 34 cholesterol reducing RCTs.

Study	Intervention type	Predominant patient group
1	Diet	Secondary
2	Drug	Secondary
3	Drug	Secondary
4	Drug	Secondary
5	Drug	Specific
6	Drug	Secondary
7	Diet	Secondary
8	Diet	Secondary
9	Diet	Secondary
10	Drug	Secondary
11	Drug	Secondary
12	Drug	Secondary
13	Drug	Secondary
14	Surgery	Secondary
15	Drug	Secondary
16	Diet	Primary
17	Diet	Secondary
18	Drug	Secondary
19	Drug	Secondary
20	Drug	Primary
21	Diet	Secondary
22	Diet	Secondary
23	Drug	Secondary
24	Diet	Primary
25	Surgery	Secondary
26	Drug	Secondary
27	Drug	Secondary
28	Drug	Primary
29	Drug	Primary
30	Drug	Primary
31	Drug	Primary
32	Drug	Specific
33	Drug	Secondary
34	Drug	Secondary

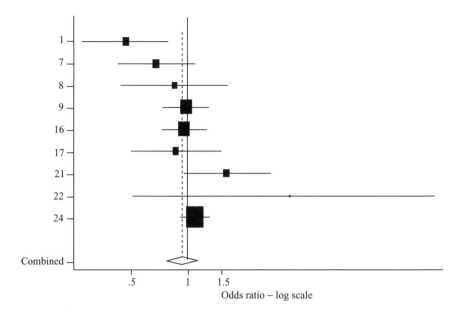

Figure 6.1   Subgroup meta-analysis combining nine dietary intervention cholesterol lowering trials.

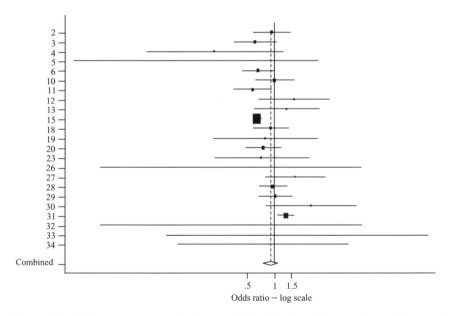

Figure 6.2   Subgroup meta-analysis combining 23 drug intervention cholesterol lowering trials.

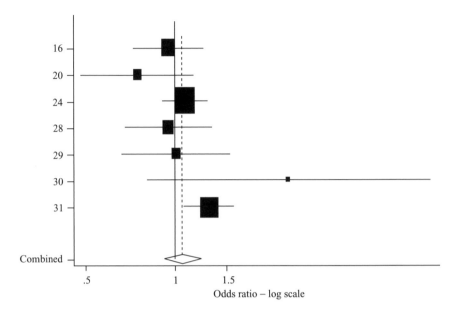

Figure 6.3 Forest plot of RCTs implementing cholesterol lowering as a primary intervention measure.

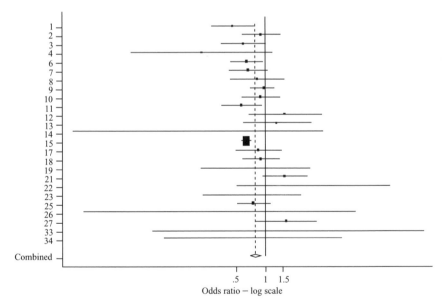

Figure 6.4 Forest plot of RCTs implementing cholesterol lowering as a secondary intervention measure.

primary prevention measure. Although some of the between study heterogeneity is explained by subdividing the trials in this way, the test for heterogeneity is still significant in the secondary studies, suggesting this is not the only factor causing variation in study results.

It should be noted that dichotomizing the trials into those administering interventions in primary and secondary situations can be seen as an (inexact) proxy subgrouping for those trials including patients at high and low baseline risks. The subgrouping is inexact because the original trialists used different definitions of primary and secondary care. In fact, some studies described as primary had higher rates of CHD mortality in the control groups than some of the secondary studies [9]. A more exact assessment of baseline risk is considered in Section 6.3.5.

6.3   Regression models for meta-analysis

Two types of regression models are possible: one is an extension of the fixed effect model, commonly known as a *meta-regression model*; and the other an extension of the random effects model, called a *mixed model* (because it includes both fixed and random terms). The fixed-effect methods are most appropriate when all variation (above that explainable by sampling error) between study outcomes can be considered accountable by the covariates included. A mixed-model is more suitable when the predictive covariates explain only part of the variation/heterogeneity, and a random effect term is used to account for the remainder. However, as with fixed and random models themselves, it has been argued that one should always include a random effect term as there will always be some degree of between study heterogeneity not captured by the covariates. It should be noted that regression models are most useful when the number of studies is large, and cannot be sensibly attempted when very small numbers of studies are being combined [10].

Modelling using regression models is not a trivial task. It is beyond the scope of this chapter to give a comprehensive introduction to regression techniques. For the reader who wants to know more about regression modelling, many introductory statistical texts cover the basics. See Collett [11] for further details on modelling binary outcomes.

6.3.1   Meta-regression model (fixed-effects regression)

A general meta-regression model, which can include continuous and discrete predictor variables, is outlined, following Hedges' account [12]. If only a limited number of categorical predictor variables are being investigated, an alternative analysis of variance (ANOVA) approach can be taken. This can be regarded as a special case of the more general model below [12].

Suppose, as before, that there are k independent effect size estimates $T_1, \ldots, T_k$, with estimated sampling variances $v_1, \ldots, v_k$. The corresponding underlying effect size parameters are $\theta_1, \ldots, \theta_k$, for each of the k studies. Suppose also that there are p known predictor variables $X_1, \ldots, X_p$ which are believed to be related to the effects via a linear model of the form

$$\theta_i = \beta_0 + \beta_1 x_{i1} + \ldots + \beta_p x_{ip}, \tag{6.1}$$

where $x_{i1}, \ldots, x_{ip}$ are the values of the predictor variables $X_1, \ldots, X_p$ for the ith study, and $\beta_0, \beta_1, \ldots, \beta_p$ are the unknown regression coefficients, to be estimated.

Recall that in the fixed effects model, $\theta_1, \ldots, \theta_k$, were all set equal, say to θ. Here they are allowed to vary (as in the random effects analysis). However, unlike the random effects model, here it is the covariate predictor variables that are responsible for the variation not a random effect; the variation is predictable not random. This model assumes approximate normality of the dependent (outcome) variable [3]. In situations when the outcomes are in the form of 2×2 tables, simulation studies indicate that such a criterion will be adequately met if the expectations of the counts contributing to the rates or ratios is four or greater [3].

The coefficients in the above model are easily calculated via weighted least squares algorithms, with the weights defined by the reciprocal of the sampling variances (i.e. weight for $T_i : w_i = 1/v_i$). Any standard statistical package that performs weighted (multiple) regression can be used.

It is important to note that the standard errors of the estimates for the coefficients, produced by standard software packages are based on a slightly different model than the above used for fixed effect meta-regression; thus, an adjustment needs to be calculated by hand:

$$S_j = SE_j / \sqrt{MS_{ERROR}}, \tag{6.2}$$

where S_j is the corrected standard error, SE_j is the standard error of coefficient b_j (the obtained estimate for β_j) as given by the computer program, and MS_{ERROR} is the 'error' or 'residual' mean square from the analysis of variance for the regression as given by the computer program [12, 13].

Approximate confidence intervals for each of the regression coefficient estimates (the b_js) are given by:

$$b_j - Z_{\alpha/2}(S_j) \leq \beta_j \leq b_j + Z_{\alpha/2}(S_j), \tag{6.3}$$

where $Z_{\alpha/2}$ is the two-tailed critical value of the standard normal distribution. The corresponding two sided significance test is $H_0 : \beta_j = 0$, and is rejected if the above confidence interval does not contain zero. If it is not rejected, one concludes there is no, or insufficient, evidence of a relationship between the jth predictor variable and outcome.

As with all modelling exercises, testing of assumptions and considering the adequacy of model fit is an important aspect of the analysis that should not be overlooked [14].

6.3.2 Meta-regression example: A meta-analysis of Bacillus Calmette-Guérin (BCG) vaccine for the prevention of tuberculosis (TB)

A meta-regression model is is illustrated in an analysis of RCTs comparing Bacillus Calmette-Guérin (BCG) vaccine for the prevention of tuberculosis (TB) against a non-vaccinated group, originally reported by Colditz *et al.* [15], with further details of the methodology discussed in Berkey *et al.* [16]. The data from the 13 RCTs included is reproduced in Table 6.2.

Note, in this example, that the (log) Relative Risk (RR) (calculated using equation (2.4)), rather than the odds ratio is used as the outcome measure. One advantage of using this outcome is that the protective effect of the vaccine can be calculated directly as $1 - RR$. Considering the standard test for heterogeneity described in Section 3.2.1, the test statistic Q for this dataset is 152.2, which is highly significant ($p < 0.001$), suggesting considerable heterogeneity between studies. A standard random effects model, including no covariates, produces a pooled RR of 0.49 (0.35 to 0.70), with the between study variation ($\hat{\tau}^2$) estimated at 0.31.

Table 6.2 Data from clinical trials of BCG vaccine efficacy (reproduced in a modified form from Berkey *et al.* [16]).

Trial	Latitude (degrees from the equator)	Vaccinated		Not vaccinated		RR (95% CI)
		Disease	No disease	Disease	No disease	
1	44	4	119	11	128	0.41 (0.13 to 1.26)
2	55	6	300	29	274	0.20 (0.09 to 0.49)
3	42	3	228	11	209	0.26 (0.07 to 0.92)
4	52	62	13 536	248	12 619	0.24 (0.18 to 0.31)
5	13	33	5 036	47	5 761	0.80 (0.52 to 1.25)
6	44	180	1 361	372	1 079	0.46 (0.39 to 0.54)
7	19	8	2 537	10	619	0.20 (0.08 to 0.50)
8	13	505	87 886	499	87 892	1.01 (0.89 to 1.14)
9	−27	29	7 470	45	7 232	0.63 (0.39 to 1.00)
10	42	17	1 699	65	1 600	0.25 (0.15 to 0.43)
11	18	186	50 448	141	27 197	0.71 (0.57 to 0.89)
12	33	5	2 493	3	2 338	1.56 (0.37 to 6.53)
13	33	27	16 886	29	17 825	0.98 (0.58 to 1.66)

It was suspected that the distance from the equator affected the efficacy of the vaccine, and hence the effect of such a covariate is investigated in a meta-regression model. The covariate latitude is centred by subtracting the mean lattitude (33.4615) from the latitude value of each study. (Note, study 9 was carried out on the opposite side of the equator, but as it is distance from the equator which we are interested in the negative sign is dropped for the analysis.) The weights required for each study, the inverse of the variance of the $\ln(RR)$, are calculated using equation (2.5). The three columns of data required for fitting the model of the form of equation (6.1) are presented in Table 6.3.

Fitting a weighted (linear) regression model to this data in a standard statistics package (see Appendix A) produces the following model:

$$\log(RR) = -0.635 - 0.029(x - 33.46),$$

where x is the distance from the equator in degrees latitude. Hence, as the distance from the equator increases the $\log(RR)$ decreases, corresponding to greater vaccine efficacy. Figure 6.5 plots the individual trials and the regression line on the same axes; the size of each plotting circle indicating the relative size of each study. The standard errors given by the package for the intercept and slope parameters are 0.074 and 0.0044, respectively, but these need adjusting, as noted previously, using equation (6.2). The residual mean square error for the model is 2.79; dividing both standard errors by the square root of this (1.67) produces lower values of 0.045 and 0.0027 for intercept and slope, respectively. Confidence intervals for both parameters can now be calculated using equation (6.3). Hence, the intercept estimate is

Table 6.3 Data required for meta-regression analysis on BCG vaccine trials.

Trial	$\ln(RR)$	Weight	Latitude (centred)
1	−0.89	3.07	10.54
2	−1.59	5.14	21.54
3	−1.35	2.41	8.54
4	−1.44	49.97	18.54
5	−0.22	19.53	−20.46
6	−0.79	144.81	10.54
7	−1.62	4.48	−14.46
8	0.01	252.42	−20.46
9	−0.47	17.72	−6.46
10	−1.37	13.69	8.54
11	−0.34	80.57	−15.46
12	0.45	1.88	−0.46
13	−0.02	14.00	−0.46

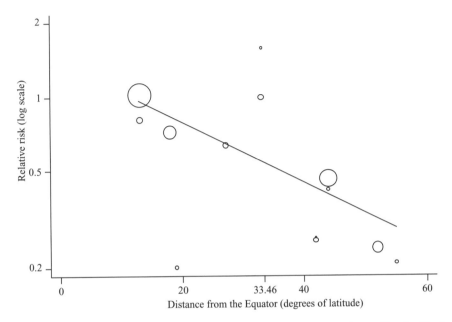

Figure 6.5 Regression line fitting distance from the equator, using a (fixed effect) meta-regression model.

−0.635(−0.722 to −0.547), and slope estimate is −0.029 (−0.034 to −0.024). The RR for the average distance from the equator observed in the trials can be calculated by simply taking the exponential of the intercept estimate, which is 0.530(0.486 to 0.578) (this is an advantage of centring the coefficient used in the model).

6.3.3 Mixed effect model (random-effects regression)

The derivation below is modified from Raudenbush [10]. As a starting point take the random effects model outlined previously (5.1): $T_i = \theta_i + e_i$, where T_i is the estimated effect size of the true effect size θ_i for each of the k studies, $i = 1, \ldots, k$. We also assume the e_i are statistically independent, each with a mean of zero and estimation variance v_i. The variance for these estimates of treatment effect, as before (5.2) can be expressed as: $\mathrm{Var}(T_i) = \tau^2\theta + v_i$, where $\tau^2\theta$ is the between study, or random effects variance and v_i is the within study variance. Now we extend this to formulate a model for the true effects depending on a set of study characteristics plus error:

$$\theta_i = \beta_0 + \beta_1 X_{i1} + \beta_2 X_{i2} + \ldots + \beta_p X_{ip} + u_i, \tag{6.4}$$

where β_0 is the model intercept, $X_{i1}, , X_{ip}$ are coded characteristics of studies hypothesised to predict the study effect size; $\beta_1, \ldots, \beta_p$ are regression coefficients capturing the association between study characteristics and effect sizes; u_i is the random effect of study i, i.e. the deviation of the true effect in study i from the value predicted on the basis of the model. Each random effect u_i is assumed independent, with a mean of zero and variance σ_i^2.

Under the fixed effects specification, the study characteristics $X_{i1}, \ldots, X_{ip}$ are presumed to account completely for variation in the true effect sizes. In contrast, the random effects specification assumes that part of the variability in these true effects is unexplainable by the model. This model is a consistent extension of the models presented previously. If the model has no predictors, i.e. $\beta_1 \ldots = \beta_p = 0$, then it reduces to that of the random effects model (Section 5.1). If the random effects variance is null, i.e. $\tau^2 \theta = 0$, then the results will be identical to that of the fixed effects meta-regression model (equation (6.1)).

Estimating the parameters

Substituting (5.1) into (6.4) gives:

$$T_i = \beta_0 + \beta_1 X_{i1} + \beta_2 X_{i2} + \ldots + \beta_p X_{ip} + u_i + e_i. \tag{6.5}$$

This equation has two components in its error term $u_i + e_i$, so the variance of T_i, controlling for the Xs, is

$$v_i^* = \mathrm{var}(u_i + e_i) = \tau_\theta^2 + v_i. \tag{6.6}$$

A weighted least squares approach is used; optimal weights are given as before by the inverse of each study's variance:

$$w_i^* = 1/v_i^* = 1/\left(\tau^2 \theta + v_i\right). \tag{6.7}$$

We can estimate the v_is from the data. We also need an estimate of τ_θ^2, which is generally unknown and must be estimated from the data. In fact, an estimate of the regression coefficients (the βs) is required to obtain estimates of τ_θ^2 and hence w_i^*. So, unfortunately, estimation of the βs is dependent upon knowing τ_θ^2, and estimation of τ_θ^2 depends upon knowing the β's. Several different approaches to solving this problem have been proposed.

First, the method of moments can be used in an iterative scheme. To do this the provisional estimates of the βs in equation (6.5) are computed (from ordinary or weighted regression). Then, based on these estimates, an estimate τ_θ^2 can be obtained and, therefore, the weights w_i calculated. These weights can then be used in a weighted least squares regression to obtain new estimates of the βs. Alternatively, the method of maximum likelihood can be used to estimate both simultaneously. This requires a further assumption that each T_i is normally distributed (see Section 6.4.1) [17].

6.3.4   Mixed model example: a re-analysis of Bacillus Calmette-Guérin (BCG) vaccine for the prevention of tuberculosis (TB) trials

In Section 6.3.2, 13 BCG vaccine trials were analysed using a (fixed effects) meta-regression model. In this section, a closer examination and re-analysis of these trials is carried out. In Section 6.3.2, no consideration was given to heterogeneity, beyond that attributable to the covariate distance from the equator. Here, a random effect term is included, producing a model of the form of equation (6.4). This produces the following regression model:

$$\log(RR) = -0.718 - 0.029(x - 33.46).$$

The point estimates of the estimated parameters are very similar to those produced by the fixed effects regression model in Section 6.3.2, but their standard errors are considerably larger at 0.101 and 0.0067 (compared to 0.074 and 0.0044) for intercept and slope, respectively. This directly translates into wider confidence intervals for the study parameters (the interval estimate for the intercept is (-0.917 to -0.520), and the slope is (-0.042 to -0.016)). The between study variance estimate for this model is 0.063, which gives an indication of the residual variation between studies having corrected for distance from the equator. This value is much lower than that for the random effects model including no covariates (0.31), but still suggests some heterogeneity is present beyond that explained by the covariate. This is reflected in the width of confidence intervals of the parameters. In instances like this, if the fixed effect regression results are reported, then it will produce overly precise estimates in the results [16]. Therefore, when any residue heterogeneity is present, above that explained by covariates, a mixed model should be used.

6.3.5   Mixed modelling extensions

*Alternative methods for obtaining mixed model solutions

Several alternative solutions to the mixed effect model of Section 6.3.3 have been proposed. These are outlined below.

Berkey et al. [16] derive an iterative random effects regression model, which uses an iterative scheme which alternates between estimating the regression coefficients via weighted least squares, where the weights incorporate the current estimate of the between study variance, and estimating the between-study variance, in a similar manner to the solution of Section 6.3.3. However, the authors found that small biases were present in the estimates of the regression coefficients and the between study variance, and that there is the potential to eliminate these using an alternative estimator. This can be considered as an empirical Bayes estimate (see Section 12.5); see Berkey et al. [16] for the formulae.

Although not straightforward, it is possible to derive a moment estimator for τ^2 – the between study variance in a mixed model when only one covariate is included. This is a direct extension of the DerSimonian and Laird weighted least squares random effect model outlined in Section 5.2. Thompson and Sharp provide equations [13]; they also discuss further ML and REML estimates [13].

Alternative formulation of the fixed and random effect regression models

The previous sections outline the most commonly used fixed and random effect regression models. Several alternative formulations of regression models have been proposed.

Probably the most significant alternative to the regression models proposed in the previous sections is the logistic, as opposed to linear, weighted regression for modelling log odds ratio outcomes. Logistic regression has several theoretical advantages over weighted regression. First, the assumption that the log odds ratios are normally distributed, which may not hold especially for studies with small numbers of events, is not required. Additionally, when there are 0 cells in studies of 2×2 tables, there is no need to use a continuity correction factor (such as adding 0.5 to every cell).

A fixed effect logistic regression model is simple to implement, and has been used by several authors [13, 18–20]. Recently, due to advances in computational power and software, it has become possible to implement random effect logistic regression models [13] which include a between study variation term, much like the mixed model of Section 6.3.

In Sections 6.3.1 and 6.3.3, it was assumed that model parameters were normally distributed when calculating confidence intervals; however, use of the t-distribution has been suggested for small samples [13].

A comparison of the results produced by many of the different model variations and solution methods discussed in this and the previous section have been performed [13]. Whilst differences in the results using these methods do exist, in many situations, these will be small.

*Generalization and extensions to fixed/random effect regression models

Stram [22] presents a very general mixed-effects regression model framework. He developed a model from which most other models used in meta-analysis can be viewed as special cases. Explicitly, this model builds on the standard random effects model [23], the mixed model, the model of Begg and Polite [24] – to incorporate single arm studies with two-arm studies (see Section 18.3), and the model of Tori et al. [25] for combining surrogate endpoints (see Section 15.10). The general form of the model is

$$\mathbf{Y}_i = \mathbf{X}_i\alpha + \mathbf{Z}_i\beta_i + \zeta_i + e_i, \qquad (6.8)$$

where there are $i = 1, 2, \ldots, K$ independent studies. Y_i is an $(n_i \times 1)$ vector of one or more related estimates of treatments or treatment comparisons of interest; X_i is an $(n_i \times p)$ matrix of known covariates related to the p vector of unknown fixed effect parameters, α; and Z_i is an $(n_i \times q)$ vector of known covariates related to a $(q \times 1)$ vector of unobserved random effects, β_i, for each study. The two remaining $n_i \times 1$ unobserved random vectors, ζ_i and e_i, specify two types of error in Y_i. The ζ_i specify the sampling errors in Y_i, and e_i specifies other sources of error or heterogeneity between studies and between arms of the same study. In this model, it is assumed that β_i, u_i and e_i are each independent multivariate normal random vectors. One of the new extensions offered by this model is the possibility for random effect covariates.

Multi-level or hierarchical models, which can also implement weighted random effect regression, have been applied to meta-analysis [26]. This is an area of current research. Such models have the potential to combine summary data with individual patient data, including both study level and patient level covariates [27, 28].

Regression models can be used to explore the impact of study quality on a meta-analysis, this is discussed in Chapter 8. Chapter 11 discusses mixed models from a Bayesian perspective. If individual patient data are available, a more highly structured model may be more appropriate [29]; regression using patient level (as opposed to study level) covariates is possible; this is also considered in Chapter 12.

In an effort to explore the effect of duration of study on outcome, an extension to the standard regression model was developed [20]. Here, subgroup data for pre-specified time intervals was obtained from the original investigators. This type of analysis could be seen as somewhere between the meta-regression model used above, and regression analysis on individual patient data (see Chapter 12).

Regression modelling when more than one outcome is being considered simultaneously is discussed in Chapter 15. Several distinct meta-regression techniques specific to observational studies, such as dose-response analysis are covered in Section 16.7.2. Further extension of mixed models are discussed in Chapter 17 for the generalized synthesis of evidence.

Methods for assessing heterogeneity of underlying patient risk – a 'special case'

Studies may appear heterogeneous because of differences in the baseline risk of the patients. If the overall effectiveness of a new treatment is related to the severity of the disease, this could affect decisions about which patients should be treated [30]. The usual way of investigating baseline risk within trials is to consider the observed risk of events in the control group (or sometimes the average risk in the control and treatment groups) [31]. This variable could be included in a regression model in an effort to explain heterogeneity between study results. However, this type of analysis can lead to bias [32], since this measure of baseline risk forms part of the definition of the treatment difference (i.e. the event risk in the control group is used in the calculation of an odds ratio, relative risk, etc.). In other words, if by chance, the risk of an event in the control group is high then the estimated treatment effect will be greater (OR will be further from 1). If, on the other hand, the event risk in the control group is low, then the OR will be nearer 1. Thus, even if there is no true relationship between baseline risk and treatment effectiveness, one is likely to be observed due to this statistical artefact – regression to the mean [30]. This problem is reduced if studys are large (leading to less random variation in control risk), and if there are a large number of studies.

Conventional approaches, which relate treatment effect to the proportion of events in the control group, have problems [33]. Alternative analyses which have been developed to avoid the problem of 'regression to the mean' are outlined below.

Thompson et al. [31] present a solution using Bayesian methods (see Chapter 11). This can be extended to include other covariates. One potential drawback is that this method only works on the log odds ratio scale; using other scales is possible in principle, but currently difficult in practice.

McIntosh [34] has presented a method to examine population risk as a source of heterogeneity by representing clinical trials in a bivariate two-level hierarchical model, and estimate model parameters by both maximum likelihood and Bayes procedures. This method assumes bivariate normality of true treatment effects and control group risks across trials, and uses a normal approximation for binary outcome data, but these assumptions need investigating [30].

Cook and Walter [35] have presented another method which does not depend upon bivariate normality assumptions, and uses an unconditional maximum likelihood approach. Thompson et al. [30] compare their Bayesian approach to this, and find the results do differ. Further research is needed to ascertain which method is superior.

One method of avoiding the use of patient risk directly would be to use a prognostic score based on patient covariates (i.e. predictors of risk) and then relate treatment effects to this score for individual patients [30]. Thompson et al. suggest the prognostic score would best be based on data other than that from the trials which form the meta-analysis for treatment effects. Note that the score of risk used should, where possible, be one which clinicians can use so as to determine which of their patients are likely to benefit sufficiently from an intervention.

Another alternative, if individual patient data are available (see Chapter 12), is to relate treatment effects to individual patient covariates in an attempt to investigate heterogeneity. This avoids the problems discussed above, and would be directly useful to the clinician considering treatment for an individual patient [30].

Example: Mixed model using patient baseline risk as a covariate

The 34 cholesterol trials are examined to see if baseline risk explains any of the heterogeneity between studies. In the original analysis [9], baseline risk in the control group was defined as

$$Baseline\ risk = \left(\frac{Coronary\ heart\ disease\ death}{\frac{years\ of}{follow\ up} \times \left(\frac{number\ alive}{at\ end\ trial} \right) + 05 \left(\frac{number\ dying}{during\ study} \right)} \right) \times 1000 \quad (6.9)$$

This definition is more desirable than using

$$Baseline\ risk = \frac{Total\ mortality\ in\ the\ control\ group}{Number\ of\ patients\ who\ survived\ in\ the\ control\ group}. \quad (6.10)$$

because it takes into account the length of follow-up of each trial in the calculation which varied considerably (less than a year to nearly 10 years) in this dataset. It provides a death rate per 1000 person years.

Unfortunately, regression analysis methods which can adjust for regression to the mean are difficult using this definition (although current research suggests a similar analysis is possible using the event rate ratio outcome measure [36]). Hence, for this illustrative example the simpler definition of (6.10) is used. By examining the raw data in Table 2.1, it can be seen that the variation in baseline risk between studies is considerable.

All of the methods proposed require specialist software routines to implement them. The results which follow were computed using the method of Thompson et al. [30], which uses Bayesian methodology. This somewhat pre-empts the consideration of a Bayesian framework for meta-analysis, and the reader should consult Chapter 11 for further details.

The coefficients obtained from fitting the model are given in Table 6.4. For comparison, the coefficients obtained from fitting a standard mixed model to the data are also contained in the table.

In both instances, the credibility interval for β (the Bayesian equivalent of a confidence interval) does not contain zero. Thus, it would appear that baseline risk explains a considerable proportion of the between study heterogeneity. This suggests that the greater at risk patients are, the more potential benefit they have to gain from treatment. For this example, the coefficient for beta from the standard (unadjusted) model is slightly larger than that from the correct adjusted model. This is a typical result, which can be explained by the fact that regression to the mean exaggerates the relationship between baseline risk and estimates of treatment effect. In some instances, the over-estimation can be considerably greater than for the

Table 6.4 Model parameter estimates fitting baseline risk as a covariate; adjusting and not adjusting for regression to the mean.

Parameter estimate (95% CrI)	Adjusted model	Un-adjusted model
ln(*baseline risk*) (β)	−0.143 (−0.24 to −0.04)	−0.153 (−0.25 to −0.05)
ln(*OR*)	−0.382 (−0.59 to −0.17)	−0.417 (−0.64 to −0.18)
Between study variance	0.040 (0.01 to 0.11)	0.034 (0.003 to 0.10)

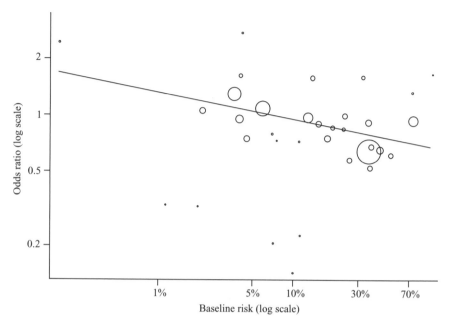

Figure 6.6 Regression line fitting baseline risk to cholesterol trials, adjusting for regression to the mean.

example presented here [30]. The estimate of between study variation has been reduced due to the inclusion of the covariate (it was calculated as 0.068 in Section 5.5 using a standard random effects model), however it is non-zero, suggesting some degree of heterogeneity remains, indicating that the inclusion of a random effect term is sensible. Figure 6.6 plots the studies together with the adjusted regression line. At around 5% baseline risk the corresponding odds ratio for the fitted line is approximately 1, suggesting that the risks of cholesterol lowering interventions used in the studies included in the meta-analysis may outweigh the benefits for people at this and lower risks. In this case, this is probably due to the fact that cholesterol lowering drugs had life-threatening side effects, which became numerically more important in patients at lower risk of dying from heart disease.

6.4   Summary/Discussion

The use of subgroup analyses and regression methods to explore and potentially explain between study heterogeneity have been described in this chapter. Subgroup analyses can be carried out on (average) patient or study characteristics, and are a simple alternative to regression models when the influence of a single categorical characteristic is being explored. Regression techniques can also be used to investigate between study heterogeneity. Two different basic models (with extensions) have been described; one an extension of the fixed effect model of Chapter 4; and the other an extension of the random effect model of Chapter 5. Special models to investigate the effect of patient baseline risk are also considered.

To-date, regression methods, and especially mixed models, have been used relatively rarely in meta-analysis; this is a potentially important approach with which increased experience in practical applications, including criteria for choice of covariates to be included with random coefficients, is needed. Caution is required, however, because aggregation bias has the potential to cause spurious results when covariates relating to average patient characteristics, such as age are used.

References

1. Gelber, R.D., Goldhirsch, A. (1987). The interpretation of results from sub-set analyses within overviews of randomized clinical trials. *Stat. Med.* **6**: 371–88.

2. Lau. J., Ioannidis, J.P., Schmid, C.H. (1998). Summing up evidence: one answer is not always enough. *Lancet* **351**: 123–7.

3. Greenland, S. (1987). Quantitative methods in the review of epidemiological literature. *Epidemiol. Rev.* **9**: 1–30.

4. Dickersin, K., Berlin, J.A. (1992). Meta-analysis: state-of-the-science. *Epidemiol. Rev.* **14**: 154–76.

5. Schneider, B. (1989). Analysis of Clinical Trial Outcomes: Alternative Approaches to Subgroup Analysis. *Controlled Clin. Trials* **10**: 176S–86S.

6. Yusuf, S., Wittes, J., Probstfield, J., Tyroler, H.A. (1991). Analysis and interpretation of treatment effects in subgroups of patients in randomized clinical trials. *J. Am. Med. Asso.* **266**: 93–8.

7. Counsell, C.E., Clarke, M.J., Slattery, J., Sanderock, P.A.G. (1994). The miracle of DICE therapy for acute stroke: Fact or fictional product of subgroup analysis? *Br. Med. J.* **309**: 1677–81.

8. Oxman, A.D., Guyatt, G.H. (1992). A consumers guide to subgroup analyses. *Annals of Internal Med.* **116**: 78–84.

9. Smith, G.D., Song, F., Sheldon, T.A., Song, F.J. (1993). Cholesterol lowering and mortality: The importance of considering initial level of risk. *Br. Med. J.* **306**: 1367–73.

10. Raudenbush, S.W. (1994). Random effects models. In: Cooper, H., Hedges, L.V., editors. *The Handbook of Research Synthesis*. New York: Russell Sage Foundation; 301–22.

11. Collett, D. (1991). *Modelling Binary Data*. London: Chapman & Hall.

12. Hedges, L.V. (1994). Fixed effects models. In: Cooper, H., Hedges, L.V., (editors). *The Handbook of Research Synthesis*. New York: Russell Sage Foundation; 1994; 285–300.

13. Thompson, S.G., Sharp, S.J. (1999). Explaining heterogeneity in meta-analysis: a comparison of methods. *Stat. Med.* **18**: 2693–708.

14. Altman, D.G. (1991). *Practical Statistics for Medical Research*. London: Chapman & Hall.

15. Colditz, G.A., Brewer, T.F., Berkey, C.S., Wilson, M.E., Fineberg, H.V., Mosteller, F. (1994). Efficacy of BCG vaccine in the prevention of tuberculosis. Meta-analysis of the published literature. *J. Am. Med. Assoc.* **271**: 698–702.

16. Berkey, C.S., Hoaglin, D.C., Mosteller, F., Colditz, G.A. (1995). A random-effects regression model for meta-analysis. *Stat Med.* **14**: 395–411.

17. Huque, M.F., Dubey, S.D. (1994). A metaanalysis methodology for utilizing study-level covariate-information from clinical-trials. *Commu. Statistics–Theory and Methods* **23**: 377–94.

18. Detsky, A.S., Naylor, C.D., O'Rourke, K., McGeer, A.J., LAbbe, K.A., O'Rourke, K., L'Abbe, K.A. (1992). Incorporating variations in the quality of individual randomized trials into meta-analysis. *J. Clin. Epidemiol.* **45**: 255–65.

19. L'Abbé, K.A., Detsky, A.S., O'Rourke, K. (1987). Meta-analysis in clinical research. *Annals of Internal Med.* **107**: 224–33.

20. Thompson, S.G. (1993). Controversies in meta-analysis: the case of the trials of serum cholesterol reduction. *Stat. Methods Med. Res.* **2**: 173–92.

21. Larholt, K.M. (1989). Statistical Methods and Heterogeneity in Meta-Analysis; Unpublished ScD thesis, Department of Biostatistics, Harvard School of Public Health, Boston.

22. Stram, D.O. (1996). Meta-analysis of published data using a linear mixed-effects model. *Biometrics* **52**: 536–44.

23. Peto R. (1987). Why do we need systematic overviews of randomised trials? *Stat. Med.* **6**: 233–40.

24. Begg C.B., Pilote L. (1991). A model for incorporating historical controls into a meta-analysis. *Biometrics* **47**: 899–906.

25. Tori, V., Simon, R., Russek-Cohen, E., Midthune, D., Friedman, M. (1992). Statistical model to determine the relationship of response and survival in patients with advanced ovarian cancer treated with chemotherapy. *J. Nat. Cancer Inst.* **84**: 407–14.

26. Lambert, P.C., Abrams, K.R. (1996). Meta-analysis using multilevel models. *Multilevel Modelling Newsletter* **7(2)**: 17–9.

27. Goldstein, H., Yang, M., Omar, R.Z., Turner, R.M., Thompson, S.G. Meta-analysis using multilevel models with an application to the study of class size effects. (Submitted.)

28. Turner, R.M., Omar, R.Z., Yang, M., Goldstein, H., Thompson, S.G. Multilevel models for meta-analysis of clinical trials with binary outcomes. (Submitted.)

29. Boissel, J.P., Blanchard, J., Panak, E., Peyrieux, J.C., Sacks, H. (1989). Considerations for the meta-analysis of randomized clinical trials: summary of a panel discussion. *Controlled Clin. Trials* **10**: 254–81.

30. Thompson, S.G., Smith, T.C., Sharp, S.J. (1997). Investigation underlying risk as a source of heterogeneity in meta-analysis. *Stat. Med.* **16**: 2741–58.

31. Brand, R., Kragt, H. (1992). Importance of trends in the interpretation of an overall odds ratio in the meta-analysis of clinical trials. *Stat. Med.* **11**: 2077–82.

32. Senn, S. (1994). Importance of trends in the interpretation of an overall odds ratio in the meta-analysis of clinical trials. *Stat. Med.* **13**: 293–6.

33. Sharp, S.J., Thompson, S.G., Altman, D.G. (1996). The relation between treatment benefit and underlying risk in metaanalysis. *Br. Med. J.* **313**: 735–8.

34. McIntosh, M.W. (1996). The population risk as an explanatory variable in research synthesis of clinical trials. *Stat. Med.* **15**: 1713–28.

35. Cook, R.J., Walter, S.D. (1997). A logistic model for trend in $2 \times 2 \times$ kappa tables with applications to meta-analyses. *Biometrics* **53**: 352–7.

36. Arends, L.R., Hoes, A.W., Lubsen, J., Grobbee, D.E., Stijnen, T. (1999). Baseline risk as predictor of treatment benefit; three clinical meta-re-analyses. (Submitted.)

CHAPTER 7

Publication Bias

7.1 Introduction

To avoid drawing unbiased conclusions from a meta-analysis it is important that all, or at the very least, the majority of the relevant primary studies need to be identified on a given subject. Unfortunately, even comprehensive searches of the literature (including grey material) and the use of other less formal methods, such as personal communication, may not produce an unbiased sample of studies. It has long been accepted that research with statistically significant results is potentially more likely to be submitted, published or published more rapidly than work with null or non-significant results [1], which leads to a preponderance of false-positive results in the literature [2]. The implications of this for meta-analysis are that, combining only the identified published studies uncritically may lead to an incorrect, usually over-optimistic, conclusion. This problem is known as *publication bias*.

Further, the biases directly associated with the reporting and publication of results have been split into three distinct mechanisms [3]: (i) publication bias (explained above); (ii) retrieval bias (bias left after trying to obtain unpublished results); and (iii) pipeline effects (effect of waiting [or not] for unpublished studies to become published).

Additionally, other related biases exist. Begg [4] and Begg and Berlin [2] comment on the effect of subjective reporting of results as a form of publication bias. They suggest that exaggerated claims based on the 'biased' opinion of the investigator(s) may effect what results are reported, and in extreme cases, only significant results are included in a report, while the non-significant ones are omitted. This phenomena has been studied [5] and an association demonstrated. Another source of possible bias is due to the duplication of reporting (publishing) results. This may occur because authors want to increase their authorship by essentially submitting the same results to different journals, or because different groups report multi-centre trials based on at least part of the same data. A further problem is that of a language bias (see Section 7.2.5). Additional reasons for results not being published include students who leave the academic arena and do not publish their PhD or MSc dissertations; or studies being suppressed by those who do not wish to have results appear that are against their own vested interests, political beliefs, or funding source's interests [6]. It is also worth being aware that the line between published and unpublished results becomes blurred for example when study results presented at conferences are not subsequently published [7].

A brief review of the evidence for the seriousness and consequences, and the predictors, of publication bias follow in Sections 7.2 through 7.4. Many methods have been suggested for identifying, estimating the impact, and adjusting a meta-analysis for publication bias. These are described in Sections 7.5 and 7.6. Broader perspective solutions to publication bias are briefly considered in Section 7.7.

7.2   Evidence of publication and related biases

Publication bias was first suspected from the observation that a large proportion of published studies had rejected the null hypotheses. In 1959, Sterling found that the results of 97% of studies published in four major psychology journals were statistically significant, concluding that studies with non-significant results might be under-represented [8]. In 1995, the same author concluded that practices leading to publication bias had not changed over a period of 30 years [9].

Since this discovery, many studies investigating the existence and magnitude of publication have been carried out using various methodologies, first in the social science, and later in the medical literature [4]. Several of these are outlined below; for a recent review, see Song *et al.* [10].

7.2.1   Survey of authors

Several studies [11–13] have surveyed authors, and found that, generally, studies with non-significant results are less likely to be submitted for publication compared to those with statistically significant results. Dickersin *et al.* [14] observed that the proportion of trials in which the new treatment were superior to the control therapy was 55% in 767 published studies versus only 14% in 178 unpublished studies.

7.2.2   Published versus registered trials in a meta-analysis

Simes [15] compared alkylating agent monotherapy with combination chemotherapy in advanced ovarian cancer. Meta-analysis of only the published trials yielded a large and significant survival advantage for combination chemotherapy. This was not substantiated when all studies in the International Cancer Research Databank, an unbiased list of unpublished and published trials, were used.

7.2.3   Follow-up of cohorts of registered studies

The existence of publication bias has been consistently confirmed by studies which have followed up cohorts of studies approved by the Research Ethics Committee or cohorts of trials registered by the research sponsors [1, 16–18].

Studies with significant results were also more likely to generate multiple publications and more likely to be published in journals with a high citation impact factor when compared with studies with non-significant results [1]. However, risk factors for publication bias were not consistently identified across these cohort studies [1, 17, 18].

Stern and Simes [18], and others [19], observed that clinical trials with statistically significant results were published much *earlier* than those with non-significant results (median 4.7 years versus 8.0 years) [18]. This finding was confirmed using a cohort of randomized controlled trials on AIDS, in which it was found that the median time from starting enrolment to publication was 4.3 years for positive trials and 6.5 years for negative trials [20].

7.2.4   Non-empirical evidence

Melton [21] states explicitly as editorial policy that the journal gives preference to study reports demonstrating statistical significance. However, contrary to previous thought, there is evidence suggesting author preferences are a more important cause of publication bias than editorial preferences [12, 22]. That is to say, authors do not submit non-significant studies to journals because they believe it to be a waste of time as they will get rejected. A large percentage of the non-significant studies that are submitted do in fact get published, so in fact, it is the authors' beliefs not journal policy that prevents many from being published.

7.2.5   Evidence of language bias

It is common to limit searching for research reports to those only published in English. Grégoire *et al.* [23] suggest this may lead to a 'Tower of Babel bias', with the rationale that authors in non-English speaking countries, having completed a clinical trial yielding negative results, might be less confident about having it published in a large diffusion international journal written in English, and would then submit it to a local journal. Hence, positive results by authors from non-English speaking countries are thus more likely to be published in English, and negative results in the investigators' language [23].

Grégoire *et al.* [23], Moher *et al.* [24] and Egger [25] have all investigated foreign language bias in meta-analysis. Grégoire's hypothesis under test was that negative results are more likely to go in smaller journals with low distribution, but this was not supported empirically.

Moher *et al.* [24] found differences in the completeness of reporting, design characteristics and analytical approaches of RCTs published in English with those published in French, German, Italian and Spanish. This provides evidence to support inclusion of all trial reports, irrespective of the language in which they are published, in systematic reviews.

7.3   The seriousness and consequences of publication bias for meta-analysis

The most serious potential consequence of publication bias is that it leads to a biased estimate (usually over-estimate) of treatment effects or risk-factor associations in published work [1]. When a meta-analysis has a large aggregated sample size, this problem is potentially more serious because the results may appear to be extremely precise and convincing, even though the observed effect or association is entirely due to bias [2]. However, it has been suggested that the problem of publication bias has been exaggerated [26, 27] as studies with negative results tend to be poorer in quality, weakened by small sample size (type II error).

Although the existence of publication bias is well demonstrated, there is little empirical evidence about its consequences. An exception is that of the trials of 1C anti-arrhythmic agents [28–30].

7.4   Predictors of publication bias (factors effecting the probability a study will get published)

Beyond the significance of the treatment effect, and the size of the study, other potential factors which may effect the chances of a study being published have been identified. Dickersin *et al.* [14], for example, found that studies which favour the new therapy are more likely to be published. In addition, Begg and Berlin [3] highlight other potentially distinguishing features as being the presence or absence of randomization, sample size, exploratory versus confirmatory studies, protocol definition, the nature of the journal, calendar time and source of funding.

7.5   Identifying publication bias in a meta-analysis

Having assembled the primary studies to be considered in a meta-analysis, it is important to assess the chances of and likely effects of publication bias [4]. The various tools for identifying publication bias are described below.

7.5.1   The funnel plot

The results from smaller studies will be more widely spread around the mean effect because of larger random error. A plot of sample size versus treatment effect from individual studies in a meta-analysis should thus be shaped like a funnel if there is no publication bias [31]. If the chance of publication is greater for studies with positive statistically significant results, or larger effect size estimates, or some other less defined mechanism, the shape of the funnel plot may become skewed.

When the true outcome effect is small but not zero, small studies reporting a small effect size will not be statistically significant, and therefore less likely to be published, while small studies reporting a large effect size may be statistically significant and more likely to be published. Consequently, there will be a lack of small studies with small effect estimates in the funnel plot, and the funnel plot will be skewed with a larger effect among smaller studies and a smaller effect among larger studies [31]. This will result in an overestimation of the treatment effect in a meta-analysis.

Example funnel plots are displayed in Figures 7.1 and 7.2. Figure 7.1 displays a set of studies with no evidence of publication bias, while Figure 7.2 displays 'classic' asymmetry with a suggestive lack of studies in the bottom left-hand corner of the plot.

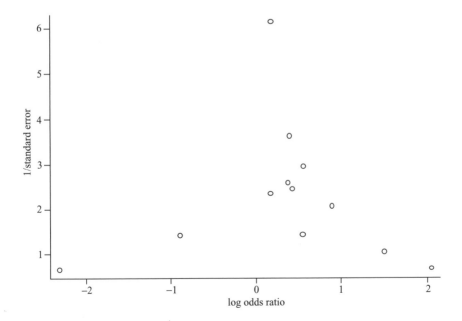

Figure 7.1   Example funnel plot showing no evidence of publication bias.

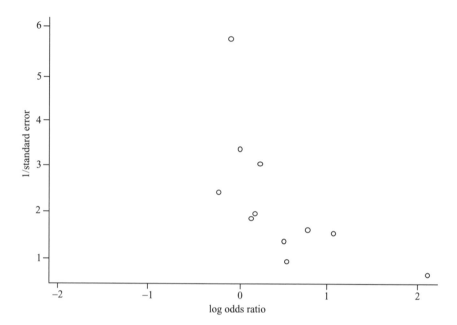

Figure 7.2 Example funnel plot showing classic asymmetry suggesting the presence of publication bias.

If, however, the true outcome effect is assumed to be zero, the results of small studies can only be statistically significant when they are far away from zero, either positive or negative. If studies with significant results are published and studies with results around zero are not published, there will be an empty area around zero in the funnel plot [31]. In this case, the funnel plot may not be obviously skewed but would have a hollow area inside. These polarized results (significant negative or positive results) may cause many debates, but the overall estimate obtained by combining all studies is unlikely to be biased. In the authors' experience to-date, however, this kind of publication bias has rarely been encountered.

In a funnel plot, the treatment effects from individual studies are often plotted against their standard errors (or the inverse of the standard error), instead of the corresponding sample sizes. Use of standard errors may have some advantages because the statistical significance is determined not only by the sample size, but also by the level of variation in the outcome measured, or the number of events in the case of categorical data. However, the visual impression of a funnel plot may change by plotting treatment effects against standard errors instead of against the inverse of standard errors [32].

There are some limitations in the use of funnel plot to detect publication bias. For a funnel plot to be useful, there needs to be a range of studies with varying sizes. The funnel plot is an informal method for assessing the potential publication bias and different people may interpret the same plot differently (more formal statistical tests are described below). It should be stressed that a skewed funnel plot may be caused by factors other than publication bias [33]. For example, it has been shown that if the quality of studies varies with the study size, a funnel plot may give the visual impression of publication bias when this is really confounded by study quality [34]. Other possible sources of asymmetry in funnel plots include different intensity of intervention, differences in underlying risk, poor methodological design of small studies, inadequate analysis, fraud, choice of effect measure and chance [33].

Example assessment of presence of publication bias

A meta-analysis of risperidone versus typical neuroleptics in the treatment of schizophrenia [35] is used to illustrate the methods for detecting publication bias throughout Section 7.5. The data for the 10 RCTs included in the analysis is presented in Table 7.1. The outcome measured was absence of clinical improvement, so an odds ratio of less than one suggests that risperidone is superior to typical neuroleptics.

Plotting the log odds ratios against the inverse of their standard errors produces the funnel plot in Figure 7.3. Visual inspection suggests that publication bias may be present in this meta-analysis, as the 'funnel' appears to be skewed with a scarcity of small studies in the bottom right-hand corner of the funnel. (Note: since failure to improve is an undesirable outcome, the funnel plot is the opposite way round to that in Figure 7.2 where the missing studies were in the bottom left-hand corner of the plot.)

Table 7.1 Data from 10 trials of risperidone versus typical neuroleptics in the treatment of schizophrenia.

Study	$\ln(OR)$	$se(\ln(OR))$	$1/se(\ln(OR))$	$\ln(OR)/se(\ln(OR))$
1	−1.04	0.72	1.39	−1.45
2	−0.46	0.39	2.53	−1.17
3	−0.18	0.48	2.07	−0.38
4	−0.46	0.68	1.46	−0.67
5	−0.28	0.30	3.31	−0.92
6	−0.41	0.40	2.49	−1.02
7	−0.66	0.41	2.44	−1.61
8	−0.66	0.30	3.38	−2.23
9	0.52	0.73	1.36	0.71
10	−0.11	0.15	6.78	−0.75

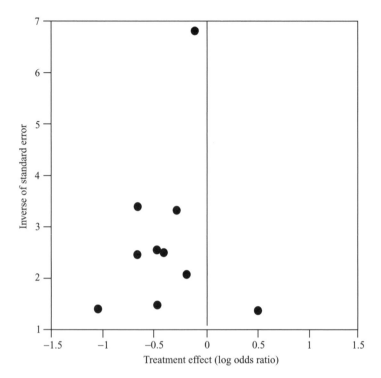

Figure 7.3 Funnel plot for 10 RCTs of risperidone versus typical neuroleptics in the treatment of schizophrenia.

7.5.2   Rank correlation test

The funnel plot is an informal, visual method. Various statistical tests have been developed. The rank correlation test, described by Begg and Mazumdar [36], examines the association between effect estimates and their variances, to exploit the fact that publication bias will tend to induce a correlation between the two factors (i.e. smaller studies (with larger variances will tend) to tend to have larger effect size estimates), and constructs the rank-ordered sample on the basis of one of them. The test is a distribution-free method, which involves no modelling assumptions, but it suffers from a lack of power, and so the possibility of publication bias cannot be ruled out even when the test is non-significant.

The formulae for the test are given below [36]. Define the standardized effect sizes of the k studies to be combined to be

$$T_i^* = (T_i - \bar{T}_\bullet)/(\tilde{v}_i^*)^{\frac{1}{2}}, \tag{7.1}$$

where

$$\bar{T}_\bullet = \left(\sum_{j=1}^{k} v_i^{-1} T_j \right) \bigg/ \sum_{j=1}^{k} v_i^{-1}, \tag{7.2}$$

and T_i and v_i are the estimated effect size and sampling variance from the ith study, and

$$\tilde{v}_i^* = v_i - \left(\sum_{j=1}^{k} v_j^{-1} \right)^{-1}, \tag{7.3}$$

which is the variance of $(T_i - \bar{T}_\bullet)$.

It is then necessary to evaluate P, the number of all possible pairings in which one factor is ranked in the same order as the other, and Q, the number in which the ordering is reversed. A normalized test statistic is obtained by calculating

$$Z = (P - Q)/[k(k - 1)(2k + 5)/18]^{\frac{1}{2}}, \tag{7.4}$$

which is the normalized Kendall rank correlation test statistic for data that have no ties, though this can be relaxed [37]. This statistic is compared to the standardized normal distribution. Any effect size scale can be used, as long as it is assumed distributed asymptotic normal.

This test can be considered complementary to the funnel plot. Begg [4] suggests using a very liberal significance level. Additionally, due to the test having very low power for meta-analyses including only a small numbers of studies, more emphasis should be given to an informal visual inspection of the funnel plot [36].

Example: Applying the rank correlation test to the risperidone for the treatment of schizophrenia meta-analysis

Applying the rank correlation test to the risperidone meta-analysis produces a test statistic, $Z = 0.09$, which is non-significant ($p = 0.93$). This result is not surprising due to the small number of studies included in the meta-analysis.

7.5.3 Linear regression test

To test the asymmetry of a funnel plot, Egger *et al.* [33] suggested a method based on a regression analysis of Galbraith's radial plot [38] (see Section 3.3.3). For $i = 1 \ldots k$ studies in the meta-analysis, let T_i and v_i be the estimated effect sizes and sample variances from each study. Define the standardized effect (z-statistic) as $T_i^* = T_i/v_i^{\frac{1}{2}}$, the precision as $s^{-1} = 1/v_i^{\frac{1}{2}}$, and the weight as normal ($w_i = 1/v_i$). To perform the test, T^* is fitted to s^{-1} using standard weighted linear regression with weights w and the equation $T^* = \alpha + \beta s^{-1}$.

The intercept $\hat{\alpha}$ is used to measure asymmetry; if it is estimated to be significantly different from 0, then it is concluded that there is evidence of publication bias in the meta-analysis dataset. A negative intercept indicates that smaller studies are associated with bigger effects. In their original paper, Egger *et al.* [33] also performed an unweighted regression, and reported the most extreme result.

By applying this method, Egger *et al.* [33] observed significant asymmetry in 38% of published meta-analyses in a selection of journals, and in 13% of Cochrane reviews.

From comparisons [33, 39] between the tests, it would appear that the linear regression test is more powerful that the rank correlation test; however, a considerable discrepancy in the results of the two tests has been observed [39]. It has been suggested that the operating characteristics of this method need to be evaluated more thoroughly [40].

Example: Applying the linear regression test to the risperidone for the treatment of schizophrenia meta-analysis

The linear regression test is applied to the meta-analysis of risperidone for schizophrenia. The corresponding Galbraith plot including the fitted line is displayed in Figure 7.4. Note the confidence interval for the intercept of the line with the *y*-axis is marked by two small circles. The intercept of the regression line is -0.70 (-1.93 to 0.52), which is not statistically significant ($p = 0.22$). Although the test is not statistically significant, the sign of the intercept coefficient (negative) indicates that the small studies are associated with a larger treatment effect.

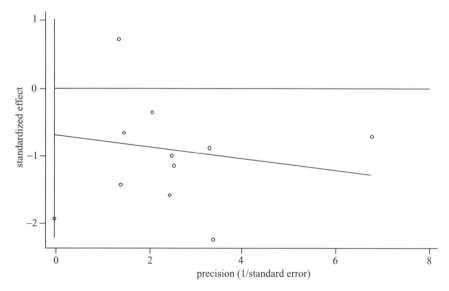

Figure 7.4 Galbraith regression plot with linear regression test for the meta-analysis of risperidone for schizophrenia.

In the original Galbraith plot, the slope of the line indicated the size and direction of effect; a greater gradient indicates a greater treatment effect. The slope coefficient for this plot is -0.09 (-0.48 to 0.30), which is an estimate for the treatment effect, which in a weak sense might be considered adjusted for the effects of publication bias [37]. This contrasts with the pooled estimate from a standard random effects analysis ($\ln OR = -0.288(-0.49$ to $-0.089)$), because not restricting the regression line to go through the origin means that the gradient of the slope will become closer to zero or positive when the estimated effect is greater in smaller studies.

7.5.4   Other methods to detect publication bias

A method used to adjust meta-analyses for publication bias, known as 'trim and fill' [32], has an associated test based on the number of studies estimated as missing. This is described in Section 7.6.5.

7.5.5   Practical advice on methods for detecting publication bias

Although the idea of the funnel plot has been around for some time, the idea of carrying out a formal significance test for the presence of publication bias is much more recent. It is difficult to offer practical advice on which test to use as they appear to come to contrasting conclusions a large proportion of the time [39], and clearly more research is required into their relative performance. Presently, we recommend a funnel plot should be inspected routinely, ideally supported by the results of several tests.

7.6   Taking into account publication bias or adjusting the results of a meta-analysis in the presence of publication bias

Several methods have been developed to assess the importance of or adjust the results of a meta-analysis for publication bias [10], if its presence is suspected following application of the methods of Section 7.5.

If publication bias is suspected, then efforts could be made to try and find the missing studies, before, or instead of, adjusting the analysis. One way of doing this would be to write to interested investigators, although this can be a painstaking process [41], or consulting registry of trials [15]. However, if one does include data from unpublished studies, which have not passed peer review, one is at risk of lowering the quality and credibility of the data [3]. Opinion seems split among researchers, whether this is a sensible thing to do.

The idea of adjusting the results of a meta-analysis is currently very controversial, and none of the methods for adjusting the results are carried out routinely. There has been little empirical research comparing the performance of the different methods. Until more research into these methods is done, we recommend that these methods should be used as a form of sensitivity analysis only (see Chapter 9). If the results of a meta-analysis change dramatically after using one or a combination of the methods described here, caution is needed when drawing conclusions from the results obtained. In some instances, a funnel plot may display 'classic' publication bias features; however, even if the missing studies had been included, the result could still be very similar, as it is usually only the smallest studies which are not published. If this can be demonstrated through a sensitivity analysis based on the use of these analytic methods, then this is valuable information; the meta-analysis of Linde *et al.* [42] is a good example of this.

It remains to be seen how useful and appropriate these methods are in practice, and whether they move from experimental to more mainstream methodology. Unfortunately many of them are computationally complex to perform, require non-standard software, and hence are currently difficult to implement.

7.6.1   Analysing only the largest studies

If a funnel plot looks skewed, suggesting that small negative studies are missing, a simple method which may reduce bias is to combine the results of only the largest studies [3]. In carrying out such an analysis a cut-off between large and small studies needs to be determined, usually arbitrarily. A related but slightly more sophisticated approach is to carry out a cumulative meta-analysis (see Chapter 19) combining the studies sequentially, ordered by their precision [43]. Like many of the methods in this section this can be useful as a form of sensitivity analysis.

Example: Risperidone for the treatment of schizophrenia meta-analysis

The risperidone for schizophrenia meta-analysis dataset highlights the problem of attempting to analyse only the largest studies. Other than study 10, they all could be considered reasonably small; no obvious cut-off exists. More sensible would be to perform a cumulative meta-analysis, ordered by precision, examples of which are given in Chapter 19.

7.6.2   Rosenthal's 'file drawer' method

In essence, this method considers the question: 'How many new studies averaging a null result are required to bring the overall treatment effect to non-significance?' [44]. It was developed by Rosenthal [44, 45], and has been referred to as the 'file drawer problem', as it could be seen as estimating the number of studies filed away by researchers' without being published.

The method is based on combining the normal z-scores (the Z_i) corresponding to the p-values observed for each study (this is covered in more detail in Section 14.7). The overall z-score can be calculated by

$$Z = \sum_{i=1}^{k} Z_i / \sqrt{k}, \qquad (7.5)$$

where k is the number of studies in the meta-analysis. This sum of z-scores is a z-score itself, and the combined z-scores are considered significant (i.e. the outcome measured in the studies is significant) if $Z > Z_{\alpha/2}$ (the $\alpha/2$ percentage point of a Standardized Normal Distribution–which is 1.96 for $\alpha = 5$). Now we determine the number of unpublished studies with an average observed effect of zero that there would need to be in order to reduce the overall z-score to non-significance. Define k_0 to be the additional number of studies required, such that

$$\sum_{i=1}^{k} Z_i / \sqrt{k + k_0} < Z_{\alpha/2}; \qquad (7.6)$$

rearranging the above gives

$$k_0 > -k + \left(\sum_{i=1}^{k} Z_i \right)^2 / \left(Z_{\alpha/2} \right)^2. \qquad (7.7)$$

After k_0 is calculated, one can judge if it is realistic to assume that this many studies exist unpublished in the research domain under investigation. The plausible number of unpublished studies may be hundreds in some areas, or only a few in others. Therefore, the estimated fail-safe N should be considered in proportion to the number of published studies (k). Rosenthal suggested that the fail-safe N may be considered as being unlikely, in reality, to exist if it is greater than a tolerance level of '$5k + 10$' [44]. If k_0 is less than this tolerance level, then one must have doubts about the validity of the meta-analysis.

This method is far from perfect, and should be considered nothing more than a crude guide [3]. Shortcomings include: (a) the fact that combining Z scores does not directly account for the sample sizes of the studies; (b) the choice of zero for the average effect of the unpublished studies is arbitrary, and certainly biased [3]; (c) it is guesswork estimating the magnitude of unpublished studies in the area; (d) the method does not adjust or deal with treatment effects; (e) heterogeneity among the studies are ignored [46]; and (f) the method is not influenced by the shape of the funnel graph [4]. In its favour, the value k_0 is easy to calculate and easily interpretable [4]. Despite all the drawbacks, this method has been used widely as a tool in meta analysis [47]. This test is generally used as a sensitivity test once a meta-analysis has been found to give a significant result.

Several extensions and variations of Rosenthal's 'file-drawer' method have been developed [46–48].

Example: Applying the file-drawer calculation to the risperidone for the treatment of schizophrenia meta-analysis

The random effect odds ratio from pooling the 10 trials of risperidone versus typical neuroleptics for the treatment of schizophrenia is 0.75 (0.62 to 0.92), implying that more patients had clinically improved on the risperidone treatment arms of the trials. Applying Rosenthal's file drawer calculation to these data estimates how many unpublished studies–which, on average, show no treatment effect–would be needed to turn this statistically significant result into a non-significant one. The Z values for each trial are calculated by dividing the log odds ratio by its standard error. These are given in the final column of Table 7.1. There appears to be some confusion in the literature as to whether a one-or two-sided critical value should be used. Rosenthal originally used a one-sided figure [44], however since the original significance test for the treatment difference was two-sided, we believe a two-sided figure should be used. Hence, for a 95% significance level $Z_{1-\alpha/2} = 1.96$. Calculating k_0 using equation (7.7) gives

$$k_0 > -10 + ((-1.45) + \ldots\ldots + (-0.75))^2/(1.96)^2$$
$$> 13.5.$$

Hence, at least 14 unpublished studies with a zero treatment effect, on average, are required in this meta-analysis to change the statistically significant result into a statistically non-significant result. Although this number sounds improbable, it is much less than 60 (i.e. $5*10 + 10$), the tolerance level suggested by Rosenthal, which may be much too conservative in this context.

*7.6.3   Models which estimate the number of unpublished studies, but do not adjust

The fail-safe N estimated by using Rosenthal's method is not necessarily related to the actual number of unpublished studies. Glesser and Olkin [49] have developed two general methods that attempt to estimate the actual number N of missing studies using the p-values reported in the published studies. N and its confidence bounds can then be evaluated for plausibility through sensitivity analysis by the meta-analyst.

The first model considers the possibility that the p-values observed are the k smallest p-values among the $N + k$ unreported and reported studies.

The second type of model is a full selection model of the form described elsewhere [3, 50], and outlined in Section 7.6.4, where the probability that a study is reported is a function of the attained p-value.

*7.6.4   Selection models using weighted distribution theory

Weight functions were first introduced into meta-analysis by Hedges [46]. Prior to this, they had been used in various disciplines, including sample surveys and ecology. They are used to adjust results where only partial information is available, and the chance of having particular data is related to a feature of the data [46]. Hence, in a meta-analysis setting, weight functions are used to model the selection process, and develop estimation procedures that take that selection process into account [51]. There are two aspects to such models: (a) the effect size model which specifies what the distribution of the effect size estimates would be if there were no selection; and (b) the selection model which specifies how this effect size distribution is modified by selection [52]. Usually, these models assume that the chance of a study being included in the meta-analysis is related to the statistical significance of its outcome (implying that journals are more likely to publish significant results than non-significant ones). In these instances, the outcome considered is the observed p-value.

*7.6.5   The 'Trim and Fill' method

The Trim and Fill method is a simple rank-based data augmentation technique to formalize the use of the funnel plot [32]. This recently-developed method can be used to estimate the number of missing studies, and more importantly, to provide an estimate of the treatment effect by adjusting for potential publication bias in a meta-analysis. The mechanics of the approach are displayed diagramatically in Figure 7.5 using a meta-analysis of the effect of gangliosides on mortality from ischaemic stroke [53]. Briefly, the number of 'asymmetric' studies on the right-hand side of the funnel is estimated: these can broadly be thought of as studies which have no left-hand-side counterpart. These studies are then removed, or 'trimmed', from the funnel, leaving a symmetric remainder from which the true 'centre' of the funnel is estimated using a standard meta-analysis procedure as in Figure 7.5(b). The 'trimmed' studies are then replaced and their 'missing counterparts' imputed or 'filled': these are mirror images of the 'trimmed' studies with the mirror axis placed along the adjusted pooled estimate as in Figure 7.5(c). This then allows an adjusted overall confidence interval to be calculated.

Three different estimators for the number of missing studies have been derived [32, 54]. The properties of these estimators have been investigated via simulation studies, and generally it is concluded that they work well in all but very extreme cases. A significance test based on the number of studies estimated as missing is also described [32, 54]. The new test would appear to be quite powerful if there are more than 5–6 missing studies [32]. See Duval and Tweedie for formulae required to use this method [32, 54].

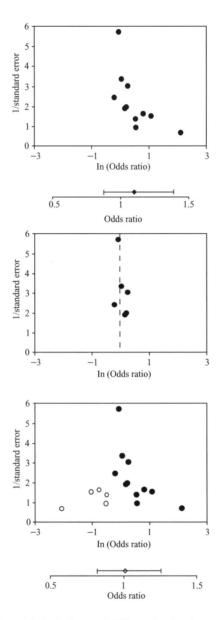

Figure 7.5 Illustration of the Trim and Fill method using a meta-analysis of the effect of gangliocides on mortality from ischaemic stroke. (a) Original funnel plot; (b) asymmetric studies 'trimmed', and centre value of remaining studies is calculated; (c) funnel is 'filled' allowing a 95% confidence interval to be calculated for the adjusted pooled estimate. (Adapted from Duval and Tweedie [71].)

Example: Applying the trim and fill adjustment methods to the risperidone for the treatment of schizophrenia meta-analysis

Although the computational details of the trim and fill approach are beyond the scope of this book, the results of applying it to the risperidone example are included so the reader can compare the results it gives with the other tools for dealing with publication bias. Consider again the funnel plot for these trials (Figure 7.3). The method of trim and fill suggests that the three trials on the left-hand-side with the most extreme results (furthest left) would need to be removed for the remainder of the funnel to be symmetric. The odds ratio pooling the remaining seven trials is 0.82, and 95% confidence calculated after reinserting the three extreme results and inputing three mirror image studies on the right-hand-side of the plot is (0.68 to 1.01). The original significant result (0.75 (0.61 to 0.92)) becomes just non-significant at the 5% level using this method to allow for publication bias. It is concluded that the benefit of risperidone over typical neuroleptics is questionable, due to the potential impact of publication bias.

Previous methods to adjust for publication bias (i.e. the selection models above) are complex and highly computer intensive to run [6]. In contrast, this method is both conceptually easy, and simpler to implement than any of the previous methods described to adjust a meta-analysis for publication bias. This method is non-parametric, and relies only on symmetry assumptions, which are satisfied by both fixed and random effect meta-analysis models. The key assumption of the method is that it is the most extreme negative studies which have not been published. In simulation studies, it was found that the method estimated the point estimate of the overall effect size approximately correctly and the coverage of the confidence interval is substantially improved, as compared to ignoring publication bias [32].

The authors do, however, stress that the main goal of their work should be seen as providing method for sensitivity analyses, rather than actually finding the values of missing studies, and that the method does seem to give good indications of which meta-analyses do not suffer from publication bias, and which need to be evaluated much more carefully.

7.6.6   The sensitivity approach of Copas

Copas [55] presents a method for adjusting for publication bias based on a method described by Copas and Li [56]. A random effects meta-analysis model is used [57], together with a separate selection equation with a single correlation parameter ρ linking selection to outcome. A likelihood approach is taken, but the model cannot be fully identified without strong and unverifiable assumptions, so a sensitivity approach based on an overall probability of study selection is adopted.

7.7   Broader perspective solutions to publication bias

The previous section has outlined several, often complex, methods of largely unknown validity, for assessing or adjusting meta-analyses results for the presence of publication bias. Clearly, if at all possible, it would be desirable not to have to use these methods at all. Discussed below are some broader measures which may help to reduce publication bias in the health and related literature.

7.7.1   Prospective registration of trials

There is little disagreement that prospective registration of all trials at their inception is the best solution to eliminating publication bias [15]. A method has been described by Simes [15, 58] which removes the possibility of publication bias, at the cost of potentially only including a selected proportion of all studies carried out. This involves limiting the meta-analysis to a subset of studies which represent an exhaustive collection from a sampling frame which is independent of the publication process [3]; that is, to restrict attention to studies that (prospectively) have been registered at some international trials registry. The analyst then follows up all registered studies, including (a cohort of) unpublished studies identified by the international register. It is important to note that studies published, but not on the register, are ignored, and not included in the meta-analysis. This method would produce a list of trials whose inclusion in the analysis would not be influenced by their results [58]. Simes [15, 58] give examples of using this methodology in meta-analyses of RCTs of cancer treatments. Further support for this idea has come from Begg and Berlin [3] who considered their construction to be seen as a policy goal, and Easterbrook [59], who has assembled a compendium of existing registries. There is also some indication that pharmaceutical companies will in future provide publicly accessible registers of trials carried out on their products.

7.7.2   Changes in publication process and journals

It has been suggested that journals could reduce publication bias by accepting manuscripts for publication based mainly on the research protocol [60]. Since the beginning of 1997, a general medical journal, *The Lancet*, has been assessing and registering selected protocols of randomized trials, and providing a commitment to publish the main clinical findings of the study.

To motivate investigators to register their trials, it was also suggested that prospective registration should be a requirement laid down by journal editors and registering agencies [61]. By disclosing 'conflict of interest' or 'competing interests', potential bias due to sources of research funding may be revealed [62].

Recently, over 100 medical journals around the world have invited readers to send in information on unpublished trials in a so-called 'trial amnesty' [63, 64]. This trial amnesty had registered 150 trials by the end of 1998 [65]. However, there is little hope that conventional paper journals can solve the problem of publication bias because of space limitation, and the requirement of newsworthy articles for maintaining or increasing the journals' circulation level.

Publication bias may be reduced by introducing new peer reviewed electronic journals without space limitation, in which papers should be accepted based only on the research methodology or validity criteria, not on the results [66–69]. Because the originality is no longer a requirement, these kind of electronic journals will encourage the submission of trials with negative or non-striking results, and trials that simply replicate previous trials. Such journals would be mainly used as medical recorders, and therefore would be most useful for investigators and people who conduct systematic reviews.

7.8   Including unpublished information

It may seem clear that, if one could identify all the unpublished studies and retrieve the relevant information, publication bias could be alleviated. However, Cook *et al.* [70] investigate attitudes towards unpublished data in a meta-analysis. They found that 46 out of 150 meta-analyses examined used unpublished results. Of editors asked, 46.9% felt that unpublished data should probably or certainly be included, and 30% would not publish an overview that included unpublished data.

This leaves the reviewer in a quandary if unpublished studies on a topic are known to exist. For a discussion on how study quality affects the results, and ways of incorporating study quality into a meta-analysis, see Chapter 8.

7.9   Summary/Discussion

In conducting a meta-analysis, researchers should always be aware of the potential for publication bias, and make efforts to assess to what extent publication bias may affect their meta-analysis. This includes use of tools to test for its presence, informally and formally, and possibly using methods to check the robustness of the pooled result using various methods as forms of sensitivity analysis. An assessment of the likely impact of including unpublished data is desirable.

The methods to 'adjust' the results of a meta-analysis for publication bias should currently be considered experimental. However, it is likely that their use will grow as more research is done and software becomes available to facilitate their routine use.

However, no method is ever going to alleviate the need for the problem of publication bias to be eliminated. With advances in worldwide communications, via the Internet, etc., the feasibility of worldwide study registries has increased, together with the potential to publish material not published in a paper format. The inception of groups such as the Cochrane Collaboration, and their retrospective search for all known RCTs, it would seem to suggest that the first advances have been made.

References

1. Easterbrook, P.J., Berlin, J.A., Gopalan, R., Matthews, D.R. (1991). Publication bias in clinical research. *Lancet* **337**: 867–72.

2. Begg, C.B., Berlin, J.A. (1989). Publication bias and dissemination of clinical research. *J. Natl. Cancer Inst.* **81**: 107–15.

3. Begg, C.B., Berlin, J.A. (1988). Publication bias: a problem in interpreting medical data (with discussion). *J. Roy. Statist. Soc. A.* **151**: 419–63.

4. Begg, C.B. (1994). Publication bias. **In**: Cooper, H., Hedges, L.V., editors. *The Handbook of Research Synthesis*. New York: Russell Sage Foundation; 399–409.

5. Chalmers, T.C. (1982). Informed consent, clinical research and the practice of medicine. *Trans. Am. Clin. Climatol Assoc.* **94**: 204–12.

6. Givens, G.H., Smith, D.D., Tweedie, R.L. (1997). Publication bias in meta-analysis: a Bayesian data-augmentation approach to account for issues exemplified in the passive smoking debate. *Stat. Sci.* **12**: 221–50.

7. Boissel, J.P., Haugh, M.C. (1993). The iceberg phenomenon and publication bias: The editors' fault? *Clinical Trials and Meta-Analysis* **28**: 309–15.

8. Sterling, T.D. (1959). Publication decisions and their possible effects on inferences drawn tests of significance – or vice versa. *Am. Stat. Assoc. J.* **54**: 30–4.

9. Sterling, T.D., Rosenbaum, W.L., Weinkam, J.J. (1995). Publication decisions revisited: the effect of the outcome of statistical tests on the decision to publish and vice versa. *Am. Statistician* **49**: 108–12.

10. Song, F., Eastwood, A., Gilbody, S., Duley, L., Sutton, A.J. (2000). Publication and other selection biases in systematic reviews. *Health Technol. Assess.* (to appear).

11. Greenwald, A.G. (1975). Consequences of prejudice against the null hypothesis. *Psychol Bull* **82**: 1–20.

12. Coursol, A., Wagner, E.E. (1986). Effect of positive findings on submission and accept-ance rates: a note on meta-analysis bias. *Professional Psychol.* **17**: 136–7.

13. Sommer, B. (1987). The file drawer effect and publication rates in menstrual cycle research. *Psychol. of Women Quart.* **11**: 233–42.

14. Dickersin, K., Chan, S., Chalmers, T.C., Sacks, H.S., Smith, H.J. (1987). Publication bias and clinical trials. *Controlled Clin Trials* **8**: 343–53.

15. Simes, R.J. (1987). Confronting publication bias: A cohort design for meta-analysis. *Stat. Med.* **6**: 11–29.

16. Dickersin, K., Min, Y.I., Meinert, C.L. (1992). Factors influencing publication of research results: follow-up of applications submitted to two ionstitutional review boards. *J. Am. Med. Assoc.* **263**: 374–8.

17. Dickersin, K., Min, Y.I. (1993). NIH clinical trials and publication bias. *Online J. Curr. Clin. Trials* Doc: No. 50.

18. Stern, J.M., Simes, R.J. (1997). Publication bias: evidence of delayed publication in a cohort study of clinical research projects. *Br. Med. J.* **315**: 640–5.

19. Misakian, A.L., Bero, L.A. (1998). Publication bias and research on passive smoking. Comparison of published and unpublished studies. *J. Am. Med. Assoc.* **280**: 250–3.

20. Ioannidis, J. (1998). Effect of the statistical significance of results on the time to completion and publication of randomized efficacy trials. *J. Am. Med. Assoc.* **279**: 281–6.

21. Melton, A. (1962). Editorial. *J. Exp. Psychol.* **64**: 553–7.

22. Gotzsche, P.C. (1987). Reference bias in reports of drug trials. *Br. Med. J.* **295**: 654–6.

23. Gregoire, G., Derderian, F., Lelorier, J., Le Lorier, J. (1995). Selecting the language of the publications included in a meta-analysis – is there a Tower-of-Babel bias? *J. Clin. Epidemiol.* **48**: 159–63.

24. Moher, D., Fortin, P., Jadad, A.R., Juni, P., Klassen, T., Le Lorier, J., Liberati, A., Linde, K., Penna, A. (1996). Completeness of reporting of trials published in languages other than English: implications for conduct and reporting of systematic reviews. *Lancet* **347**: 363–6.

25. Egger, E., ZellwegerZahner, T., Schneider, M., Junker, C., Lengeler, C. (1997). Language bias in randomised controlled trials published in English and German. *Lancet* **350**: 326–9.

26. Freirnan, J.A., Chalmers, T.C., Smith, H.J., Kuebler, R.R. (1978). The importance of beta, the type II error and sample size in the design and interpretation of the randomized controlled trial: survey of 71 'negative' trials. *New Engl. J. Med.* **299**: 690–4.

27. Angell, M. (1989). Negative studies. *New Engl. J. Med.* **321**: 464–6.

28. Cowley, A.J., Skene, A., Stainer, K., Hampton, J.R. (1993). The effect of lorcainide on arrhythmias and survival in patients with acute myocardial-infarction–an example of publication bias. *Int. J. Cardiology* **40**: 161–6.

29. The Cardiac Arrhythmia Suppression Trial (CAST) (1989). Investigators. Preliminary report: effect of encainide and flecainide on mortality in a randomised trial of arrhythmia suppression after myocardial infarction. *New Engl. J. Med.* **321**: 406–12.

30. The Cardiac Arrhythmia Suppression Trial II Investigators. (1992). Effect of the antiarrhythmic agent moricisine on survival after myocardial infarction. *New Engl. J. Med.* **327**: 227–33.

31. Light, R.J., Pillemar, D.B. (1984). *Summing Up: The science of Reviewing Research.* Cambridge, MA: Harvard University Press.

32. Duval, S., Tweedie, R. (2000). Trim and fill: A simple funnel plot based method of testing and adjusting for publication bias in meta-analysis. *Biometrics* (to appear).

33. Egger, M., Smith, G.D., Schneider, M., Minder, C. (1997). Bias in meta-analysis detected by a simple, graphical test. *Br Med J* **315**: 629–34.

34. Petticrew, M., Gilbody, S., Sheldon, T.A. (1999). Relation between hostility and coronary heart disease. Evidence does not support link [letter]. *Br. Med. J.* **319**: 917.

35. Kennedy, E., Song, F., Hunter, R., Gilbody, S. (1998). Risperidone versus 'conventional' antipsychotic medication for schizophrenia. The Cochrane Library (Issue 3) Oxford: Update Software.

36. Begg, C.B., Mazumdar, M. (1994). Operating characteristics of a rank correlation test for publication bias. *Biometrics* **50**: 1088–101.

37. Steichen, T.J. (1998). Tests for publication bias in meta-analysis. *Stata Technical Bull.* **41**: sbe20: 9–15.

38. Galbraith, R.F. (1988). A note on graphical presentation of estimated odds ratios from several clinical trials. *Stat. Med.* **7**: 889–94.

39. Sutton, A.J., Duval, S.J., Tweedie, R.L., Abrams, K.R., Jones, D.R. The impact of publication bias on meta-analyses within the Cochrane Database of Systematic Reviews (submitted).

40. Naylor, C.D. (1997). Meta-analysis and the meta-epidemiology of clinical research: meta-analysis is an important contribution to research and practice but it's not a panacea. *Br. Med. J.* **315**: 617–9.

41. Yusuf, S., Peto, R., Lewis, J., Collins, R., Sleight, P. *et al.* (1985). Beta blockade during and after myocardial infarction: an overview of the randomised trials. *Progress in Cardiovascular Dis.* **27**: 335–71.

42. Linde, K., Clausius, N., Ramirez, G., Melchart, D., Eitel, F., Hedges, L.V., Jonas, W.B. (1997). Are the clinical effects of homoeopathy placebo effects? A meta-analysis of placebo-controlled trials. *Lancet* **350**: 834–43.

43. Lau, J., Schmid, C.H., Chalmers, T.C. (1995). Cumulative meta-analysis of clinical trials: Builds evidence for exemplary medical care. *J. Clin. Epidemiol.* **48**: 45–57.

44. Rosenthal, R. (1979). The file drawer problem and tolerance for null results. *Psychol. Bull.* **86**: 638–41.

45. Rosenthal, R. (1978). Combining the results to independent studies. *Professional Psychol.* **17**: 136–7.

46. Iyengar, S., Greenhouse, J.B. (1988). Selection models and the file drawer problem. *Stat. Sci.* **3**: 109–35.

47. Orwin, R. (1983). A fail-safe N for effect size in meta-analysis. *J. Ed. Statist.* **8**: 157–9.

48. Spiegelhalter, D.J., Myles, J.P., Jones, D.R., Abrams, K.R. (2000). Bayesian methods in health technology assessment. *Health Technology Assess.* (to appear).

49. Gleser, L.J., Olkin, I. (1996). Models for estimating the number of unpublished studies. *Stat. Med.* **15**: 2493–507.

50. Dear, K.B.G., Begg, C.B. (1992). An approach for assessing publication bias prior to performing a meta-analysis. *Stat. Sci.* **7**: 237–45.

51. Hedges, L.V. (1992). Modeling publication selection effects in meta-analysis. *Stat. Sci.* **7**: 246–55.

52. Hedges, L.V., Vevea, J.L. (1996). Estimating effects size under publication bias: small sample properties and robustness of a random effects selection model. *J. Educ. Behav. Stat.* **21**: 299–333.

53. Duval, S., Tweedie, R. (1998). Practical estimates of the effect of publication bias in meta-analysis. *Australian Epidemiologist* **5**: 14–17.

54. Duval, S., Tweedie, R. (2000). A non-parametric 'trim and fill' method of assessing publication bias in meta-analysis. *J. Am. Stat. Assoc.* (to appear).

55. Copas, J. (1999). What works?: selectivity models and meta-analysis. *J. Roy. Stat. Soc. A* **161**: 95–105.

56. Copas, J.B., Li, H.G. (1997). Inference for non-randon samples. *J. Roy. Stat. Soc. B* **59**: 55–95.

57. DerSimonian, R., Laird, N. (1986). Meta-analysis in clinical trials. *Controlled Clin. Trials* **7**: 177–88.

58. Simes, R.J. (1986). Publication bias: the case for an international registry of clinical trials. *J. Clin. Oncol.* **4**: 1529–41.

59. Easterbrook, P.J. (1992). Directory of registries of clinical trials. *Stat. Med.* **11**: 345–423.

60. Newcombe, R.G. (1988). Discussion of the paper by Begg and Berlin: Publication bias: a problem in interpreting medical data. *J. Roy. Stat. Soc. A* **151**: 448–9.

61. Julian, D. (1998). Meta-analysis and the meta-epidemiology of clinical research. Registration of trials should be required by editors and registering agencies [letter]. *Br. Med. J.* **316**: 311.

62. Smith, R. (1998). Beyond conflict of interest: transparency is the key. *Br. Med. J.* **317**: 291–2.

63. Horton, R. (1997). Medical editors trial amnesty. *Lancet* **350**: 756.

64. Smith, R., Roberts, I. (1997). An amnesty for unpublished trials: send us details on an unreported trials. *Br. Med. J.* **315**: 622.

65. CCTA. Cochrance Controlled Trials Register. The Cochrane Library (1999). Oxford: Update Software.

66. Chalmers, I. (1990). Underreporting research is scientific misconduct. *J. Am. Med. Assoc.* **263**: 1405–8.

67. Berlin, J.A. (1992). Will publication bias vanish in an age of online journals. *Online J. Current Clin. Trials* doc 12.

68. Song, F., Eastwood, A., Gilbody, S., Duley, L. (1998). Reducing publication bias: is electronic journal an answer? *MEDNET98: Third Annual World Congress on the Internet in Medicine.* London: Educational Technology Research Papers Series, The University of Birmingham, 1998.

69. Chalmers, I., Altman, D.G. (1993). How can medical journals help prevent poor medical research? Some opportunities presented by electronic publishing. *Lancet* **353**: 490–3.

70. Cook, D.J., Guyatt, G.H., Ryan, G., Clifton, J., Buckingham, L., Willan, A., McIlroy, W., Oxman, A.D. (1993). Should unpublished data be included in metaanalyses– current convictions and controversies. *J. Am. Med. Assoc.* **269**: 2749–53.

71. Duval, S., Tweedie, R. (1998). Practical estimates of the effect of publication bias in meta-analysis. *Austr. Epidemiol.* **5**: 14–17.

CHAPTER 8

Study Quality

8.1 Introduction

The subject of judging research quality in synthesis dates back to Glass in 1976 [1]. The primary concern is that combining study results of poor quality may lead to biased, and therefore misleading, pooled estimates being produced. Sophisticated analyses will not eliminate the limitations of poor data [2] '... in some respects, the quantitative methods used to pool the results from several studies in a meta-analysis are arguably of less importance than the qualitative methods used to determine which studies should be aggregated' [3]. However, assessment of quality is controversial; Greenland [4] has indicated that quality assessment is the most insidious form of bias in the conduct of meta-analysis.

There are at least three approaches for assessing research quality. The first system [5] applies the validity framework developed by Cook and Campbell [6], and focuses on non-randomized studies often found in the social science literature [this will not be pursued further here, and the interested reader is referred to the above cited papers and Wortman [7]). The second is via a quality scoring system, the first of which was developed by Chalmers *et al.* [8, 9] for assessing RCTs exclusively, although checklists were available before this [10]. The objective of these scales is to provide an overall index of quality (a comparison with the validity framework approach can be found in Wortman [7]). Since these first attempts, many different scales and checklists have been developed; for a review of these for RCTs, see Moher *et al.* [10] and for those for observational studies, see Deeks *et al.* [11]. Recently, a scale which could assess the quality of both randomized and observational studies has been developed [12]. Finally, using individual markers of quality (e.g. randomization procedure) can be considered as a third alternative [13].

An appealing feature about using a scale is that it provides an overall quantitative estimate of quality. However, the validity of many of the present scales has been criticized. It has been suggested [14] that most scales have been developed in an arbitrary fashion with no attention to accepted methodological standards. Additionally, many scales are not truly measuring quality, but focus on extraneous factors more related to the adequacy of reporting or generalizability [14].

Unfortunately, no clear association between study quality and study results exists consistently across all trials. However, an empirical study has shown [15] that over a large number of RCTs, encompassing many subject areas, inadequate methodological approaches, particularly those representing poor allocation concealment, are associated with bias. Dickersin and Berlin [16] review meta-analyses of both RCTs and observational studies which have addressed this issue.

While a detailed review of the specific scales and checklists and their pros and cons is beyond the scope of this book, it is important to note that large differences in results can be observed by using different scales, at least those for RCTs, which is where empirical investigations have been focused [13, 14, 17]. The next section does consider factors which affect the quality of a study, which can be used in an individual component assessment. The remainder of this chapter then focuses on how quality score assessments and individual markers of quality can be incorporated into a meta-analysis.

There would appear to be little consensus concerning the optimum way of dealing with study quality in meta-analysis (although there is broad agreement that a quality assessment should always be carried out). There is perhaps growing support for using individual indicators, rather than an overall quality score, following a comparative assessment [17, 18]. Perhaps the best path to take is to consider the methods outlined below as part of a sensitivity analysis, (Chapter 9), and assess the influence adjustment for quality has on the results of an analysis unmodified by study quality. The key problems are that: (a) in each type of study, the factors influencing internal validity are likely to differ, and so standard scores will not be generically relevant; (b) the influence of factors affecting validity will differ, depending on the question and context of the trial; and (c) studies do not uniformly report sufficient details of the methods used in their design, conduct and analysis, to be able to accurately measure these factors in the same way in each study.

8.2   Methodological factors that may affect the quality of studies

Table 9.1 suggests a hierarchy of the sources of best evidence. The reasoning behind this was is different study designs are susceptible to biases in varying degrees, and thus vary in the reliability of the results. It has become accepted that large Randomized Controlled Trials (RCTs) are the 'gold standard' source of evidence of efficacy, giving potentially unbiased estimates of intervention effects if well conducted, because of the comparability of the groups at baseline. However, no empirical measure of the amount of bias that other study designs are susceptible

Table 8.1 An example of a hierarchy of evidence (adapted from Deeks *et al.* [11]).

I	Well-designed randomized controlled trials
	Other types of trial:
II-1a	Well-designed controlled trial with pseudo-randomization
II-1b	Well-designed controlled trials with no randomization
	Cohort studies:
II-2a	Well-designed cohort (prospective study) with concurrent controls
II-2b	Well-designed cohort (prospective study) with historical controls
II-2c	Well-designed cohort (retrospective study) with concurrent controls
II-3	Well-designed case-control (retrospective) study
III	Large differences from comparisons between times and/or places with and without intervention (in some circumstances, these may be equivalent to level II or I)
IV	Opinions of respected authorities based on clinical experience; descriptive studies and reports of expert committees

to is available. Despite this, on the grounds of specific features which are known to increase bias, such as collecting the data retrospectively or using a historical comparison group, an argument can be made for the superiority of evidence from one study design over another. As one can only determine the methodologic quality of a study to the extent that study design and analytic methods are reported [19], we restrict ourselves to only the reported factors for the remainder of this chapter. A distinction that can be made is between experimental and observational studies, so we focus initially on clinical trials.

8.2.1 Experimental studies

The assessable design features of trials which affect quality can be split into four areas: assignment; masking and concealment of allocation; patient follow-up; and statistical analysis [10].

- **Assignment:** this could well be the single most important design feature of a study. As randomized controlled trials provide the most valid basis for the comparison of interventions in health care [20], randomization is clearly a desirable feature, and thus RCTs are considered the most reliable method on which to assess the efficacy of treatments [21].
  Despite this, the details of randomization are not often reported [20]. Even if a study is described as randomized, there may be doubts about

the care or probity with which it has been performed. Bias has also been detected in trials not reporting adequate allocation concealment [22], but this is thought not to be as important as the generation of allocation procedure details from investigators [20, 23].

- **Masking:** this is also known as 'blinding'. A trial is said to be blind if the patient does not know what intervention arm of the study they are on. A study is said to be 'double blind' if, in addition, the assessor of outcome is also unaware of the treatment each patient is on. This source of bias may be as important as assignment biases when the outcome measurement of interest involves some subjective judgement.
- **Patient follow-up:** patients drop out of trials for several reasons. Patients may also switch to other arms in some instances (for example, if the patients are allergic to the original treatment). How these events are documented, and subsequently dealt with in the analysis, can affect the overall treatment estimate.
- **Statistical analysis:** obviously if an inappropriate statistical analysis of trial results is carried out, (or a correct type of analysis, but with mistakes) misleading results may be produced.

8.2.2   Observational studies

When investigating the effectiveness of an intervention, observational studies are more prone to bias than RCTs, primarily because treatment allocation may be related to prognosis or prejudgement of likely appropriateness of treatment [11]. Thus, establishing that differences observed between groups of patients in observational studies are the effect of the interventions is a far harder exercise than it is in experimental studies. Similarly, studies which are planned prospectively are also less likely to be biased than studies which are undertaken retrospectively [11]. Cohort studies, in which groups receiving the different interventions being compared are evaluated concurrently, are regarded as more valid than studies which make comparisons with 'historical' controls [11]. It is also worth being aware that treatment effects could be under-estimated due to over-matching on factors which are related to allocation of the intervention [11].

In addition to study design, factors such as the conduct of the study and its analysis affects the quality of a study. In this way, a poor RCT may be less valid than a well conducted observational study [11]. It is for this reason that scales and checklists were developed; to make the assessment of such factors more systematic. Currently, it is uncommon for randomized and observational studies to be combined in the same meta-analysis, largely due to concerns that the observational studies potentially add too much bias to the pooled results, but this potentially useful application of meta-analysis is discussed in Chapter 17.

8.3 Incorporating study quality into a meta-analysis

Once a formal assessment of study quality has been made, using a measurement scale, or individual quality markers, a decision has to be made how to use this information. The methods which have been proposed to incorporate an assessment of study quality into a meta-analysis are described below. Incorporating study quality into a meta-analysis can be considered a special case of exploring heterogeneity (Chapter 3), i.e. to what extent does variation in measures of quality between studies explain variation in estimates of treatment or exposure effects?

8.3.1 Graphical plot

A plot of the point estimate and 95% confidence interval for each study's outcome estimate against quality score (derived from a scoring system) or individual study marker can be investigated to explore whether the two variables are related [24].

Example

This example uses data from a meta-analysis in the Cochrane Database of Systematic Reviews [25] investigating the effect of Continuous Electronic Heart Rate Monitoring (CEHRM) versus Intermittent Auscultation (IA) for assessment during labour. The outcome considered here is the risk difference for caesarean deliveries for mothers given CEHRM versus IA. The quality scores recorded by the original investigators were measured using the scale devised by Chalmers *et al.* [8]. These, together with outcome data, are presented in Table 8.2.

Figure 8.1 plots these five studies results against their quality scores. In this example, no clear relationship between quality score and outcome exists, although there is a slight suggestion that the risk difference reduces in magnitude the better the quality of the study.

Table 8.2 Outcomes and quality scores from five RCTs of continuous electronic heart rate monitoring versus intermittent auscultation for assessment during labour.

Study id	RD	se(RD)	Quality score (QS_i)(%)
1	0.030	0.0101	60
2	0.062	0.0183	45
3	0.065	0.0173	54
4	0.026	0.0323	71
5	0.004	0.0104	57

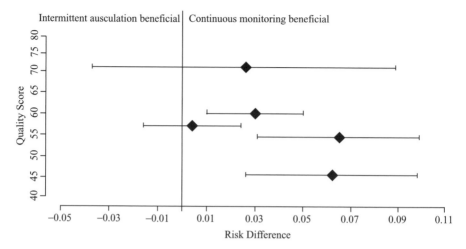

Figure 8.1 Plot of RCT outcomes versus quality score for the risk difference of caesarean deliveries for CEHRM vs. IA.

8.3.2 Cumulative methods

A cumulative meta-analysis (described in Chapter 19) based on a quality score can be conducted. In such an analysis, trials are combined starting with the highest quality study and adding in studies one by one in descending order of quality score. A pooled estimate is calculated after each additional study [24]. Inspection of this graph can provide an opportunity to discern the effect of quality on outcome [24].

8.3.3 Regression model

If there appears to be an association between outcome and quality, each study's quality score can be included as a continuous covariate in a regression model (see Chapter 6). If individual markers associated with study quality are being investigated, then these can be entered singularly, or simultaneously in a multi-factorial regression model. Either a fixed effect regression or a mixed model can be used. As with the use of regression techniques generally in meta-analysis, this approach often lacks power due to the small numbers of studies included in many meta-analyses [24].

Example

Sowden *et al.* [26] present a meta-analysis investigating whether there is a relationship between numbers of coronary artery bypass grafts performed in a hospital and hospital death rates. The analysis presented follows the original very closely, except a random effect regression model is used here (see Section 6.3), whereas a (fixed effect) logistic regression model (see Section 6.4.2) was used in the original analysis. Six observational studies were included in the original analysis. The outcome measure used to combine studies was the (log) odds ratio for mortality in high versus low volume hospitals. The degree of adjustment of this estimate for prognostic variables, and the actual variables used for adjusting, varied greatly between the primary studies. This lead to concern that the relationship between bypass graft surgery and hospital death rates could be due to confounding because of differences in case mix. To examine this, a novel (though arbitrarily scaled) scoring system was devised to measure how well each study had adjusted for case mix; this is reproduced in Table 8.3. This score, together with outcome estimate and standard error for each of the primary studies, is given in Table 8.4.

The Q statistic from the test for heterogeneity for these studies is 20.1 (5 df) which is highly significant ($p = 0.001$). Pooling the studies using a random effects model, ignoring study quality produces an odds ratio of 0.68 (0.57 to 0.80), and estimate of between study variance of 0.03. This suggests that there is a significant reduction in mortality in the hospitals with a high volume of coronary artery bypass

Table 8.3 Scoring of adjustment for case mix in the coronary artery bypass graft studies. (Reproduced by permission of BMJ Publishing Group from Sowden *et al.* [26].)

Adjustment score	Criteria
0	No adjustment for case mix
1	Adjustment for age, sex, and whether patients had multiple diagnosis
2	Adjustment for age, sex, and nature of other heart and medical conditions as described in discharge abstracts
3	Adjustment for age, sex, and nature of other heart and medical conditions as described in clinical databases

Table 8.4 Data including quality adjustment score, on six primary studies investigating the relationship between volume and outcome in coronary artery bypass graft surgery.

Study	OR (95% CI)	ln(OR)	se(ln(OR))	adjustment score
1	0.44 (0.29 to 0.65)	−0.82	0.21	1
2	0.58 (0.52 to 0.65)	−0.54	0.06	1
3	0.64 (0.53 to 0.77)	−0.45	0.10	0
4	0.83 (0.69 to 0.99)	−0.19	0.09	2
5	0.74 (0.58 to 0.94)	−0.30	0.12	2
6	0.84 (0.66 to 1.07)	−0.17	0.12	3

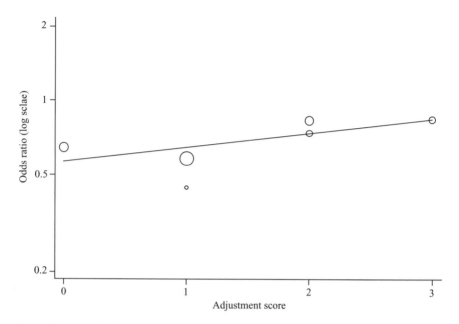

Figure 8.2 Analysis incorporating adjustment quality score for meta-analysis of relationship between volume of coronary artery bypass graft surgery and hospital death rates.

graft surgery. However, if the adjustment score coefficient is included as a coefficient in a mixed model, the resulting model is

$$\ln(OR) = -0.585 + 0.137 \times (\text{adjustment score}).$$

This regression line, together with the results of the six studies, is plotted in Figure 8.2. A visual inspection of this plot suggests a trend with quality score is apparent, and the confidence interval for the adjustment score coefficient does not contain zero (0.08 to 0.27). It appears that as quality score increases the odds ratio decreases, indicating a reduction in beneficial effect of increased volume. The odds ratio estimate when the quality score is 0 is 0.56; this increases to 0.84 when the quality adjustment score is 3. Hence, it would appear that a large proportion of the difference observed between low and high volume hospitals is due to lack of adjustment for case mix.

8.3.4 Weighting

Rather than weight each study by a measure derived from its precision, as is normally done in a meta-analysis, each of the individual study estimates could be weighted by a variable which measures the perceived quality of the study [24]. In doing this, it should be noted that, although actual estimates are affected only by the relative weights used, the width of the confidence intervals is affected by the absolute weights used [24]. To avoid this problem, each study's score can be divided by the mean score, to leave the width of confidence intervals unchanged.

A further possibility is to multiply the precision of the study by its quality score, and to use this product as the weighting for each study [18, 27]. In such a way, both the size and quality of the study are incorporated into the weighting calculation. This could be done using either a fixed or random effects model [27]. For a fixed effect model, the new weights become

$$w_i' = (QS_i)(w_i), \qquad (8.1)$$

and similarly for the random effects model,

$$w_i' = (QS_i)(w_i^*), \qquad (8.2)$$

where QS_i is the quality score allocated to the ith trial in the analysis, and w_i and w_i^* are the original weights given to the ith trial in fixed and random effects analysis, respectively, as defined in equations (4.2) and (5.7). The pooled estimate can now be calculated as before using the new weights. The variance of the pooled estimate can be calculated using [27]

$$\mathrm{var}(\overline{T}\cdot) = \frac{\sum_{i=1}^{k} QS_i w_i'}{\left(\sum_{i=1}^{k} w_i'\right)^2}. \qquad (8.3)$$

Caution has been expressed at incorporating or using quality score in the weighting given to studies, because, while weighting study estimates by the precision has desirable statistical properties, quality scores are not direct measures of precision, and this approach lacks statistical or empirical justification [24].

Example: Incorporating study quality into study weighting

This example uses data from a meta-analysis published by Berlin *et al.* [28], investigating the effect of physical activity in the prevention of coronary heart disease. Specifically, Table 8.5 gives results for coronary heart disease mortality for occupational studies which compared a high to a low level exercise group. The quality scores allocated to each study are from an assessment in the original meta-analysis. The fourth column of Table 8.5 provides the inverse-variance weights, the w_i, which can be used to combine the $\ln(RR)$ in a fixed effect model. Such an analysis (using equations outlined in Section 4.2) produces a pooled relative risk of 1.87 (1.62 to 2.15). Column 6 of Table 8.5 shows the adjusted weights created by multiplying the original weights by the quality score as outlined in equation (8.1).

Table 8.5 Results of occupational studies reporting CHD mortality comparing a high to a low level activity group.

Study	$\ln(RR)$	$se(\ln(RR))$	$1/\mathrm{var}(\ln(RR)) = w_i$	Quality score (QS_i)	$QS_i \times w_i = w_i'$
1	0.69	0.18	31.98	14	447.76
2	0.69	0.10	103.31	18	1859.64
3	0.47	0.14	49.07	17	834.14
4	0.10	0.52	3.66	14	51.31
5	0.88	1.05	0.90	14	12.66

The calculation required to combine the studies using the new variance-weighted fixed effect model of Section 4.2 is outlined below:

$$\ln(RR_{Pooled}) = \frac{[(447.76 \times 0.69) + \ldots\ldots + (12.66 \times 0.88)]}{(447.76 + \ldots\ldots + 12.66)} = 0.626.$$

The variance of this pooled estimate is calculated using equation (8.3):

$$\mathrm{var}(\ln(RR_{Pooled})) = \frac{[(14 \times 477.76) + \ldots\ldots + (14 \times 12.66)]}{(447.76 + \ldots\ldots + 12.66)^2} = 0.0053.$$

Calculating 95% confidence intervals and taking anti-logarithms produces a relative risk of 1.87 (1.62 to 2.16), which is almost identical to the non-quality adjusted result. The reason for the similarity in the result is largely due to the fact that the range of quality scores among the studies was not very large, and hence the relative weighting did not change very much on introduction of the quality weighting. It should be noted that adjusting for quality can sometimes have a much larger effect on a pooled result than in this example. A fixed effects model was considered here to simplify the example; it should be noted that a random effect model may be more appropriate for combining these studies.

8.3.5 Excluding studies

Another approach is to exclude the studies of poor(est) quality altogether. This can be viewed as an extreme form of weighting – giving the poorest studies zero weight [29]. To determine what classifies as an unacceptably poor quality study, a threshold value needs to be defined in the inclusion criteria for the systematic review. If a scoring system is used, figures such as the mean, the mean plus one standard deviation or the median can be used as this threshold (assuming a high score on the scale indicates good quality) [24]. Alternatively, if one is considering individual markers of quality simpler criteria can be used, such as whether randomisation was adequately performed, or whether there were blinded outcome assessments [24]. Unsurprisingly, there does not appear to be a consensus as to the criteria to be adopted in deciding whether to reject studies. Some authors recommend inclusion of all but the very worst of studies [30], while others advocate the exclusion of all but the 'best' studies [31]. For an alternative approach to combining studies which takes this latter approach to its logical extreme, see Section 20.2.3, describing best evidence synthesis. An assessment [17] of 25 different quality scales concluded that use of such scales to identify trials of high quality is problematic, as the results obtained differed considerably depending on the particular scale being used.

Example: Excluding the poorest quality studies

We return to the meta-analysis investigating whether there is a relationship between volume of coronary artery bypass graft surgery and hospital death rates described in Section 8.3.3. A case could be made for excluding the studies in which adjustment for fewest prognostic variables had been made; for example, those scoring less than two on the scoring system outlined in Table 8.3. The pooled random effect odds ratio combining the remaining three studies (numbers 4, 5 and 6) is 0.81 (0.71 to 0.91). (The pooled random effect estimate when combining all six studies was 0.68 (0.57 to 0.80).) Hence, excluding the poorest quality studies has diminished the effect of volume, although a significant difference between low and high volume groups remains. This result is consistent with the regression analysis described in Section 8.3.3. The problem with these approaches is that it is not clear the threshold above which quality is sufficiently high as to protect against significant bias or confounding.

8.3.6   Sensitivity analysis

It has already been suggested that perhaps the best way of dealing with the quality of studies in a meta-analysis is through the use of sensitivity analysis. In such a way, the robustness of the results of a meta-analysis can be assessed through changing the studies included, and the weights allocated to each study using the methods described above.

8.4   Practical implementation

There are practical issues to consider when assessing the quality of studies as part of a meta-analysis. The first issue is whether to blind the assessors to aspects of the studies. There have been suggestions that assessors should only have the methods and results sections, with the authors and setting masked, and even the names of the treatment groups deleted to reduce assessor bias [24]. Jadad [32] has recently investigated the effects of blinding, and found evidence to suggest that blinded assessment produced significantly lower and more consistent scores than open assessment. This is the first piece of evidence to support what was previously seen as a purely speculative and possibly over-elaborate precaution [24]. It should be noted that implementing such a masking scheme can be problematic due to difficulty in distinguishing between 'inputs' and 'results'. For example, study attributes and baseline data are proposed to be available to the assessors, although they could be considered as results, and in such a way influence the quality assessment. It is also likely that the degree of assessor bias in study evaluation will depend on the type of assessors, their involvement in the area and knowledge of the field. Assessment by 'methodologists' with little prior interest in or knowledge of the field is probably less susceptible to this source of bias.

Another problem is that some large and complex trials report the details of study methodology in separate earlier publications. Detsky *et al.* [24] argue that looking at this material would probably increase quality score of the trial above the score it would achieve if considered in isolation. It is important to note that when reporting the results of a meta-analysis, a log of rejected trials, should be given [9].

8.5　Summary/Discussion

This chapter has considered the use of quality assessments in meta-analysis. There is general agreement that a quality assessment of the primary studies should be carried out routinely, possibly using a scale, checklist or individual components, but no consensus exists on whether it should remain descriptive or be incorporated at the analysis stage [27]. Further, whilst a number of methods have been proposed for incorporating study quality into a meta-analysis, little information concerning the merits of each approach is currently available. As far as the use to which such quality scores can be put, a number of possibilities exist, but in specific situations the analysist should not be totally reliant on any one quality-adjusted method (in addition to an unadjusted analysis). It is recommended that reviewers should perform sensitivity analysis using at least two different methods of incorporating quality.

Finally, it should be remembered that one of the roles of meta-analysis is to clarify weaknesses in the existing data on a particular subject and to encourage better quality future studies, and possibly by setting guidelines for such studies [16].

References

1. Glass, G.V. (1976). Primary, secondary and meta-analysis of research. *Educ. Res.* **5**: 3–8.

2. Thacker, S.B. (1988). Meta-analysis. A quantitative approach to research integration. *J. Am. Med. Assoc.* **259**: 1685–9.

3. Naylor, C.D. (1988). Two cheers for meta-analysis: problems and opportunities in aggregating results of clinical trials. *Can. Med. Assoc. J.* **138**: 891–5.

4. Greenland, S. (1994). Invited commentary: a critical look at some popular meta-analytic methods. *Am. J. Epidemiol.* **140**: 290–6.

5. Wortman, P.M. (1983). Evaluation research: A methodological perspective. *Ann. Rev. Psychol.* **34**: 223–60.

6. Cook, T.D., Campbell, D.T. (1979). *Quasi-experimentation: Design & Analysis Issues for Field Settings.* Boston: Houghton Mifflin.

7. Wortman, P.M., Cooper, H., Hedges, L.V., (editors). (1994). Judging research quality. In: *The Handbook of Research Synthesis.* New York: Russell Sage Foundation; 97–110.

8. Chalmers, T.C., Smith, H. Jr., Blackburn, B., Silverman, B., Schroeder, B., Reitman, D., Ambroz, A. (1981). A method for assessing the quality of a randomized control trial. *Controlled Clin. Trials* **2**: 31–49.

9. Sacks, H.S., Berrier, J., Reitman, D., Ancona-Berk, V.A., Chalmers, T.C. (1987). Meta-analysis of randomized controlled trials. *New Engl. J. Med.* **316**: 450–5.

10. Moher, D., Jadad, A.R., Nichol, G., Penman, M., Tugwell, P., Walsh, S. (1995). Assessing the quality of randomized controlled trials – an annotated bibliography of scales and checklists. *Controlled Clin. Trials* **12**: 62–73.

11. Deeks, J., Glanville, J., Sheldon, T. (1996). Undertaking systematic reviews of research on effectiveness: CRD guidelines for those carrying out or commissioning reviews. Centre for Reviews and Dissemination. York: York Publishing Services Ltd. Report #4.

12. Downs, S.H., Black, N. (1998). The feasibility of creating a checklist for the assessment of the methodological quality both of randomised and non-randomised studies of health care interventions. *J. Epidemiol. & Community Health* **52**: 377–84.

13. Moher, D., Jadad, A.R., Tugwell, P. (1996). Assessing the quality of randomised controlled trials: current issues and future directions. *Int. J. Technol. Assess. in Health Care* **12**: 195–208.

14. Moher, D., Cook, D.J., Jadad, A.R., Tugwell, P., Moher, M., Jones, A. *et al.* (1999). Assessing the quality of randomised controlled trials: implications for the conduct of meta-analyses. *Health Technol. Assess.* 3(12).

15. Schulz, K.F., Chalmers, I., Hayes, R.J., Altman, D.G. (1995). Empirical evidence of bias: Dimensions of methodological quality associated with estimates of treatment effects in controlled trials. *J. Am. Med. Assoc.* **273**: 408–12.

16. Dickersin, K., Berlin, J.A. (1992). Meta-analysis: state-of-the-science. *Epidemiol. Rev.* **14**: 154–76.

17. Juni, P., Witschi, A., Bloch, R., Egger, M. (1999). The hazards of scoring the quality of clinical trials for meta-analysis. *J. Am. Med. Assoc.* **282**: 1054–60.

18. Moher, D., Pham, B., Jones, A., Cook, D.J., Jadad, A.R., Moher, M., Tugwell, P., Klassen, T.P. (1998). Does quality of reports of randomised trials affect estimates of intervention efficacy reported in meta-analysis? *Lancet* **352**: 609–13.

19. Cho, M.K., Bero, L.A. (1994). Instruments for assessing the quality of drug studies published in the medical literature. *J. Am. Med. Assoc.* **272**: 101–4.

20. Schulz, K.F., Chalmers, I., Grimes, D.A., Altman, D.G. (1994). Assessing the quality of randomization from reports of controlled trials published in obstetrics and gynecology journals. *J. Am. Med. Assoc.* **272**: 125–8.

21. Cook, D.J., Guyatt, G.H., Laupacis, A., Sackett, D.L. (1992). Rules of evidence and clinical recommendations in the use of antithrombotic agents. *Antithrombotic Therapy Consensus Conference. Chest* **102**: 305S–11S.

22. Chalmers, T.C., Celano, P., Sacks, H.S., Smith, H. (1983). Bias in treatment assignment in controlled clinical-trials. *New. Engl. J. Med.* **309**: 1358–61.

23. Chalmers, T.C., Levin, H., Sacks, H.S., Reitman, D., Berrier, J., Nagalingam, R. (1987). Meta-analysis of clinical trials as a scientific discipline. I: control of bias and comparison with large co-operative trials. *Stat. Med.* **6**: 315–25.

24. Detsky, A.S., Naylor, C.D., O'Rourke, K., McGeer, A.J., L'Abbe, K.A. (1992). Incorporating variations in the quality of individual randomized trials into meta-analysis. *J. Clin. Epidemiol.* **45**: 255–65.

25. Thacker, S.B., Stroup, D.F. (1999). Continuous electronic heart rate monitoring versus intermittent auscultation for assessment during labour (Cochrane Review). The Cochrane Libary, Issue 3. Oxford: Update Software.

26. Sowden, A.J., Deeks, J.J., Sheldon, T.A. (1995). Volume and outcome in coronary artery bypass graft surgery: True association or artefact? *Br. Med. J.* **311**: 151–5.

27. Berard, A., Bravo, G. (1998). Combining studies using effect sizes and quality scores: application to bone loss in postmenopausal women. *J. Clin. Epidemiol.* **51**: 801–7.

28. Berlin, J.A., Colditz, G.A. (1990). A meta-analysis of physical activity in the prevention of coronary heart disease. *Am. J. Epidemiol.* **132**: 612–28.

29. Light, R.J. (1987). Accumulating evidence from independent studies – what we can win and what we can lose. *Stat. Med.* **6**: 221–31.

30. Abramson, J.H. (1990). Meta-analysis: a review of pros and cons. *Public Health Rev.* **9**: 149–51.

31. Slavin, R.E. (1995). Best evidence synthesis: an intelligent alternative to meta-analysis. *J. Clin. Epidemiol.* **48**: 9–18.

32. Jadad, A.R., Moore, R.A., Carroll, D., Jenkinson, C., Reynolds, D.J.M., Gavaghan, D.J., McQuay, H. (1996). Assessing the quality of reports of randomized clinical trials: Is blinding necessary? *Controlled Clin. Trials* **17**: 1–12.

CHAPTER 9

Sensitivity Analysis

9.1 Introduction

Sensitivity analysis:

> ... provides reviewers with an approach to testing how robust the results of the review are, relative to key decisions and assumptions that were made in the process of conducting a review. Each reviewer must identify the key decisions and assumptions that are open to question, and might conceivably have affected the results, for a particular review. [1, p83].

It has been argued that sensitivity analysis should in fact be carried out to reflect decisions made at all stages of a meta-analysis [2]. Indeed, in many facets of meta-analysis, covered in previous sections, it has been indicated that sensitivity analysis should be used to assess the robustness of the results to specific methods used and decisions made. The more the result obtained is materially unchanged by sensible sensitivity analyses, the more confident we will be in the final results of the meta-analysis. This chapter provides a summary of sensitivity analyses previously suggested and adds some new ones. Section 9.2 considers the sensitivity of results to the inclusion/exclusion of studies, while Section 9.3 considers sensitivity of results to the meta-analytic methods used.

9.2 Sensitivity of results to inclusion criteria

This section broadly follows the advice given in the Cochrane Collaboration handbook [1]:

- *Changing the inclusion criteria*. The impact on the results of criteria used to select the studies to be included in a meta-analysis can be examined. Re-analysis can be carried out varying the types of patients included,

147

interventions administered, and the outcome measure definitions, where this is thought to be important.

- *Including or excluding studies where there is some ambiguity as to whether they meet the inclusion criteria.* There may be disagreement between researchers as to whether certain studies meet the inclusion criteria. The effect of including or excluding these studies should be examined.

- *Including or excluding unpublished studies.* Section 7.8 considered whether is was appropriate to include studies which had been identified but which had not been formally peer reviewed or published. Concerns exist to whether such studies may be of inferior quality and bias the meta-analyses findings. For this reason, exploring the influence these studies has on the analysis through sensitivity analysis is valuable.

- *Impact of studies of lower methodological quality.* Chapter 8 considered in detail the issue of the quality of the primary studies to be analysed. Several methods were described which could be used to carry out a sensitivity analysis into the effects of taking into account quality on outcome.

- *Re-analysing the data where uncertainties concerning values extracted exist.* Extraction of data from primary studies is not covered in this book. However, differences in the way outcomes are defined and measured, or inconsistencies in how the results are reported, may make it impossible directly to extract all relevant information. In some situations, assumptions may have to be made to allow for the transformation of outcomes on to one scale. The effect of using a reasonable range of values (possibly the upper and lower limits) where uncertainty exits about the results should be explored through sensitivity analysis.

- *Publication bias assessment.* The issue of publication bias and methods to assess its likely influence on the conclusions of a meta-analysis were the topic of Chapter 7. Methods for testing and/or adjusting the pooled result if publication bias is suspected were described there, and should be implemented as part of a sensitivity analysis.

- *Re-analysing the data where missing values exist.* Due to imperfections in original study reports, or otherwise, missing values may be a feature of the dataset to be analysed. The issue of missing values in meta-analysis has, up to now, been relatively neglected with little literature existing on the subject. Perhaps the simplest way to deal with missing values is to impute a reasonable range of values for them and assess the influence on results. Chapter 13 outlines more sophisticated methods which could be adopted for missing data in a meta-analysis.

- *Simulation of extra trials.* Simulations of extra trials can be carried out to assess the robustness of the results. These may be particularly useful if one knows of trials currently underway, and can estimate likely outcomes of these trials, so that the effect of their addition to the meta-analysis can be checked. Further, this sort of analysis can also inform the design of future trials if a meta-analysis is currently inconclusive, by considering the size of the trial needed to produce a decisive answer when included in the meta-analysis [4].

Example: Investigating the impact of each individual study on the pooled result

A simple but informative sensitivity assessment can be performed by repeating the meta-analysis systematically excluding each individual study in turn. This assessment can indicate which studies are most influential, and help inform the judgement as to whether conclusions rely on the inclusion/exclusion of any particular study. Perhaps the most effective way of displaying the results of such an assessment is to create a plot which shows each pooled result, having excluded a study, compared to the pooled result including all studies. This plot has been generated for the meta-analysis of fluoride preparations to prevent dental caries first described in Section 2.4.1 and is presented in Figure 9.1. Studies have been pooled using a fixed effect model, for this example.

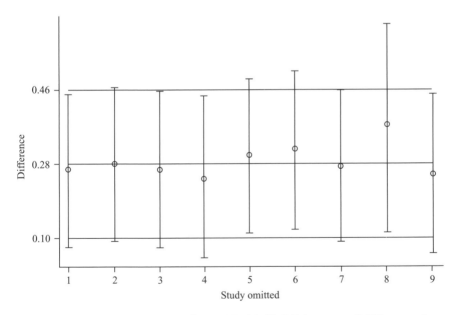

Figure 9.1   Meta-analysis comparing NaF with SMFP in terms of differences from baseline in DMFS dental index. Plot indicating the influence of each study on the overall pooled result.

In this example, the point estimate and confidence interval does not change materially with the exclusion of any individual study. However, it does suggest that study 8 is the most influential.

9.3   Sensitivity of results to meta-analytic methods

Re-analysing the data using different statistical approaches. The two types of model commonly used in meta-analysis, namely fixed and random effects, were described in Chapters 4 and 5. Since there is dispute over the most appropriate model, analysing the data using both methods is recommended, to ascertain how much difference the choice of model makes. Further, if extensions to these models have been used, for example if covariates have been included in a regression model (see Chapter 6), an assessment of their impact is recommended.

9.3.1   Assessing the impact of choice of study weighting

Thompson [3] describes a plot with is helpful in considering the impact of the choice of statistical methods. A random effects analysis can be viewed as simply changing the percentage of weight allocated to each trial, compared to a fixed effect analysis. The random effect analysis gives each study more equal weighting than the fixed effect model, how much more equal depending on the value for the between-study variance.

It is possible to determine the pooled outcome as a function of the between study variance. This means that a value of zero corresponds to a fixed effect analysis and a value of infinity gives the trials an equal weighting. The random effects analysis falls somewhere between these two extremes. Thus, a graph of the pooled odds ratio over this range of values for the between study variance can be constructed, to assess how the overall result varies as the relative study weightings change.

Example: Assessment of the influence of weighting scheme for the cholesterol lowering trials

The 34 cholesterol lowering trials were first analysed using a random effects model in Chapter 5. The reader may recall that there was considerable heterogeneity between studies and the estimate for between study variation was quite large (0.068). Figure 9.2 displays the influence weighting scheme has on the pooled odds ratio using the method described above.

The x-axis has been scaled so that the value zero corresponds to a fixed effect analysis; the value one is equal to pooling giving each study equal weighting; and one half to a random effect analysis (the between study variance/(between study

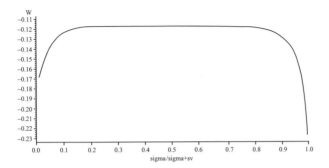

Figure 9.2 Influence of weighting scheme on outcome, pooling cholesterol lowering RCTs.

variance + actual estimated between study variance) is plotted on the *x*-axis) [3]. It can be seen that over a broad range of weighting schemes, the pooled odds ratio is very robust and approximately equal to 0.89 (ln(OR) = 0.12). Only at each extreme does the estimate vary, and then only modestly (note the scale on the *y*-axis). The largest difference from the random effects estimate occurs when studies are given equal weighting, but even then the odds ratio has only reduced to approximately 0.80. Hence, it would appear that analyses of this dataset are robust to the choice of weighting scheme used.

9.4   Summary/Discussion

The importance of carrying out a sensitivity analysis cannot be over-emphasised, and in its various guises it should be done routinely. Assumptions which may need testing, and the methods for doing so are described throughout this book. This chapter is intended to summarize these and to describe some further methods.

References

1. Oxman, A.D. (editor). (1996). *The Cochrane Collaboration handbook: preparing and maintaining systematic reviews. Second ed.* Oxford: Cochrane Collaboration.

2. Olkin, I. (1994). Re: 'A critical look at some popular meta-analytic methods'. *Am. J. Epidemiol* **140**: 297–9.

3. Thompson, S.G. (1993). Controversies in meta-analysis: the case of the trials of serum cholesterol reduction. *Stat. Meth. Med. Res.* **2**: 173–92.

4. Rushton, L., Jones, D.R. (1992). Oral contraceptive use and breast cancer risk: a meta-analysis of variations with age at diagnosis, parity and total duration of oral contraceptive use. *Br. J. Obstet. Gynaecol* **99**: 239–46.

CHAPTER 10

Reporting the Results of a Meta-analysis

10.1   Introduction

This section discusses ways in which results of a meta-analysis can be reported and presented. In previous chapters of this book, results of various methods have already been presented for the illustrative examples, so the reader that has reached this point from the beginning will already be familiar with several of the methods discussed here.

It has been reported that the quality of the reporting of meta-analyses has been generally poor [1]. Over the past 15 years or so, approximately 20 scales and checklists have been developed to evaluate the quality of systematic review reports [2]. Very recently, a new checklist and flow diagram were created at a conference on the Quality Of Reporting Of Meta-analyses (QUORUM) [3]. This includes probably the most comprehensive checklist, including 18 items, nine of which are evidence-based. This has become known as the QUORUM Statement, and addresses primarily the content of the abstract, introduction, methods and results section of a systematic review of randomized trials. This statement is in the same spirit to the CONSORT statement for reporting RCTs [4]. It is recommend reading for those preparing reports of meta-analyses of RCTs.

At another conference, Blair *et al.* [5] attempted to reach a consensus on how the results of meta-analyses in environmental epidemiology should be reported. Their discussion considers the presentation, interpretation and communication of results. Many of the issues covered could relate to meta-analyses in other fields, and this work is recommended as further reading for researchers carrying out meta-analyses in epidemiology, generally.

10.2   Overview and structure of a report

Deeks *et al.* [6] report the NHS Centre for Reviews and Dissemination (CRD) guidelines for the format their systematic review reports should take in medicine and related fields. Although the format of a report for other sources may vary, not least due to space constraints, it is a very good starting reference. Halvorsen [7] follows a very similar structure, and gives many more details. A suggested report outline based on these is presented below. Examples of the application of this report structure is given in the Database of Abstracts of Reviews of Effectiveness (DARE), which is a database of structured abstracts of systematic reviews in the health field (DARE website: *http://nhscrd.york.ac.uk/welcome.html*).

1. *Abstract or executive summary.* Ideally, this should be structured under the headings: objectives, data sources, study selection, data extraction, data synthesis, results and conclusions [6].

2. *Background information.* 'The need for the report should be justified by clearly describing the problem for which evidence of effectiveness is required, and describing the needs of the health care professionals and consumers who are to benefit from the report' [6].

3. *Hypotheses tested/question to be addressed in the review.*

4. *Review methods.* A section on each of the following should be included [6]:
   - Search strategy
   - Inclusion criteria
   - Assessments of relevance and validity of primary studies
   - Data extraction
   - Data synthesis
   - Investigations of differences between studies.

5. *Details of studies included in the review.* A section on each of the following should be included [6]:
   - Details relating to the patient groups included
   - Mode of intervention and the outcomes assessed in each study
   - Details of study results, study design and other aspects of study quality and validity.

   Generally, sufficient information should be provided to allow replication of the analysis [6].

6. *Details of studies excluded from the review.* Give reason for exclusion and details of the specific studies excluded.

7. *Results of the review.* 'The estimates of efficacy from each of the studies should be given, together with the pooled effect if this has been calculated. All results should be expressed together with confidence intervals. The table or diagram should indicate the relative weight that each study is given in the analysis. The test for heterogeneity of study results should be given if appropriate and all investigations of the differences between the studies

should be reported in full. As well as reporting the results in relative terms the impact of the results in absolute terms (such as absolute risk reduction (ARR) and number needed to treat (NNT) (see Section 2.3.4)) should be given. This permits the clinical significance and possible impact of the intervention to be assessed' [6].

8. *Analysis of the robustness of the results.* 'Sensitivity analyses should be performed and documented to investigate the robustness of the results where there is missing data, uncertainty about study inclusion, or where there are large studies which dominate the data synthesis' [6] (see Chapter 9 for more details on sensitivity analysis).

9. *Discussion.* 'A discussion of the strength of the causal evidence, potential biases in both the primary studies and the review, and the limitations they place on inferences, should be given' [6].

10. *Implications of the review.* 'The practical implications of the results both for health care and future research should be discussed. This section should take account the needs of the target audience' [6].

11. *Reference lists.* Three lists of studies should be given [6]:
    • Studies included in the review
    • Studies excluded from the review
    • Any other literature which is referred to in the report.

12. *Dissemination and further research.* 'Suggestions of the main messages for dissemination and the important target audiences should be discussed. Implications for further research should be outlined with a discussion of lessons of the review for the research methods that may be useful' [6].

10.3 Graphical displays used for reporting the findings of a meta-analysis

10.3.1 Forest plots

This type of plot does not appear to have a standard name, however it is often referred to as a 'forest' plot. Much information is succinctly conveyed in such a figure [8], with point estimates and 95% Confidence Intervals (CI) for each study, along with the final combined result and confidence interval all being displayed. In addition, the size of the study may be represented by the size of the box, indicating the estimated treatment effect. This plot has its drawbacks, however, since the reader's eyes are often drawn to the least significant studies, because these have the widest confidence intervals and are graphically more imposing [9]. A point that needs considering is which scale to use on the horizontal axis. If the results of trials are presented in the form of odds ratios, then a log scale may be more appropriate as confidence intervals will be symmetrical. Several forest plots for different meta-analysis examples are presented throughout this book. It is possible to include further information on such a plot, including the name of each study, the raw data and results of subgroup analyses. Producing such a plot may be especially valuable if report space is limited, as it provides a very succinct way of displaying much essential information.

Example forest plots including extra information

Two examples of such plots are given; Figure 10.1 displays the results of the meta-analysis of clozapine vs 'typical' drugs for schizophrenia, first described in Section 2.4.1 using the standardized mean difference outcome scale. Here a study identifier, the numbers in each group of each study, their mean response and its standard deviation are included on the plot, as well as the pooled random effect result, expressed graphically and numerically.

Figure 10.2 displays the results from a meta-analysis of the cholesterol lowering interventions introduced in Section 2.3.1. This provides an example of the extra information which can be included on a forest plot when combining binary data. The total number of patients in each group are displayed, together with the number of events in each group. Additionally, the results of the subgroup analyses pertaining to the different intervention methods, as described in Section 6.2, are given as, well as an overall pooled result.

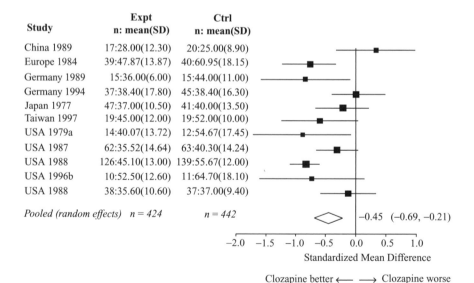

Figure 10.1 Annotated forest plot for meta-analysis of Clozapine vs. 'typical' drugs for schizophrenia. Taken from Higgins JPT. (1999). *ci. plot: Confidence interval plots using S-PLUS, Mannal version 2*, London: Institute of child Health.

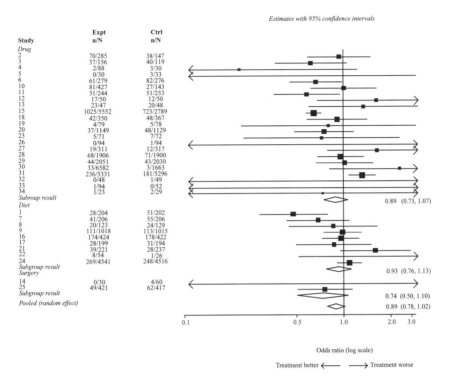

Figure 10.2 Annotated forest plot, including subgroup analyses for the Cholesterol lowering meta-analysis.

10.3.2 Radial plots

These were first described as an informal method to assess heterogeneity in Section 3.3.3. Information regarding the point estimate and precision of each trial is displayed. The gradient of the line of best fit, constrained to go through the origin, indicates the pooled fixed effect result; confidence intervals for this pooled result can also be constructed. When aspects of these studies need to be considered, these plots can provide a useful focus for discussion, enabling differences between subgroups (e.g. plotted using different symbols) and exceptions to be seen easily [10].

10.3.3 Funnel plots

These plots can be used to assist in the assessment of the presence of publication bias, and was described in detail in Section 7.5.1.

10.3.4   Displaying the distribution of effect size estimates

Light *et al.* [8] state that it is important to display the distribution of the effect sizes estimates from the individual studies. Simple boxplots and stem and leaf plots are advocated for doing this.

10.3.5   Graphs investigating length of follow-up

Light *et al.* [8] present a display used when studies have different follow up times. Here time is plotted on the horizontal axis and treatment effect on the vertical. Each study estimate is plotted along with vertical 95% confidence intervals. A running mean line is plotted through these points in addition to a horizontal median line. This graph could help determine if follow up time had an effect on the treatment estimate.

10.4   Summary/Discussion

This section provides an overview of what should be included in the report of a systematic review and meta-analysis. It is important to make methods explicit, and include details of the data combined. This should in principle allow a competent reader to replicate the meta-analysis, and update the analysis if new studies are published.

References

1. Sacks, H.S., Berrier, J., Reitman, D., Ancona-Berk, V. A., Chalmers, T.C. (1987). Meta-analysis of randomized controlled trials. *New Engl. J. Med.*, **316**: 450–5.

2. Shea, B., Dube, C., Moher, D. (2000). Assessing the quality of reports of systematic review: The QUORUM statement compared to other tools. In: Egger, M., Davey Smith, G., Altman, D.G. editors. *Systematic Reviews. 2nd ed.* London: BMJ Publishing Group, (to appear).

3. Moher, D., Cook D.J., Eastwood, S., Olkin, I., Rennie, D., Stroup, D., for the QUORUM group (1999). Improving the quality of reporting of meta-analysis of randomised controlled trials: the QUORUM statement. *Lancet* (to appear).

4. Begg, C., Cho, M., Eastwood, S., Horton, R., Moher, O., Olkin, I. (1996). Improving the quality of reporting of randomised controlled trials: the CONSORT statement. *J. An. Med. Assoc.* **276**: 637–9.

5. Blair, A., Burg, J., Foran, J., Gibb, H., Greenland, S., Morris, R., Raabe, G., Savitz, D., Teta, J., Wartenberg, D., *et al.* (1995). Guidelines for application of meta-analysis in environmental epidemiology. ISLI Risk Science Institute. *Regul Toxicol Pharmacol* **22**: 189–97.

6. Deeks, J., Glanville, J., Sheldon, T. (1996). *Undertaking Systematic Reviews of Research on effectiveness: CRD guidelines for those carrying out or commissioning reviews.* Centre for Reviews and Dissemination. York: York Publishing Services Ltd. Report #4.

7. Halvorsen, K.T. (1994). The reporting format. In: Cooper H., Hedges L.V., (editors). *The Handbook of Research Synthesis.* New York: Russell Sage Foundation; 425–38.

8. Light, R.J., Singer, J.D., Willett, J.B. (1994). The visual presentation and interpretation of meta-analysis. In: Cooper H., Hedges L.V., (editors). *The Handbook of Research Synthesis.* New York: Russell Sage Foundation; 439–54.

9. Galbraith, R.F. (1988). A note on graphical presentation of estimated odds ratios from several clinical trials. *Stat. Med.* **7**: 889–94.

10. Galbraith, R.F. (1994). Some applications of radial plots. *J. Am. Stat. Assoc.* **89**: 1232–42.

11. Oxman, A.D. (editor). (1996). *The Cochrane Collaboration Handbook: Preparing and maintaining systematic reviews. Second ed.* Oxford: Cochrane Collaboration.

Part B:

Advanced and Specialized Meta-analysis Topics

CHAPTER 11

Bayesian Methods in Meta-analysis

11.1 Introduction

Over the last few years, Bayesian methods have been used more frequently in a number of areas of health care research, including meta-analysis [1–3]. Though much of this increase in their use has been directly as a result of advances in computational methods, it has also been partly due to their intuitively appealing and interpretable nature for decision making [4], and also specifically the fact that they overcome some of the difficulties encountered by other methods traditionally used.

Bayesian methods are conceptually quite different from the classical methods described and discussed in Part A of this book. For this reason, a brief overview of the nature and principles underlying Bayesian methods is given before their use in a meta-analysis context is considered.

11.2 Bayesian methods in health research

11.2.1 General introduction

Bayesian methods can be considered as an alternative to the Classical approach to statistical analysis. The name originates from the Rev. Thomas Bayes (1702–1761), who in papers published posthumously [5] outlined an alternative approach for making statements regarding probabilities and random phenomena.

Though this book is primarily concerned with meta-analysis, it is perhaps instructive to consider how Bayes' ideas relate to a single study before generalizing it to the meta-analytical context. A Classical statistical analysis of a single RCT would make use of *only* the data contained in the trial. By contrast, a Bayesian analysis would proceed by first summarizing the evidence external to trial, perhaps from laboratory, animal or non-randomized studies, or based on subjective beliefs. This external evidence is then combined with the observed data to arrive at the *current* state of knowledge regarding the intervention in question. Obviously, this in itself raises a number of issues; for example, when extrapolating from animal studies to humans, different people will hold different beliefs about how animal results will carry over to humans. They might also hold differing beliefs about how reliable the evidence was from, say, a number of perhaps small observational studies. The issue here is that different people will interpret the evidence which was available, external to the RCT, differently. This is a key element of the Bayesian approach, namely that different individuals have their own view of the world, and this introduces the idea of *subjective probability* [6], in contrast to the objective probabilities traditionally attached to specific, often repeatable, events.

Returning to the trial example, assuming that an individual has been able to summarize quantitatively their beliefs prior to the RCT being performed, then the question which the Bayesian approach addresses is *how do these beliefs change in the light of the evidence generated by the trial?* The answer to such a question is that the prior beliefs of the individual are combined with the evidence generated by the trial using *Bayes' Theorem*, which forms the basis of the Bayesian approach.

The quantity of interest, for example the log odds ratio for mortality, is denoted by θ. The prior beliefs are specified as a probability density function $P(\theta)$. The actual data collected in the trial is summarized via a *likelihood function*, denoted $P(Data|\theta)$. Bayes' Theorem can now be used to update the prior beliefs regarding θ in the light of the data, in order to obtain the *posterior density function*, $P(\theta|Data)$, by multiplying the prior density by the likelihood function. Thus, algebraically,

$$P(\theta|Data) \propto P(\theta)P(Data|\theta). \tag{11.1}$$

So that $P(\theta|Data)$ in equation (11.1) is a proper probability density function (i.e. so that it integrates to one), the constant of proportionality is also required, which is given by $\int P(\theta)P(Data|\theta)d\theta$. All inferences regarding θ then use the posterior density function $P(\theta|Data)$. For example, summary measures of location (e.g. mean), and dispersion (e.g. variance) may be obtained. *Credibility Intervals (CrI)*, which are intervals within which θ lies with a specific probability, can also be obtained. These are similar to Confidence Intervals (CI) in a Classical analysis, but they have a more intuitive interpretation that many wrongly ascribe to CIs [7]. The posterior density may also be used to make direct probability statements regarding θ; for example, the probability that θ is less than a specific value, say 0, which may indicate the probability that an intervention has some beneficial effect. Finally, the posterior density may also be used to obtain the *predictive distribution*.

From this, predictive statements regarding future observations, conditional upon the data collected so far and the prior beliefs, may be made.

The resulting beliefs *a posteriori* to the trial are then the beliefs an individual would hold if they updated their prior beliefs in the light of the trial evidence in a rational and coherent manner. A number of points should be noted. First, the prior beliefs need not necessarily be *a priori* in a temporal sense, but rather that they are *external* to the study in question. Secondly, the posterior beliefs obtained by the application of Bayes' Theorem to an individuals prior beliefs may not in fact be the posterior beliefs held by that individual, since s/he may not be rational and coherent in their probabilistic reasoning. Thirdly, there is the important issue of whose prior beliefs to use. The actual numerical application of the Bayesian approach to this issue of the first RCT in the cholesterol meta-analysis is considered again in Section 11.2.3.

The application of Bayes' Theorem to problems when there is only a single parameter is often relatively straightforward, and this is particularly so when the prior density is such that, when it is combined with a specific likelihood function, the resulting posterior density belongs to the same family of distributions as the prior. Such models are termed *conjugate models*, and although they can be restrictive in terms of the distributions that can be used, they often serve as a suitable initial analysis. An example of a Normal conjugate model is considered below in the analysis of the first cholesterol RCT.

When there are numerous parameters, for example in regression settings, conjugate models are not always feasible. In these situations, rather than θ being a single unknown parameter it is now a vector of, say, k unknown parameters, $\mathbf{\theta}$. Bayes' Theorem proceeds exactly as in equation (11.1), but this time it yields a *joint* posterior density for all k parameters. To obtain *marginal posterior* densities for specific individual parameters of interest, the joint posterior density has to be integrated over the remaining *nuisance* parameters. It is has been this need to evaluate often high dimensional integrals which has hampered, until recently, the widespread practical use of Bayesian methods in health care research.

A number of approaches to this problem, including the use of asymptotic approximation methods, numerical integration and simulation [8], have been developed. The use of such numerical methods means that models which use non-conjugate prior distributions can also be considered. *Gibbs Sampling*, a particular type of *Markov Chain Monte Carlo* (MCMC) method [9], has been increasingly used in applied Bayesian analyses within a health care research setting [2, 9, 10]. Gibbs sampling generates samples from the conditional posterior densities which will eventually converge to the desired marginal posterior densities [9, 11, 12]. Although assessment of convergence can be difficult, one advantage of Gibbs sampling is its simplicity, and that it can be performed in any programming environment. The development of a specific package *BUGS* [13], and more recently *WinBUGS* [14], has greatly increased its appeal and use. However, this approach still has some way to go before it is accepted as common-place in the health research literature. Strong polar reactions are still to be found among many statisticians, who are either strong advocators or opponents of the Bayesian approach.

The role of the prior distribution is an important aspect of a Bayesian analysis, and indeed one of the criticisms of the Bayesian approach is its dependence on such prior distributions. The specification or elicitation of prior beliefs, especially in a multi-parameter setting, is a non-trivial task. Therefore, a number of approaches have been developed in which *vague prior distributions* have been used, so that the data, through the likelihood function, effectively dominate the prior distribution. What is considered to be a vague prior distribution is very much dependent upon the context. In some situations, uniform prior distributions may be used so that the posterior density becomes the *standardized likelihood*. In other contexts, prior distributions which have particular properties may be appropriate, for example a *Jeffreys' prior* [15]. Whichever prior distributions are used, a crucial aspect of any Bayesian analysis is the sensitivity of the conclusions to the choice of prior distribution(s), and it is important that a *sensitivity analysis* is performed [7].

A group of methods termed *Empirical Bayes* have become increasingly used in health care research, though there is also a considerable body of literature on these methods generally [16–22]. Such methods have acquired the name Empirical Bayes because they make use of some aspects of the (fully) Bayesian approach, but not that of subjective probability and inclusion of subjective beliefs. They use the idea of a prior distribution to impose structure on a problem, but they do not use subjective *a priori* beliefs to derive or elicit actual numerical values for the hyper-parameters of the prior distributions, for example the mean and variance in the case of a Normal prior distribution. Instead, they estimate the most plausible values of the hyper-parameters from the data. The key issue is that empirical Bayes methods only use the observed data, though some element of subjective judgement does have a role to play in the choice of the actual *form* of the prior distribution. In a meta-analysis context, empirical Bayes methods will generally produce very similar results to the random effect models described in Chapter 5 [23, 24].

More detailed accounts of the application of Bayesian methods, generally [25–32], within a health care research context [3, 7], and specifically in a Randomized Controlled Trial (RCT) setting [33–44], are available elsewhere.

So far, this discussion has been somewhat abstract, but in Section 11.2.3 a Bayesian analysis of the first trial of the cholesterol meta-analysis using a Normal conjugate model is presented; later in Section 11.3, Bayesian methods are specifically applied to the problem of meta-analysis.

11.2.2   General advantages/disadvantages of Bayesian methods

Whilst there are specific advantages to adopting a Bayesian approach, there are also a number of disadvantages. Some of the main advantages and disadvantages are listed below:

Advantages

- Probability statements may be made directly regarding quantities of interest, e.g. the probability that patients receiving drug A have better outcome than those receiving drug B.
- All evidence regarding a specific problem can be taken into account rather than (a) in the context of a single study, just the information contained in that study, or (b) in the context of a meta-analysis, just the studies whose results are being combined directly. Thus, this allows a summary of the current state of knowledge.
- Predictive statements can be made easily, conditional on the current state of knowledge.
- Elicitation of prior beliefs (or expectations) of investigators about, say, the size of treatment effect, combined with the elicitation of *demands* – the minimum effect that would be considered clinically worthwhile – requires explicit consideration of the rationale for the initiation, design, monitoring and stopping of studies.
- In a Bayesian meta-analysis, individual studies *borrow strength* from one another, which can contribute to the estimation of individual study effects (see Section 11.6).
- Facilitates a decision theoretic framework which can also take into account costs/utilities when making health care or policy decisions.

Disadvantages

- Use of prior beliefs undermines the notion (or illusion) of objectivity.
- Elicitation of prior beliefs is non-trivial, and at present there are few guidelines.
- There is no automatic or conventional single measure of statistical significance.
- Bayesian analyses can be computationally complex to implement, and thus time consuming to perform.
- At present there are some software limitations, although availability of suitable software is improving rapidly.

11.2.3 Example: Bayesian analysis of a single trial using a normal conjugate model

Consider the first cholesterol lowering trial in Table 2.1. As previously mentioned, one has to express prior beliefs regarding the relative effectiveness of cholesterol reduction in terms of a probability density function. For illustration, assume that the data are viewed as a random sample of size n from a Normal distribution with unknown mean θ and known standard deviation σ, and the aim is to assess uncertainty about θ in light both of the data and available prior information.

Assuming that the summary statistic for the data x_n can be assumed to be normally distributed, then

$$x_n \sim N[\theta, \sigma^2/n]. \tag{11.2}$$

Assuming further that the *a priori* beliefs regarding θ can be expressed as a Normal distribution with mean θ_0 and variance σ^2/n_0, thus

$$\theta \sim N[\theta_0, \sigma^2/n_0], \tag{11.3}$$

then the resulting posterior distribution, combining the likelihood (11.2) with the prior distribution (11.3), is given by

$$\theta|x_n \sim N[(n_0\theta_0 + nx_n)/(n + n_0), \sigma^2/(n + n_0)], \tag{11.4}$$

where x_n is the log odds ratio $\sigma^2 = 4$ [43]. Hence, in this simple application of Bayesian methods, the posterior distribution of interest can be written down 'directly', without the need of more complex numerical methods.

In terms of the first primary study in the cholesterol meta-analysis, the observed data were 28 deaths out of 204 patients on the treatment arm, and 51 deaths out of 202 patients on the control arm. Thus, the observed log odds ratio is -0.74, and $n = 79$.

If we wish to represent vague *a priori* beliefs regarding θ, then as n_0 tends towards zero in equation (11.4), i.e. *a priori*, we are less certain regarding beliefs about θ, the posterior distribution approaches the standardized likelihood, i.e. $N[x_n, \sigma^2/n]$. Figure 11.1(a) shows graphically the posterior distribution in this case. Although such a prior distribution has the advantage that it allows the data to dominate, it could be considered unrealistic, since it effectively assumes that all values of θ are equally likely. If instead we assumed an *a priori* sceptical stance (i.e. not believing there is a positive effect), we could set $\theta_0 = 0$ and $n_0 = 40$, which assumes that the uncertainty associated with our beliefs that $\theta_0 = 0$ is equivalent to a hypothetical trial in which 40 events had been observed, i.e. approximately half the number in the first primary study. Such a prior distribution would correspond with a prior probability of $\theta < -0.5$, i.e. an odds ratio of 0.6, of less than 5%. Thus, the posterior distribution in this case is given by

$$\theta|x_n \sim N[(40 \times 0 + 79 \times -0.74)/(79 + 40), 4/(79 + 40)]$$
$$\sim N[-0.049, 0.03].$$

We can see from Figure 11.1(b) that the posterior mean for θ has been shifted slightly towards zero as a result of the prior beliefs, compared to the observed log odds ratio of -0.79, but that the amount by which is has been modified is in proportion to the ratio of the *a priori* and observed variances. The other point to notice is that the posterior variance 0.03 is smaller than the observed variance 0.05, reflecting the fact that there has been an increase in the amount of evidence on which the analysis has been based (i.e. reduced uncertainty). The posterior probability that the log odds ratio is less than -0.5 is now 48%. Alternatively, an

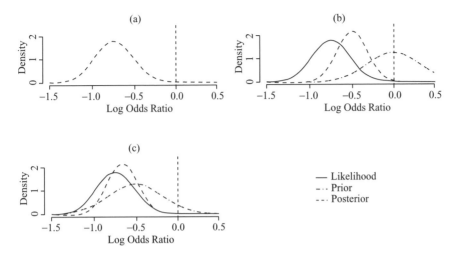

Figure 11.1 (a) Standardized likelihood, (b) sceptical prior, likelihood and posterior, (c) enthusiastic prior, likelihood and posterior, for first primary RCT.

enthusiastic prior distribution could be considered as one which gives a small probability that $\theta > 0$, i.e. that cholesterol reduction is harmful. Again assuming that uncertainty is represented by $n_0 = 40$, then letting $\theta_0 = -0.5$ would ensure that *a priori* the probability of a harmful effect of cholesterol reduction, i.e. $\theta > 0$, is less than 5%. Figure 11.1(c) shows the posterior distribution in this case, which has a mean of -0.66, and our updated beliefs are now such that the probability of cholesterol reduction having a harmful effect, i.e. $\theta > 0$, has fallen to less than 0.1% as a result of our beliefs being updated in the light of the first primary study.

11.3   Bayesian meta-analysis of normally distributed data

Many of the authors who have considered a Bayesian approach to meta-analysis have implemented a hierarchical model, which models both within-and between-study heterogeneity, making various assumptions of Normality [45–51]. In other areas of statistical science, such Bayesian hierarchical models have been used for some considerable time [52, 53]. Before considering specific approaches that have been taken, a basic hierarchical model is outlined.

Consider again the random effects model (Chapter 5) where the ith study in a meta-analysis, of size n_i can be summarized by an outcome measure T_i, and associated with the outcome measure is a within-study variance σ_i^2. Then the first level of a two level hierarchical model relates the observed outcome measure T_i to the underlying effect in the ith study θ_i. At a second level of the model, the θ_is are in turn related to the overall effect μ, in the population from which all the studies are assumed to have been sampled. τ^2 is the between-study variance or the variance of the effects in that population. From a Bayesian perspective, a number of unknown parameters σ_i^2, μ and τ^2, have to be estimated, and therefore prior distributions must be specified for them. Thus, denoting an as yet unspecified prior distribution by $[-, -]$ the model has the following form:

$$
\begin{aligned}
T_i &\sim N[\theta_i, \sigma_i^2/n_i], \ldots, \quad \sigma_i^2 \sim [-, -] \quad i = 1, \cdots, k \\
\theta_i &\sim N[\mu, \tau^2] \\
\mu &\sim [-, -] \quad \tau^2 \sim [-, -].
\end{aligned}
\tag{11.5}
$$

Having specified the three required prior distributions, estimation can then proceed using the various computational approaches mentioned in Section 11.2.1. However, the hierarchical nature of the model, together with the assumptions of Normality, and combined with the fact that there are often a considerable number of studies in any specific meta-analysis, make MCMC methods particularly suited to such models. Estimation aside specification of the prior distributions is not a trivial task, and the choice of which prior distribution to choose has received considerable attention recently [54, 55].

Choice of prior distributions for σ_i^2, μ and τ^2

Whether the σ_i^2s are assumed known and hence replaced by the observed within-study variances, or whether they are assumed random and therefore have a prior distribution specified, makes little practical difference, apart from when there are a number of small studies [23]. If the σ_i^2s are considered random, and therefore require specification of a prior distribution, a number of possibilities exist. The most frequently used prior distribution is $P(\sigma_i^2) \propto 1/\sigma_i^2$ which corresponds to a Jeffreys' prior. Although appealing for theoretical reasons, such a prior distribution is not always feasible in practice, and a commonly used alternative, distributional-based prior, is an Inverse Gamma distribution. This distribution is particularly flexible, and can accommodate a number of possible scenarios. It also has the benefit of only being defined on the positive real line.

   In terms of a prior distribution for μ, we sometimes wish to remain relatively objective about the overall population effect, and it is common practice either to assume a particularly vague proper prior distribution, or to use a Uniform distribution over the whole real line. Alternatively, when there is pertinent information external to the meta-analysis regarding the overall effect, for example from observational studies, then this may be incorporated into the prior distribution for μ, or alternatively, it could be modelled explicitly using the generalized synthesis methods of Chapter 17. The use of *any* prior distribution should, though, be subjected to a sensitivity analysis.

11.3.1 Example: Combining trials with continuous outcome measures using Bayesian methods

The meta-analysis dataset comparing clozapine to 'typical' drugs for schizophrenia, provided in Table 2.5, which was originally analysed using a standard classical random effect model and ones incorporating the uncertainty induced by estimating the between study variance in Section 5.6.1, is now analysed using Bayesian methodology.

Considering the model defined by equation (11.5), it is necessary to consider prior distributions for σ_i^2, μ and τ^2. In this example, it is assumed that the σ_i^2 s are known, and hence no prior distribution is required, reasonably vague priors are set for the remaining two parameters.

For μ, the overall population effect, a Normal prior distribution is used. This is centred at 0, and has a standard deviation of 5, hence approximately 95% of the probability mass lies within ± 10 standardized units. This is justified because it is assumed that an effect size of greater magnitude than 20 (i.e. four standard deviations in either direction) for a new depression drug for schizophrenia would be very unlikely indeed.

For τ^2, an Inverse Gamma prior distribution is used. If it is assumed that there is a one order of magnitude spread in effect sizes between studies, and secondly that it is highly unlikely that there are two orders of magnitude variability between studies, this translates approximately into an Inverse Gamma [3, 1] distribution (see Smith *et al.* [56] for further details). However, as has been commented on earlier, in practical situations the sensitivity of conclusions to the choice of particular prior distributions should be assessed via a sensitivity analysis.

Posterior estimates of the model parameters were obtained using Markov Chain Monte Carlo (MCMC) methods implemented in WinBUGS. A detailed explanation of how this program can be used for meta-analysis is given elsewhere [56]. The Bayesian model produces a pooled standardized mean difference of -0.48 (95% CrI -0.78 to -0.17), which is close to the results achieved using the two classical random effect models that incorporate the uncertainty in estimating τ^2 from the data, of -0.46 (95% CI -0.70 to -0.22) and -0.47 (95% CI -0.72 to -0.22). The estimate for τ^2 is 0.26 (95% CrI 0.12 to 0.54), and is larger than the classical estimates of 0.09 (95% CI 0.03 to 0.30) and 0.10 (95% CI 0.01 to 0.51). This can be attributed to the prior distribution placed on it, which was slightly informative, thus causing the slight increase in the width of the credibility interval for the pooled standardized mean difference.

11.4 Bayesian meta-analysis of binary data

All the model derivation in Section 11.3 has assumed that the outcome measure for each study can be assumed to be normally distributed. While making such an assumption facilitates estimation, this might not be tenable from a practical point of view.

Two possible formulations exist for modelling binary data. Consonni and Veronese [57] consider the modelling of binary outcome data in meta-analyses directly in a hierarchical model. The observed responses in a single arm of the trial are modelled using binomial distributions, with conjugate Beta distributions at the further levels of the model. Though such an approach is computationally attractive, due to the conjugate nature of the model, it is of limited value in comparative experiments.

An alternative model formulation, which has been adopted by a number of authors [56, 58–63], is briefly described in a general form below. In this approach, although the observed responses in each arm of a trial are assumed to follow a binomial distribution, a suitable transformation is then applied, frequently logit in nature, to the risk parameters. Following such a transformation the model formulation proceeds as in Section 11.3, though parameter estimation requires some form of numerical, simulation, or approximation method, to be employed.

Consider a two-arm study in which r_1 and r_2 are the observed number of responses out of n_1 and n_2, respectively. Then the first level of the model is

$$r_1 \sim \text{Bin}[\pi_1, n_1] \quad r_2 \sim \text{Bin}[\pi_2, n_2], \tag{11.6}$$

where π_1 and π_2 are the two unknown risk parameters for the two arms of the study. Consider now the logit transformation of each of the two risk parameters, such that

$$\log(\pi_1/1 - \pi_1) = \mu_i - \delta_i/2 \quad \log(\pi_2/1 - \pi_2) = \mu_i + \delta_i/2; \tag{11.7}$$

δ_i is now the parameter of interest, being the log odds ratio. This is often then assumed to be approximately Normally distributed, and the second level of the model becomes

$$\delta_i \sim \text{N}[\phi, \tau^2], \tag{11.8}$$

where ϕ represents the overall pooled effect, on a log odds ratio scale, and τ^2 is a measure of the between-study heterogeneity. As in Section 11.3, in a fully Bayesian analysis prior distributions have to be specified for both ϕ and τ^2. Thus, as before the final level of the model involves specification of prior distributions for ϕ and τ^2,

$$\phi \sim [-, -] \quad \tau^2 \sim [-, -]. \tag{11.9}$$

The key difference between this model and that of equation (11.5) is the assumption that at the lowest level of the model the responses in each arm a study are modelled directly. In equation (11.5), calculation of the log odds ratio when there are zero or complete responses in either arm of a study requires various assumptions to be made, usually the addition of 'small' constants to the responses frequencies. It is this assumption of Normality of the log odds ratio, or other transformed measures of binary data, in models such as equation (11.5) that is frequently not valid.

11.4.1   Example: Combining binary outcome measures using Bayesian methods

In Table 2.1 the 34 cholesterol trials report results for both all-cause mortality and CHD mortality. Here we consider CHD mortality. Note that there were no deaths from CHD mortality in either arm of trial number 32. This trial adds no information about this outcome, and is hence not included in the analysis. The remaining 33 cholesterol trials are now combined using Bayesian methods. The model used, is that in equations (11.6)–(11.9) described above, which assumes that the observed responses in each arm of each study follow a binomial distribution. Hence, the data are modelled directly, and no continuity correction factor is required for trials with zero response in a single arm (see Section 2.2). In addition to the 33 trials, six cohort studies have also investigated the association between cholesterol reduction and CHD mortality. Figure 11.2 shows the effects in the individual cohort studies together with 95% confidence intervals, and an overall pooled estimate using the Bayesian model of Section 11.2 for the log odds ratio. The estimate of the pooled log odds ratio is obtained as -0.28, with standard deviation 0.06 and a 95% CrI from -0.42 to -0.14, which is wider than any of the confidence intervals for the individual studies, due to large between study heterogeneity.

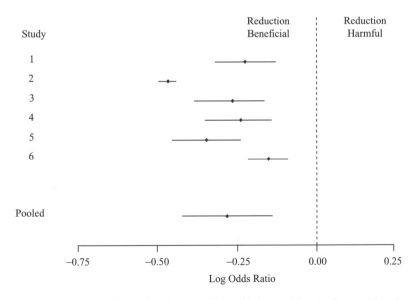

Figure 11.2 Meta-analysis of cohort studies which consider cholesterol levels and CHD mortality.

Initially, prior distributions are placed on ϕ (the log odds ratio parameter) and τ^2 in a very similar manner to the example of section 11.3.1. For ϕ, if it is assumed that the odds ratio does not exceed 500 in favour of either arm, then this equates to a log odds ratio of between -6.2 and 6.2. Hence, an approximate standard deviation can be obtained by dividing these limits by two (as for normally distributed data 95% of the distribution mass lies between two standard deviations of the centre, zero). Hence, a Normal prior distribution with mean 0 and standard deviation 10 is used. An Inverse Gamma [3, 1] distribution is used as the prior for τ^2, using the same reasoning as the Example in Section 11.3.1. Additionally, it is necessary to set priors for each μ_i in equation (11.6), these are set to be Normally distributed with a mean of 0 and a standard deviation 2 on the grounds that we wish the probability of an event (i.e. π) to lie between 0.02 and 0.98 with 95% probability (see Smith *et al.* [56] for further details).

Again, the model is implemented in WinBUGS, which uses MCMC methods. Details of how this can be achieved have been reported elsewhere [56]. This model produces an estimate of the pooled log odds ratio of -0.19 (95% CrI -0.39 to $+0.01$). These results are in broad agreement with those obtained using the classical random effects model of Chapter 5, which yielded an estimate of the log odds ratio of -0.16 (95% CI -0.31 to -0.01). The between-study variance is estimated as 0.42 (95% CrI 0.31 to 0.57) using the Bayesian approach, whilst the Classical moment-based estimate is 0.08. The slightly larger estimate of the between-study variance using the Bayesian approach thus gives rise to the wider 95% interval for the log odds ratio. This initial posterior density for the pooled log odds ratio can be seen in Figure 11.3(a). As with the analysis of the single trial in Section 11.2.3, we can make direct probability statements from the posterior distribution regarding the effectiveness of cholesterol reduction in terms of CHD mortality. For example, the posterior probability that cholesterol reduction would lead to a relative risk reduction of more than 20%, i.e. corresponding to a log odds ratio of -0.22, is 38%.

This initial Bayesian analysis has not made use of the results from the six cohort studies in Figure 11.2. A Bayesian analysis might use the overall pooled result for the cohort studies to derive a prior distribution for ϕ, the log odds ratio. Alternatively, evidence external to the meta-analysis in question may be used to derive prior distributions for other model parameters, especially τ^2 [50, 62, 64]. A *naïve* approach may use the pooled result directly, such that the mean of the prior distribution is -0.28, and the standard deviation is 0.06. Therefore, repeating the meta-analysis of the 33 trials using this prior distribution for ϕ, but retaining the same prior distributions as before for all the other parameters, the overall log odds ratio is now estimated as -0.26 (95% CrI -0.36 to -0.16), whilst the between-study variance is estimated as 0.42 (95% CrI 0.30 to 0.57). We can see from Figure 11.3(b) that by accepting the observational results at 'face value', the posterior mean has shifted towards the prior mean quite considerably as the prior standard deviation, 0.06, is smaller than the standard deviation of the log odds ratio in the 33 trials, i.e. 0.08.

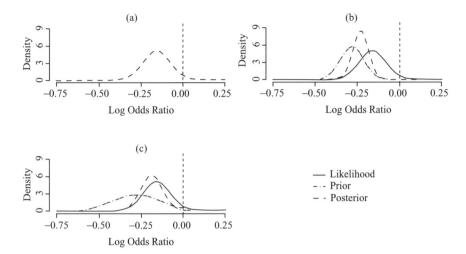

Figure 11.3 (a) Initial posterior, (b) naïve prior, likelihood and posterior, (c) down-weighted prior, likelihood and posterior, for meta-analysis of cholesterol reduction and CHD mortality.

A more appropriate analysis might be to 'down-weight' the relative importance of the observational evidence, perhaps due to the various biases we think it might contain (65). For example, if we repeated the analysis retaining the prior mean as -0.28, but this time increasing our uncertainty surrounding this value by a factor of 4 (i.e. using a new standard deviation of $0.06 \times 2 = 0.12$), then the overall pooled log odds ratio is estimated as -0.23 (95% CrI -0.38 to -0.08), and the between-study variance is estimated to be 0.42 (95% CrI 0.30 to 0.56). In Figure 11.3(c) we can see that this time, because we have down-weighted the prior evidence the posterior estimate of the effect of cholesterol reduction, is nearer to that observed in the trials. In each case, the posterior estimates for the between-study variance is similar, as we have retained the same prior distribution. Whilst the use of a weighting factor of 4 was somewhat arbitrary, it highlights the flexibility of the Bayesian approach in allowing an exploration of the sensitivity of the conclusions of a meta-analysis to the evidence which it uses, and perhaps more importantly, the relative credibility of *different* sources of evidence.

11.5   Empirical Bayes methods in meta-analysis

The use of Empirical Bayes (EB) approaches has received much attention in the literature, as until recently the use of a fully Bayes approach has been hampered by computational difficulties [17, 18, 61, 66–70]. However, the EB methods that have been used have almost exclusively assumed that the 'prior distribution' has been at the second level of the Bayesian hierarchical model (11.5), and that the hyper-parameters, in this case μ and τ^2, have then been estimated from the data. Such an approach is analogous to assuming instead a three level model as equation (11.5) and assuming uniform prior distributions for both μ and τ^2 as used by Carlin [23]. In addition, Smith *et al.* [56] and Biggerstaff *et al.* [71] explain that the random effects model of DerSimonian and Laird [72] (Chapter 5) could in fact be viewed as an empirical Bayes approach. The main drawback with this, however is that no allowance is made from the fact that τ^2 is estimated from the data, using either Maximum likelihood or moment estimation methods (however, Section 5.6.1 does describe classical methods which do this). Indeed, Carlin [23] goes on to show that, to take account of this, some form of quadrature or simulation method is required.

11.6   Advantages/disadvantages of Bayesian methods in meta-analysis

11.6.1   Advantages

Unified modelling approach

By using a Bayesian modelling approach for combining studies, the debate over the appropriateness of fixed and random effect models is overcome, whilst at the same time including the possibility of regression models [56] and the extension to more complicated hierarchical models which can accommodate different types of studies [54, 73] (see Chapter 17 for further details).

Borrowing strength

Borrowing strength can be seen as a byproduct of a fully Bayesian meta-analysis approach. When study estimates are combined, the model updates estimates of the individual studies, taking into account the results from *all* the other studies in the analysis. Thus, narrower confidence intervals will be obtained for each individual study, by *borrowing strength* from all other studies. As well as reducing the width of the confidence interval, the point estimates of the individual studies will also be affected, moving them closer together towards the overall pooled estimate. The degree to which individual study estimates will be shrunk will depend upon the relative magnitude of the within-study variances. Gaver *et al.* [74] report that a variety of statistical ideas and terms are used to describe this concept, including *shrinkage*, *empirical Bayes* and *hierarchical Bayesian* modelling.

These concepts are particularly useful if one is interested, not in some overall, 'average', of the study results, but instead about making inferences about any particular treatment effect. Then results from the other studies can be used to, 'improve', this estimate. This leads to better point estimates and narrower interval estimates of any particular effect. Gaver *et al.* [74] note that approaches to borrowing strength, with applications to medicine and health, are much rarer than the meta-analytic approach that focuses on estimating population parameters. However, DuMouchel and Harris [47] elaborate such a method to improve estimation of cancer effects in humans, by borrowing strength from experimental data on laboratory animals in experiments using the same carcinogens (see Section 17.7.5). In addition, Laird and Louis [75] give parametric and bootstrap methods for constructing empirical Bayes confidence intervals which may be applied to obtain individual study estimates. Indeed, recently interest in the concept of borrowing strength has increased with respect to institutional comparisons, see Spiegelhalter and Goldstein [76].

Example

The meta-analysis comparing clozapine to 'typical' drugs for schizophrenia, analysed using Bayesian methods in Section 11.3.1, is used to illustrate how shrinkage works. A plot of the original study estimates, and the shrunken ones produced by this analysis, is provided in Figure 11.4. Notice how the shrunken estimates (the dotted lines) have narrower confidence intervals than the original ones, and that they are 'shrunk' towards the pooled result.

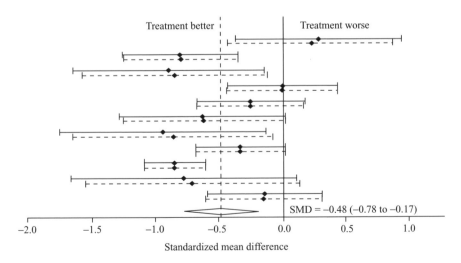

Figure 11.4 Shrinkage plot for the meta-analysis comparing clozapine to 'typical' drugs for schizophrenia.

Allowing for all parameter uncertainty

Empirical Bayes (simple models) and simple Classical approaches do not allow for the fact that both μ (mean) and τ^2 (the between-study variance) have both been estimated from the data. The Bayesian approach allows for the greater uncertainty.

Allowing for other sources of evidence

Often, meta-analyses are conducted in substantive areas in which evidence is available from sources other than RCTs, for example, the majority of evidence is from RCTs, but other evidence exists in the form of observational studies. Such external information can be used to either construct prior distributions for parameters (see the example in Section 11.4.1), or be included in a more direct manner (see Chapter 17).

Ability to make direct probability statements

Bayesian meta-analysis can give a probability that the effect is above (or below) a particular value. For example, the probability that an odds ratio is greater or less than one can be calculated, indicating the chance an intervention is greater or less than zero. In this way, Bayesian methods can give a direct answer to a question of interest [77].

Prediction

By adopting a Bayesian approach, it is possible to incorporate evidence from previous trials into the design of a new one [78]. Using this approach, one can take the result of a meta-analysis and calculate the probability that a planned or current study, with a fixed sample size, will produce conclusive results, or alternatively, estimate what the effect observed in a future study of 'average' size is likely to be. For example, in Section 11.4.1 we can estimate that the effect of cholesterol reduction in a future trial would lead to a risk reduction of 17% for CHD mortality, with a 95% CrI from a 64% risk reduction to a 95% increase in risk, based on the results of the meta-analysis.

11.6.2   Disadvantages

Specification of prior distributions

When performing a fully Bayesian meta-analysis, one needs to specify prior distributions for the population effect size μ and the between-study variance τ^2. Specification of these prior distributions, even when there is information external to the meta-analysis available can be difficult. Indeed, Louis and Zelterman [79] comment:

> Although we may 'all think like Bayesians', it can be extremely difficult to evaluate and communicate prior opinions, and considerable research continues on this aspect of Bayesian analysis.

Sensitivity to prior distributions

It should be remembered that a meta-analysis is not conducted to inform a single individual, but rather to communicate the current state of evidence to a broad community of consumers. If the prior distributions differ substantially for different consumers, then the related Bayesian analyses can produce qualitatively, as well as quantitatively, different conclusions. Therefore, it is important to perform a sensitivity analysis over the range of opinions. If conclusions are stable, then we have 'findings'. If they are not, then the collection of Bayesian analyses underscores the finding that the data are not sufficiently compelling to bring a group of relevant consumers to consensus. This situation should then motivate additional primary studies, and indeed provide a focus for the design of such studies.

Estimation of posterior distribution

Estimating the parameters of the posterior distribution can be difficult. Until recently, more complex, but yet still quite basic models, were dealt with by approximating the posterior mean and variance. However, recent advances in computational approaches allow the analyst to produce full posterior distributions for even complex models.

Comparison of Classical and Bayesian approaches

A number of authors have compared Classical and Bayesian approaches to meta-analysis [23, 56, 59, 69, 71, 80]. The conclusions of these are consistent with the descriptions in this book, i.e. that standard Classical and Empirical Bayes methods do not take into account all parameter uncertainty, and hence produce results which can be too precise. However, differences in the parameter estimates obtained using Classical, Empirical Bayes, and fully Bayesian methods, the latter using non-informative prior distributions, will generally be small, especially when the number of studies in a meta-analysis is large. Obviously, if informative prior distributions are placed on model parameters this can have considerable impact on the results, and the use of sensitivity analyses to explore fully the impact of prior distributions on overall conclusions cannot be stressed too much.

11.7   Extensions and specific areas of application

11.7.1   Incorporating study quality

The assessment of, and inclusion of, study quality generally was considered in Chapter 8. Clearly, the quantitative assessment of a measure of study quality may be included in a Bayesian analysis in the same way that it may be in a Classical analysis.

One further approach is to include prior opinions, relating quality and study bias, provided by one or more 'experts', into the model. Analyses can be performed for each prior distribution to assess the dependence of conclusions on individual opinion (regarding the quality of the studies). If the conclusions are stable over this 'community of opinion', then the meta-analysis is likely to have substantial impact. If conclusions are sensitive to the range of opinions, then no consensus has been reached [79].

This idea raises a number of questions, the key one of which is what are data and what are prior beliefs [79]. In practice, relatively little work has been conducted in this field. Prevost *et al.* [54] considered the inclusion of quality assessments in relationship to the credibility of different research designs in a generalized synthesis of evidence approach (see Chapter 17).

11.7.2   Inclusion of covariates

To-date, many of the applications of Bayesian methods in meta-analysis have been to mirror Classical random effects models. This has been partly due to the computational difficulties in applying fully Bayesian models, and partly due to the fact that the use of Bayesian methods in meta-analysis is still relatively new. In theory, extension of the model in equations (11.5) and (11.8) poses no difficulties, with μ being replaced by $\beta^T x_i$, where β is a vector of regression coefficients and x_i is a vector of study-level covariates. In a Bayesian setting just as a prior distribution was required for μ, one also needs to be specified for β. In such settings, it would appear that the use of MCMC methods is particularly appealing, since the inclusion of extra parameters will almost certainly preclude the use of other numerical methods [79].

11.7.3   Model selection

As with any modelling exercise, the eventual selection of a 'final model' is a difficult task, and one which, in the meta-analysis literature, has received relatively little attention. This is partly as a result of the relative lack of use of regression models generally, both Bayesian and Classical. That having been said, as previously noted, the choice between fixed and random effect models has received considerable attention and aroused heated debate. From a Bayesian perspective, this is almost irrelevant, since exploration of the marginal posterior distribution for τ^2 will yield an assessment of the between-study heterogeneity present. Abrams *et al.* [64, 81] have considered the question of model selection in a meta-analysis setting from a Bayesian perspective using Bayes Factors (BFs) to either discriminate between the competing models, or to average over them using the posterior model probabilities.

11.7.4 Hierarchical models

The use of Bayesian methods that mirror the Classical random effects models continue to make the assumption that all studies can be considered on an equal footing, and that any one study can be used to help inform inferences made about any other study – the notion of *exchangeability*. This may not be a reasonable assumption for a variety of reasons, and when it is not, a hierarchical modelling framework allows some relaxation of this assumption by grouping studies that can be considered exchangeable. In essence, the two level model described in equation (11.5) can be extended, so that individual studies are nested within an intermediate level. For example, in Section 17.6.2 studies with different designs are synthesized within a three level hierarchical model.

The development of hierarchical models has been considered within the mainstream Bayesian literature by a number of authors [26, 52, 82]. Use of such models has been relatively limited due to computational complexities, though developments in MCMC methods have meant that they are now more feasible in practice. From a Classical perspective, parallel models often referred to as *multi-level models* [83] or *random component models* [84] have been increasingly used in a variety of health care settings, with the advent of relatively user-friendly software [85].

In a meta-analysis context, Veronese [86] considered the case when the assumption of exchangeability might not be reasonable for all studies in a meta-analysis, but that uncertainty exists as to the optimum partition of the studies within a hierarchical model, whilst Higgins and Whitehead [62] consider the use of a Bayesian hierarchical approach to the situation when there are both patient level and study level covariates.

11.7.5 Sensitivity analysis

Sensitivity analysis is just as important when using Bayesian methods as it is when combining using those which are classically derived, and the methods described in Chapter 10 are still applicable. However, there are additional issues which require exploration when carrying out a Bayesian meta-analysis. Perhaps most important is the sensitivity of the results on the prior distributions placed on the model parameters; it is always wise to check the results over a plausible range of prior beliefs. Additionally, a trace plot [46] can be constructed which displays how the pooled estimate and the shrunken individual study estimates change over the range of values contained in the posterior distribution for τ^2, the between-study variance. Figure 11.5 shows such a plot for the clozapine versus 'typical' drugs for schizophrenia meta-analysis described in Section 11.4.1. The posterior distribution of τ is represented by the histogram. Twelve lines are also displayed; the one labelled A (and listed in the key as the intercept) shows how the pooled estimate changes over the values of τ, and the other 11, labelled B through to L, show the change in the shrunken estimates for the individual studies. In this example, the pooled estimate

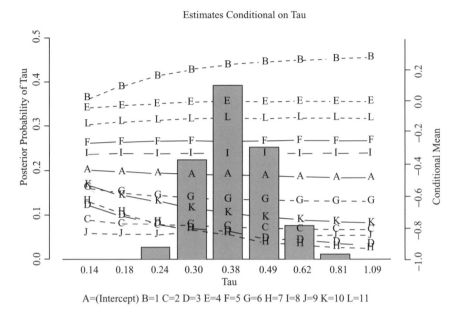

Figure 11.5 Trace plot for the clozapine versus 'typical' drugs for schizophrenia meta-analysis.

remains stable over all values in the posterior. Notice how the estimates for the individual studies diverge as τ increases – this is because as the between-study heterogeneity increases less 'borrowing of strength', and hence 'less shrinkage', occurs and the shrunken estimates remain close to the observed estimates. This plot is very similar to that described in Section 10.3.1, which considers all possible values of τ, rather than the most likely.

11.7.6 Comprehensive modelling

Meta-analysis is primarily concerned with estimating the relative effectiveness of various interventions. However, in terms of health care decision making, such estimates are required to be combined with other pieces of information, for example on costs and utilities. This is frequently done within a decision model [87]. In combining these various forms of information within a single (comprehensive) model, it is important that the uncertainty associated with each of them is also properly taken into account. The adoption of a Bayesian framework would appear to be one particularly attractive way of doing this [32].

11.7.7 Other developments

Other aspects of Bayesian methodology for meta-analysis have been developed. Some of these have already been discussed in previous chapters, and more are discussed in later chapters where they fit more naturally. For instance, Section 6.3.5 considered a Bayesian method for appropriately fitting baseline risk as a covariate within a meta-regression model. In future chapters, Bayesian approaches to missing data (Section 13.5), cumulative meta-analysis (Section 19.4), the generalized synthesis of evidence (Chapter 17), and combining diagnostic test results (Section 14.4.4) are all considered.

11.8   Summary/Discussion

This chapter has summarized the general use of empirical and fully Bayesian methods with respect to meta-analysis, and in particular, a number of specific areas in which there has been considerable research over the last few years, and in which Bayesian methods would appear to have a potential role to play. Although currently there is much methodological development, so far their use in practice would appear far from routine. A distinct advantage of the Bayesian approach is the ability to incorporate information, external to the meta-analysis, which would otherwise be excluded in a Classical analysis. However, when such external evidence is based on subjective beliefs, the issue of whose beliefs to use is raised. Though many of the computational difficulties that have plagued the application of Bayesian methods in practice have been partially solved by recent developments in Markov Chain Monte Carlo methods, these should not be seen as a 'black box' solution, since they raise a number of issues especially with regard to convergence. In general, the use of Bayesian methods in meta-analysis, like many other statistical techniques, requires *careful* application combined with *critical* assessment [7].

References

1. Breslow, N.E. (1990). Biostatisticians and Bayes (with discussion). *Stat. Sci.* **5**: 269–98.

2. Gilks, W.R., Clayton, D.G., Spiegelhalter, D., Best, N.G., McNeil, A.J., Sharples, L.D., Kirby, A.J. (1993). Modelling complexity: Applications of Gibbs sampling in medicine. *J Roy. Stat. Soc. B* **55**: 39–52.

3. Berry, D.A., Stangl, D.K. (1996). *Bayesian Biostatistics*. New York: Marcel Dekker.

4. Berger, J.O. (1980). *Statistical Decision Theory and Bayesian Analysis. 2nd ed.* New York: Springer-Verlag.

5. Bayes, T.R. (1763). An essay towards solving the doctrine of chances. *Phil. Trans. Roy. Soc. London* **53**: 370.

6. O'Hagan, A. (1988). *Probability: Methods and Measurement*. London: Chapman & Hall.

7. Spiegelhalter, D.J., Miles, J.P., Jones, D.R., Abrams, K.R. (2000). Bayesian methods in Health Technology Assessment. *Health Technol. Assess.* (to appear).

8. Thisted, R.A. (1988). *Elements of Statistical Computing – Numerical Computation*. New York: Chapman & Hall.

9. Gilks, W.R., Richardson, S., Spiegelhalter, D.J. (1996). *Markov Chain Monte Carlo in Practice*. London: Chapman & Hall.

10. Best, N.G., Spiegelhalter, D.J., Thomas, A., Brayne, C.E.G. (1996). Bayesian analysis of realistically complex models. *J. Roy. Stat. Soc. A* **159**: 323–42.

11. Gelfand, A.E., Smith, A.F.M. (1990). Sampling based approaches to calculating marginal densities. *J. Am. Stat. Assoc.* 85: 398–409.

12. Smith, A.F.M, Roberts, G.O. (1993). Bayesian computation via the Gibbs sampler and related Markov Chain Monte Carlo methods. *J. Roy. Stat. Soc. B* **55**: 3–23.

13. Gilks, W.R., Thomas, A., Spiegelhalter, D.J. (1994). A language and program for complex Bayesian models. *The Statistician* **43**: 169–78.

14. Spiegelhalter, D.J., Thomas, A., Best, N.G. (2000). *WinBUGS Version 1.2. User manual*. Cambridge: MRC Biostatistics Unit.

15. Jeffreys, H. (1961). *Theory of Probability*. Oxford: Oxford University Press.

16. Casella, G. (1985). An introduction to empirical Bayes data analysis. *Am. Statistician* 39: 83–7.

17. Vanhouwelingen, H.C., Stijnen, T. (1993). Monotone empirical bayes estimators based on more informative samples. *J. Am. Statist. Assoc.* **88**: 1438–43.

18. Efron, B. (1996). Empirical Bayes methods for combining likelihoods. *J. Am. Stat. Assoc.* **91**: 538–50.

19. Smith, A.F.M. (1973). Bayes estimates in one-way and two-way models. *Biometrika* **60**: 319–29.

20. Kass, R.E., Steffey, D. (1989). Approximate bayesian inference in conditionally independent hierarchical models (parametric empirical bayes models). *J. Am. Stat. Assoc.* **84**: 717–26.

21. Louis, T.A. (1984). Estimating a population of parameter values using bayes and empirical bayes methods. *J. Am. Stat. Assoc.* **79**: 393–8.

22. Morris, C.N. (1983). Parametric empirical Bayes inference: theory and applications. *J. Am. Stat. Assoc.* **78**: 47–65.

23. Carlin, J.B. (1992). Meta-analysis for 2×2 tables: a Bayesian approach. *Stat. Med.* **11**: 141–58.

24. Sutton, A.J, Abrams, K.R., Jones, D.R., Sheldon, T.A., and Song, F. (1998). Systematic reviews of trials and other studies. *Health Technol. Assess.* **2**(19).

25. Lindley, D.V. (1965). *Introduction to Probability and Statistics from a Bayesian Viewpoint. Part 2 – Inference*. Cambridge: Cambridge University Press.

26. Berger, J.O. (1985). *Statistical Decision Theory and Bayesian Analysis. 2nd ed*. New York: Springer-Verlag.

27. Lee, P.M. (1997). *Bayesian Statistics: An Introduction. 2nd ed*. London: Edward Arnold.

28. Press, S.J. (1989). *Bayesian Statistics: Principles, models and applications*. New York: Wiley.

29. Bernardo, J.M., Smith, A.F.M. (1993). *Bayesian Theory*. Chichester: John Wiley & Sons.

30. O'Hagan, A. (1994). *Bayesian Inference*. London: Edward Arnold.

31. Berry, D.A. (1996). *Statistics: A Bayesian perspective*. Belmont: Duxbury Press.

32. Parmigiani, G., Samsa, G.P., Ancukiewicz, M., Lipscomb, J., Hasselblad, V., Matchar, D.B. (1997). Assessing the uncertainty in cost-effectiveness analyses: application to a complex decision model. *Med. Decision Making* **17**: 390–401.

33. Spiegelhalter, D.J., Freedman, L.S. (1988). Bayesian approaches to clinical trials. In: Bernardo, J.M., DeGroot, M.H., Lindley, D.V., Smith, A.F.M., (editors). *Bayesian Statistics 3*. Oxford: Oxford University Press; 453–77.

34. Berry, D.A. (1985). Interim analyses in clinical trials: Classical versus Bayesian approaches. *Stat. Med.* **4**: 521–6.

35. Berry, D.A. (1987). Statistical inference, designing clinical trials and pharmaceutical company decisions. *The Statistician* **36**: 181–9.

36. Berry, D.A. (1989). Monitoring accumulating data in a clinical trial. *Biometrics* **45**: 1197–211.

37. Berry, D.A. (1991). Bayesian methods in Phase III trials. *Drug Infor. J.* **25**: 345–68.

38. Hughes, M.D. (1991). Practical reporting of Bayesian analysis of clinical trials. *Drugs Infor. Bull* **25**: 381–93.

39. Fletcher, A.E., Spiegelhalter, D.J., Staessen, J.A., Thijs, L., Bulpitt, C. (1993). Implications for trials in progress of publication of positive results. *Lancet* **342**: 653–6.

40. Hughes, M.D. (1993). Reporting Bayesian analyses of clinical trials. *Stat. Med.* **12**: 1651–63.

41. Spiegelhalter, D.J., Freedman, L.S., Parmar, M.K.B. (1993). Applying Bayesian thinking in drug development and clinical trials. *Stat. Med.* **12**: 1501–11.

42. Berry, D.A. (1994). Scientific inference and predictions, multiplicities and convincing stories: a case study in breast cancer research. In: Bernardo, J.M., Berger, J.O., Dawid, A.P., Smith, A.F.M., (editors). *Bayesian Statistics 5*. Oxford: Oxford University Press.

43. Spiegelhalter, D.J., Freedman, L.S., Parmar, M.K.B. (1994). Bayesian analysis of randomised trials (with discussion). *J. Roy. Stat. Soc. A* **157**: 357–416.

44. Lilford, R.J., Thornton, J.G., Braunholtz, D. (1995). Clinical trials and rare diseased: a way out of a conundrum. *Br Med J.* **311**: 1621–5.

45. DuMouchel, W. (1989). Bayesian Metaanalysis. In: Berry, D.A. (editor). *Statistical Methodology in the Pharmaceutical Sciences*. New York: Marcel Dekker; p. 509–29.

46. DuMouchel, W. (1994). Predictive cross-validation in Bayesian meta-analysis. In: Bernardo *et al.* (editors). *Proceedings of Fifth Valencia International Meeting*, Valencia, Spain.

47. DuMouchel, W.H., Harris, J.E. (1983). Bayes methods for combining the results of cancer studies in humans and other species (with comment). *J. Am. Stat. Assoc.* **78**: 293–308.

48. Waternaux, C., DuMouchel, W. (1993). Combining information across sites with hierarchical Bayesian linear models. *Proceedings of the Section on Bayesian Statistics*, San Francisco.

49. DuMouchel, W. (1994). Hierarchical Bayes linear models for meta-analysis. Technical Report #27, Research Triangle Park, NC 27709, National Institute of Statistical Sciences.

50. Abrams, K.R., Sanso, B. (1998). Approximate Bayesian inference in random effects meta-analysis. *Stat. Med.* **17**: 201–18.

51. Verdinelli, I., Andrews, K., Detre, K. *et al.* (1995). The Bayesian approach to meta-analysis: a case study. Technical Report, Department of Statistics, Carnegie Mellon University.

52. Box, G.E.P., Tiao, G.C. (1973). *Bayesian Inference in Statistical Analysis*. Reading, MA: Addison-Wesley.

53. Raiffa, H., Schlaifer R. (1961). *Applied Statistical Decision Theory*. Boston: Harvard Business School.

54. Prevost, T.C., Abrams, K.R., and Jones, D.R. (2000). Using hierarchical models in generalised synthesis of evidence: an example based on studies of breast cancer screening. *Stat. Med.* (to appear).

55. Smith, T.C., Abrams, K.R., and Jones, D.R. (1996). Assessment of prior distributions and model parameterisation in hierarchical models for the generalised synthesis of evidence. Department of Epidemiology and Public Health, University of Leicester, England, Report 96–01.

56. Smith, T.C., Spiegelhalter, D.J., Thomas, A. (1995). Bayesian approaches to random-effects meta-analysis: A comparative study. *Stat. Med.* **14**: 2685–99.

57. Consonni, G., Veronese, P. (1993). A Bayesian method for combining results from several binomial experiments. Technical Report; Studi Statistici 40, L. Bocconi University, Milan, Italy.

58. Skene, A.M., Wakefield, J.C. (1990). Hierarchical models for multicentre binary response studies. *Stat. Med.* **9**: 919–29.

59. Rogatko, A. (1992). Bayesian-approach for metaanalysis of controlled clinical-trials. *Comm. Statistics–Theory and Methods* **21**: 1441–62.

60. Smith, T.C., Spiegelhalter, D., Parmar, M.K.B. (1995). Bayesian meta-analysis of randomized triles using graphical models and BUGS. *Bayesian Biostatistics* **15**: 411–27.

61. Waclawiw, M.A., Liang, K.Y. (1994). Empirical bayes estimation and inference for the random effects model with binary response. *Stat. Med.* **13**: 541–51.

62. Higgins, J.P.T., Whitehead, A. (1996). Borrowing strength from external trials in a metaanalysis. *Stat. Med.* **15**: 2733–49.

63. Spiegelhalter, D., Thomas, A., Gilks, W. (1994). *BUGS Examples 0.30.1.* Cambridge: MRC Biostatistics Unit.

64. Abrams, K.R., Lambert, P.C., Sanso, B. *et al.* (2000). Meta-analysis of heterogeneously reported study results – a Bayesian approach. In: Stangl, D.K., Berry, D.A., editors. *Meta-Analysis in Medicine and Health policy.* New York: Marcel Dekker.

65. Hornbuckle, J., Vail, A., Abrams, K.R., Thornton, J.G. (2000). Bayesian interpretation of trials: the example of intrapartum electronic fetal heart rate monitoring. *Br. J. Obstet Gynaecol* **107**: 3–10.

66. Li, Z. (1995). A multiplicative random effects model for meta-analysis with application to estimation of admixture component. *Biometrics* **51**: 864–73.

67. Raudenbush, S.W., Bryk, A.S. (1985). Empirical Bayes meta-analysis. *J. Educ. Statist* **10**: 75–98.

68. Zhou, X.H. (1996). Empirical Bayes combination of estimated areas under ROC curves using estimating equations. *Med. Decision Making* **16**: 24–8.

69. Morris, C.N. (1992). Hierachical models for combining information and for meta-analysis. *Bayesian Statistics* **4**: 321–44.

70. Stijnen, T., Van Houwelingen, J.C. (1990). Empirical Bayes methods in clinical trials meta-analysis. *Biometrical J.* **32**: 335–46.

71. Biggerstaff, B.J., Tweedie, R.L., Mengersen, K.L. (1994). Passive smoking in the workplace: classical and Bayesian meta-analyses. *Int Arch Occup Environ Health* **66**: 269–77.

72. DerSimonian, R., Laird, N. (1986). Meta-analysis in clinical trials. *Controlled Clin Trials* **7**: 177–88.

73. Larose, D.T., Dey, D.K. (1997). Grouped random effects models for Bayesian meta-analysis. *Stat. Med.* **16**: 1817–29.

74. National Research Council. (1992). *Combining Information: Statistical Issues and Opportunities for Research*. Washington, D.C.: National Academy Press.

75. Laird, N., Louis, T.A. (1989). Empirical Bayes confidence intervals for a series of related experiments. *Biometrics* **45**: 481–95.

76. Spiegelhalter, D.J., Goldstein, H. (1996). League tables and their limitations – statistical issues in comparisons of institutional performance. *J. Roy. Stat. Soc. A* **159**: 385–409.

77. Tweedie, R.L., Scott, D.J., Biggerstaff, B.J., Mengersen, K.L. (1996). Bayesian meta-analysis, with application to studies of ETS and lung cancer. *Lung Cancer* **14**: S171–94.

78. DerSimonian, R. (1996). Meta-analysis in the design and monitoring of clinical trials. *Stat. Med.* **15**: 1237–52.

79. Louis, T.A., Zelterman, D. (1994). Bayesian approaches to research synthesis. In.: Cooper, H., Hedges, L.V., editors. *The Handbook of Research Synthesis*. New York: Russell Sage Foundation; **26**, 411–22.

80. Su, X.Y., Po, A.L.W. (1996). Combining event rates from clinical trials: Comparison of Bayesian and classical methods. *Annals of Pharmacotherapy* **30**: 460–5.

81. Abrams, K.R., Sanso, B. (1995). Model discrimination in meta-analysis – a Bayesian perspective. Technical Report 95–03, Department of Epidemiology and Public Health, University of Leicester.

82. Lindley, D.V., Smith, A.F.M. (1972). Bayes estimates for the linear model. *J. Roy. Stat. Soc. B* **34**: 1–41.

83. Goldstein, H. (1995). *Multi-level Statisatical Models. 2nd ed.* London: Arnold.

84. Longford, N. (1993). *Random Coefficient Models*. Oxford: Clarendon Press.

85. Veronese, P. (1994). Mutually compatible hierarchical priors for combining information. Technical Report, L. Bocconi University, Milan.

86. Rice, N., Leyland, A. (1996). Multilevel models: applications to health data. *J. Health Services Res. and Policy* **1**: 154–64.

87. Smith, J.Q. (1988). *Decision Analysis – A Bayesian approach*. London: Chapman & Hall.

CHAPTER 12

Meta-analysis of Individual Patient Data

12.1   Introduction

Until recently, it was customary for meta-analyses to be carried out using the aggregated summary results of studies, which were obtained from journal articles. Where these were not sufficient, or if the study was not published, the necessary summary data were requested directly from the original research group. With continually improving technology and communications, formal prospective registration of RCTs, and increasing awareness of the benefits of meta-analysis, it is becoming more feasible to obtain the full study datasets from the original researchers, making a synthesized overview using information at the *patient* level possible. This has become known as meta-analysis of Individual Patient Data (IPD), although other terms such as 'mega-analysis' [1] have also been used, and the term 'pooled analysis' *sometimes* implies use of IPD. To-date, these have largely been carried out on RCT data, but Sections 12.3 and 12.5 consider its potential for combining observational data, and Section 17.3 describes a method to combine IPD data from matched and un-matched case control studies.

There are several motivating reasons for carrying out an ambitious analysis of this type, including [2, 3]:

(a) feasibility of detailed data checking and ensuring the quality of randomisation and follow-up;

(b) ensuring the appropriateness of the analyses;

(c) updating follow-up information (with further follow-up after original publication);

(d) undertaking subgroup analyses for important hypotheses about differences in effect, more easily than with aggregate data;

191

(e) carrying out survival and other time-to-event analyses more satisfactorily (see Chapter 18);
(f) standardizing inclusion/exclusion criteria, which is not always possible using summary data;
(g) checking whether the treatment effect is constant over time, an assumption which may not be true, but which is necessary to calculate an overall estimate when the treatment effect is reported at different time points from one study to another; and
(h) identifying interactions between the treatment effect and patient profiles [3].

Chalmers [4] regards IPD meta-analyses are yardsticks against which the quality of other systematic reviews of randomized controlled trials should be measured.

To obtain data at the patient level, it will usually be necessary to contact all groups of investigators who carried out the original trials to be combined. These necessary collaborations may have several other or 'knock-on' benefits [2]:

(a) more complete identification of relevant trials,
(b) better compliance with providing missing data,
(c) more balanced interpretation of the results,
(d) wider endorsement and dissemination of the results,
(e) better clarification of further research, and
(f) collaboration on further research–perhaps multi-centre trials to address the remaining research questions identified by the meta-analysis.

There are, however, costs. IPD meta-analyses are time consuming to carry out [5]. As for resource costs of performing IPD meta-analyses, the Cochrane Working Group on IPD meta-analysis suggested around £1000 per trial, or £5–10 per patient (in 1994), whichever was less. These costs, whilst substantial, are small relative to the total amount invested in health care, and are sometimes a necessary investment [6].

Additionally, this method relies heavily on the international co-operation between the individuals and groups who have conducted relevant trials [2]. To run and report a trial takes much hard work, and thus objections to data sharing from trialists whilst problematic, are understandable. However, as Oxman et al. [6] point out, patients do not consent to participate in trials for the benefit of researchers or corporate profit, and hence it is unethical for trialists, pharmaceutical companies, or others to withhold data for private interests. There is a problem of how to proceed if all persuasion to collaborate fails, or the data for the study has been lost or destroyed (see Section 12.3).

The meta-analyses carried out by the early breast cancer trialists collaborative group [7–9] provide good examples of the use of IPD meta-analyses, and are recommended reading for those contemplating carrying out similar analyses. Generally, these use survival endpoints, and often detailed subgroup analyses are possible due to the many thousands of randomized patients included and the availability of individual patient covariates.

12.2 Procedural methodology

The focus of this book is the statistical methods used to combine information, and little consideration has been given to pre-synthesis issues such as protocol specification and data collection, which are considered elsewhere [10–12]. However, procedural issues specific to analysis of individual patient data are briefly considered here. These summarize guidelines derived by the Cochrane Working Group on meta-analysis using individual patient data. A fuller account can be found elsewhere [2].

12.2.1 Data collection

The following have been identified as the data that should be collected to carry out an IPD meta-analysis. Patient identifier, treatment allocated and outcome(s) are the essential minimum, together with the date of randomization and date of outcome if time to event is to be calculated. It is usually important to collect additional baseline variables, even when subgroup analysis are not planned, because these data are extremely useful in checking the integrity of the randomization process.

Collecting old datasets can be a difficult and slow process. The working group concluded that when a large proportion of the total randomized evidence (perhaps 90–95%) has been collected, the missing data are unlikely to alter the meta-analysis results. This can be checked by using a sensitivity analysis, such as examining the effect of including extreme results for the missing data.

12.2.2 Checking data

It is very important that the analysis should be based on the 'intention-to-treat' principle (as is the case for RCTs), and therefore that data should be collected, and analysis based on all randomized patients. Indeed, if the original study did not do this and excluded patients from the results, they can now be reintroduced into the analysis [13]. Simple procedures for checking correct randomization and follow-up have been proposed [2].

12.3 Issues involved in carrying out IPD meta-analyses

The statistical methods for synthesis at the patient level are similar to those used to analyse multicentre clinical trials. This literature is beyond the scope of this book, and the interested reader is referred, for example, to Pocock [14] or Piantadosi [15]. The main feature that differentiates this kind of analysis from that of a single trial is that a covariate can be included to indicate which study from which each patient comes. This covariate could be regarded as either a fixed or random effect.

A variant on this general approach is used in the Early Breast Cancer Trialists' Collaborative Group meta-analyses [7–9]. Generally, in these meta-analyses each trial is analysed separately, which results in one logrank statistic per trial being produced. These logrank statistics are then combined using a weighted average method. Using this approach women in one trial are compared directly only with women in the same trial, and never with women in another trial.

The Cochrane Working Group on meta-analysis using individual patient data has published guidelines for those carrying out IPD meta-analysis [2].

The introduction to this chapter outlined many of the advantages and disadvantages of using IPD data over aggregated data to perform a meta-analysis. A problem exists if a significant proportion of the IPD data cannot be obtained. In such situations, it may be possible to combine a combination of aggregated summary data from reports, where no IPD data is available, and the available IPD. Currently, it is unclear whether this is desirable, and it has been suggested, in such situations, the aggregated data should only be included in a sensitivity analysis of the IPD results [2]. This is an area of active research, and is considered further in Section 12.5.

Chapter 16 deals with meta-analysis methodology for observational studies. Analysis of individual patient data may have rewards for observational studies, as well as RCTs. For instance, uniform adjustment for confounding on a patient level could be made using IPD covariates, if consistently collected in the primary studies. For example, consider two observational studies, one which originally adjusted for patients age, and one that did not. If age is available in both studies, using individual patient data–a meta-analysis could be carried out either adjusting (data from both studies) by age or combining both studies unadjusted (see Sections 16.7.1 and 16.7.2 for further details).

12.4   Comparing meta-analysis using IPD or summary data?

The potential advantages and disadvantages of obtaining IPD for a meta-analysis have been considered in previous sections. It is unclear, however, what the potential gains are, in terms of precision and bias, for the pooled results, by using IPD over summary data.

Several studies [16–18] have been carried out to compare the results of a standard meta-analysis using aggregated data to data which use IPD; these have found discrepant results between the methods, but no clear general systematic differences (although the improved quality of IPD data is likely to lead to more valid results). Stewart and Parmars' [16] assessment used IPD which had been updated from that analysed in the published report, and hence did not compare like with like.

From a theoretical viewpoint, Olkin and Sampson [19] consider the situation in which multiple treatments are compared to a control, and the relative effect of each treatment is measured using a continuous outcome. (The model used is a simplified version of that developed by Gleser and Olkin [20] described in Section 15.4.2.) They demonstrate that standard meta-analysis estimates of the treatment contrasts are identical to the analysis of individual patient data when homogeneity is assumed for treatment contrasts among studies. Hence in this situation, no statistical advantages exist when individual patient data are pooled over standard summary methods.

Debate currently exists on when IPD meta-analyses are required. It could be argued that, although costly, IPD meta-analysis is still cheaper than carrying out a new large study in many instances, and this will often be worthwhile. However, since the advantages of IPD meta-analysis over meta-analysis of summary data are largely unestablished empirically, it has been suggested that they should not be carried out until more research into the benefits of IPD over meta-analysis of aggregate data has been done. A further pragmatic suggestion is to carry out a meta-analysis using summary data first to assess whether a meta-analysis of IPD would be beneficial, the judgement being based on factors such as the significance of the result (marginal results may require further investigation), and the potential gains of examining subgroups of patients.

For this reason, the Cochrane Working Group on individual patient data meta-analyses puts forward a research agenda to generate additional empirical evidence on the relative value of the different techniques [2].

12.5   Combining individual patient and summary data

In some contexts, IPD will be available only from a proportion of all known studies. Ways of combining IPD with summary estimates, where IPD are not available is the subject of current research. Goldstein *et al.* [21] consider a general multilevel model for combining IPD with summary data. Collette *et al.* [22] consider the issue of including summary data in an IPD with survival endpoints. (See Chapter 18). On a closely related topic, Higgins and Whitehead [23] propose a method of including both patient level and study-level covariates in an IPD meta-analysis.

12.6   Summary/Discussion

There are several advantages of carrying out a meta-analysis of individual patient data over a standard meta-analysis using aggregated data. These include the ability to: i) carry out detailed data checking; ii) ensure the appropriateness of the analyses; and iii) update follow-up information. This has led to the comment that meta-analysis of individual patient data is the yardstick against which the quality of other systematic reviews of randomized controlled trials should be measured [4].

However, IPD meta-analyses are very time consuming and resource intensive. Currently, there is little empirical evidence regarding the actual magnitude of the benefits, and it is yet to be established when the extra effort is always worthwhile.

References

1. Fortin, P.R., Lew, R.A., Liang, M.H., Wright, E.A., Beckett, L.A., Chalmers, T.C., Sperling, R.I. (1995). Validation of a meta-analysis: the effects of fish oil in rheumatoid arthritis. *J. Clin. Epidemiol.* **48**: 1379–90.

2. Stewart, L.A., Clarke, M.J. (1995). Practical methodology of meta-analyses (overviews) using updated individual patient data. Cochrane Working Group. *Stat. Med.* **14**: 2057–79.

3. Gueyffier, F., Boutitie, F., Boissel, J.P., Coope, J., Cutler, J., Ekbom, T., Fagard, R., Friedman, L., Perry, H.M., Pocock, S. *et al.* (1995). INDANA: a meta-analysis on individual patient data in hypertension. Protocol and preliminary results. *Therapie* **50**: 353–62.

4. Chalmers, I., (1993). The Cochrane collaboration: Preparing, maintaining, and disseminating systematic reviews of the effects of health care. *Annals New York Acad. Sci.* **703**: 156–65.

5. Pignon, J.P., Arriagada, R., Ihde, D.C., Johnson, D.H., Perry, M.C., Souhami, R.L., Brodin, O., Joss, R.A., Kies, M.S., Lebeau, B. *et al.* (1992). A meta-analysis of thoracic radiotherapy for small-cell lung cancer. *New Engl. J. Med.* **327**: 1618–24.

6. Oxman, A.D., Clarke, M.J., Stewart, L.A. (1995). From science to practice. Meta-analyses using individual patient data are needed. *J. Am. Med. Assoc.* **274**: 845–6.

7. Early Breast Cancer Trialists' Collaborative Group. (1990). *Treatment of Early Breast Cancer. Volume 1: Worldwide Evidence 1985–1990.* Oxford: Oxford University Press.

8. Early Breast Cancer Trialists' Collaborative Group. (1996). Effects of radiotherapy and surgery in early breast cancer: an overview of the randomized trials. *New Engl. J. Med.* **333**: 1444–55.

9. Early Breast Cancer Trialists' Collaborative Group. (1988). Effects of adjuvant tamoxifen and of cytotoxics on mortality in early breast cancer: an overview of 61 randomized trials among, 28, 896 women. *New Engl. J. Med.* **319**: 1681–92.

10. Deeks, J., Glanville, J., Sheldon, T. (1996). Undertaking systematic reviews of research on effectiveness: CRD guidelines for those carrying out or commissioning reviews. Centre for Reviews and Dissemination. York: York Publishing Services Ltd., Report #4.

11. Oxman, A.D. (editor). (1996). The Cochrane Collaboration handbook: preparing and maintaining systematic reviews. *Second ed.* Oxford: Cochrane Collaboration.

12. Sutton, A.J., Abrams, K.R., Jones, D.R., Sheldon, T.A., Song, F. (1998). Systematic reviews of trials and other studies. *Health Technol. Assess.* 2(19).

13. Clarke, M.J., Stewart, L.A. (1994). Systematic reviews – obtaining data from randomized controlled trials – how much do we need for reliable and informative meta-analyses. *Br Med. J.* **309**: 1007–10.

14. Pocock, S.J. (1983). *Clinical Trials: A Practical Approach.* Chichester: Wiley.

15. Piantadosi, S. (1997). *Clinical Trials: A Methodologic perspective.* New York: Wiley.

16. Stewart, L.A., Parmar, M.K. (1993). Meta-analysis of the literature or of individual patient data: is there a difference? *Lancet* **341**: 418–22.

17. Jeng, G.T., Scott, J.R., Burmeister, L.F. (1995). A comparison of metaanalytic results using literature vs individual patient data – paternal cell immunization for recurrent miscarriage. *J. Am. Med. Assoc.* **274**: 830–6.

18. Pignon, J.P., Arriagada, R. (1993). Meta-analysis. *Lancet* **341**: 418–22.

19. Olkin, I., Sampson, A. (1998). Comparison of meta-analysis versus analysis of variance of individual patient data. *Biometrics* **54**: 317–22.

20. Gleser, L.J., Olkin, I. (1994). Stochastically dependent effect sizes. In: Cooper, H., Hedges, L.V., editors. *The Handbook of Research Synthesis.* New York: Russell Sage Foundation; 339–56.

21. Goldstein, H., Yang, M., Omar, R.Z., Turner, R.M., Thompson, S.G. (1999). Meta-analysis using multilevel models with an application to the study of class size effects (submitted 1999).

22. Collette, L., Suciu, S., Bijnens, L., Sylvester, R. (1997). Including literature data in individual patient data meta-analyses for time-to-event endpoints [abstract only]. *Controlled Clin. Trials* **18**(3S):188S.

23. Higgins, J.P.T., Whitehead, A. (1997). Inclusion of both patient level and study-level covariates in a meta-analysis [abstract only]. *Controlled Clin. Trials* **18**(3S):84S.

CHAPTER 13

Missing Data

13.1 Introduction

The issue of missing data in meta-analysis can be considered under three broad categories:

(i) *Whole studies missing*: Chapter 7 discussed publication bias, a situation where not all the relevant studies carried out in an area have been reported, and hence included, in a meta-analysis. This is not considered further here.

(ii) *Data missing at the study level*: It may be that no treatment effect size estimate is reported, although a significance level or *p*-value is. Study level covariates may also be missing, either completely, or only partially reported (due to missing data at the patient level; for example, a study may only report the mean age of 80% of the people in a study). This type of missing data can present problems with are unique to meta-analysis.

(iii) *Data missing at the individual patient level*: Chapter 12 discusses meta-analysis of individual patient data. Here, the original data from the primary studies is collected on all patients, this is then merged into one large dataset. The analyses possible are similar in nature to those carried out in multi-centre trials. When data are missing at this level – standard techniques for missing data used in trials and observational studies can be employed in the meta-analysis.

Thus, this chapter will concentrate on situation (ii), where data are missing at the study level, although many of the techniques could also be used in situation (iii), at the individual patient level.

13.2   Reasons for missing data

Data may be missing for any of the following reasons [1]:

(a) Missing due to the *influences of research reporting practices*. The amount of information given in a study report may be limited by the nature of the publication. This can be viewed as one aspect of publication bias (see Chapter 7).

(b) Missing for *reasons unrelated to the data*. Little and Rubin [2] use the term 'missing completely at random' to make it distinct from the situation given below. In a meta- analysis context, this could relate to certain (non-crucial) patient characteristics, which may or may not have been reported. The remaining cases with complete information can be treated as a random sample of the original set of studies (and unbiased inferences can be made from them).

(c) Missing for *reasons related to completely observed variables*. In this instance, missing values occur because of the value of another completely observed variable, (e.g. data on several variables may be missing from patients who were discharged to a nursing home, as opposed to their usual home), and not because of the value of the missing variable itself. Little and Rubin [2] use the term 'missing at random'. An example in a meta-analysis context could be the reporting of a confounding factor in an observational study. If a certain factor had only been recently identified as being a confounder, then studies carried out prior to this discovery would be less likely to report it. Analysing only complete cases in this situation may not provide unbiased/generalizable results. Methods described by Rubin [3] and Little and Rubin [2], which are outlined below, are appropriate here.

(d) Missing for *reasons related to the missing values themselves*. Observations can be missing because of the value of the variable itself, or because of other unobserved variables. This can be caused by censoring mechanisms. An example of this situation are missing effect sizes not reported *because* they were *not* statistically significant. In this instance, the data are most definitely not missing at random. This situation poses one of the most difficult problems in dealing with missing data.

13.3   Categories of missing data at the study level

Missing study level effect sizes

The effect size estimate may be completely missing for particular studies. If the actual magnitude of the effect is not given when not significant (a common policy until recently), this will have an impact on the meta-analysis similar to publication bias; since treatment estimates from significant studies are more likely to be given, this will lead to a systematic over estimation of the pooled effect.

When the direction, but not the magnitude, of the effect size estimate is known, vote counting analysis can be carried out (see Section 14.7). Pigott [1] warns that replacing missing values with conservative estimates, such as zero, for missing effect sizes could lead to biased results.

Missing measures of precision

When combining continuous effect sizes, even if the outcome measure, such as a treatment difference, is reported, it is not uncommon for no measure of the precision of the estimate, such as the standard deviation, to be given.

Missing study level characteristics/covariates

Most quantitative research synthesis may only conveniently include studies with complete information on both outcomes and predictors when building models for effect size, since standard regression programs drop cases missing any variable required for modelling [1].

13.4   Analytic methods for dealing with missing data

13.4.1   General missing data methods which can be applied in the meta-analysis context

Very little has been written about methods for dealing with data missing at the study level in meta-analysis. Piggott [1] gives details of the application of standard methods for dealing with missing data in a meta-analysis context. Briefly, these include: (a) analysing only complete cases, which is straightforward but restrictive if the number of missing values is considerable; (b) single-value imputation, where missing values are replaced by 'reasonable' estimates, allowing studies with one or more missing values to be retained within the analysis; and (c) regression imputation (also known as Buck's method [4]), where it is assumed that missing variables are linearly related to other variables in the data, and regression techniques are used to estimate the missing values. Additionally, more advanced methods are also considered, including the use of maximum likelihood models which do not require any adjustments to the data, and multiple imputation, where multiple values are imputed for each missing observation, hence avoiding the problem of having to assign a single value to the missing data.

13.4.2   Missing data methods specific to meta-analysis

Recently, some research has been carried out into methods for missing data specific to meta-analysis. Bushman and Wang [5, 6] describe methods to combine studies where data to calculate an effect size and its standard deviation are not available from all the studies. These procedures combine information about study size, and the direction of the observed effect from the studies where no effect size is given, with effect sizes from studies from which they are known. Hence, the methods combine sample effect sizes and vote counts (see Section 14.7) to estimate the population effect size.

Methods to do this have been described for the standardised mean difference [6] and correlation coefficient [5] outcomes only, using fixed effect methods, however they could be adapted to other outcome indexes and random effect models [7]. These methods allow all the data available from each study to be used in the final analysis, and hence are superior to simpler approaches such as omitting studies where no effect size is given, or inputing values such as zero, or the mean effect for the outcome in studies where no outcome is reported. It can be shown that this approach produces consistent estimates, and the procedure gives weight to all studies proportional to the Fisher information they provide.

In addition to these, due to the nature of problems encountered, *ad hoc* methods are sometimes required to deal with problems with missing data in meta-analysis; an example of where this was necessary is given below.

13.4.3   Example: Dealing with missing standard deviations of estimates in a meta-analysis

A meta-analysis of selective serotonin reuptake inhibitors compared to tricyclic antidepressants in the first line treatment of depression [8] provides an illustration of the problem of missing data and how it can be dealt with, using a simple *ad hoc* approach, within a meta-analysis context. Of 53 RCTs which had measured depression change using some version of the Hamilton depression rating scale (the 17 or 21 item versions of the scale were usually used, but others were occasionally used also), only 20 of them presented a standard deviation for these scores. This meant the weights required to combine studies could not be calculated directly for 33 RCTs which used the 17 or 21 item Hamilton scale, and otherwise satisfied the inclusion criteria for the meta-analysis. There was concern that not considering over half the known studies in the analysis may produce a biased result.

It was observed that the standard deviations of the scores from the trials reporting using the 17 or 21 item scale were similar, therefore, as a sensitivity analysis, pooled estimates of the standard deviations of the treatment difference were calculated individually for the 17 and 21 item scales, using the trials where these had been reported. These values were then imputed as the standard errors in the 33 trials, where no such figure had been reported. This enabled all the trials to be combined.

The primary analysis, focusing on the 20 trials which reported standard deviations, produced a small but non-significant benefit in efficacy of tricyclic and related antidepressants compared to serotonin reuptake inhibitors. The analysis which included all trials, using input standard deviations, where required, produced a very similar result, strengthening the conclusion of the primary analysis.

13.5   Bayesian methods for missing data

There has been considerable research into missing data generally in health care settings; the research from a Bayesian perspective has concentrated on data within individual RCTs, and in the area of longitudinal data analysis [9, 10] using data augmentation techniques [11]. By comparison, there has been relatively little work in a meta-analysis setting, though Lambert *et al.* [12] considered the case when mortality data from five different centres was pooled, with both random and non randomly missing tumour marker covariate data, using the methods of Best *et al.* [10].

Additionally, Abrams *et al.* [13] considered the situation where results from studies are reported inconsistently, creating the problem that a standard deviation for the effect size estimate cannot be calculated for some studies. Specifically, mean responses and standard deviations are reported for baseline and follow-up in some trials. The desired outcome, change, can be calculated from this data, but not the standard deviation of the change because the correlation between individuals responses is required. The authors [13] show how a sensitivity analysis, imputing a range of values for the unknown correlation can be carried out, as well as describing a Bayesian solution (see Chapter 11) to this problem.

13.6   Summary/Discussion

Very little work has been done on the problem of missing data specifically in meta-analysis. The methods which have been considered have not been used extensively in a meta-analysis setting. It has been suggested that missing data is perhaps the most pervasive practical problem in research synthesis, and that new methods will evolve [14]. However at present, only methods developed in other settings are available. If such methods are used, a careful sensitivity analysis examining the impact of the assumptions these methods require should be performed.

References

1. Pigott, T.D. (1994). Methods for handling missing data in research synthesis. In: Cooper, H., Hedges, L.V., editors. *The Handbook of Research Synthesis*. New York: Russell Sage Foundation. 163–76.

2. Little, R.J.A., Rubin, D.B. (1987). *Statistical Analysis with Missing Data*. New York: Wiley.

3. Rubin, D.B. (1987). *Multiple Imputation for Non Response in Surveys*. New York: Wiley.

4. Buck, S.F. (1960). A method of estimation of missing values in multivariate data suitable for use with an electronic computer. *J. Roy. Stat. Soc.* **22**: 302–3.

5. Bushman, B.J., Wang, M.C. (1995). A procedure for combining sample correlation coefficients and vote counts to obtain an estimate and a confidence interval for the population correlation coefficient. *Psychol. Bull.* **117**: 530–46.

6. Bushman, B.J., Wang, M.C. (1996). A procedure for combining sample standardized mean differences and vote counts to estimate the population standardized mean difference in fixed effect models. *Psychol. Meth.* **1**: 66–80.

7. Wang, M.C., Bushman, B.J. (1999). Integrating results through meta-analytic review using SAS(R) software. Cary, NC, USA: SAS Institute Inc.

8. Song, F., Freemantle, N., Sheldon, T.A., House, A., Watson, P., Long, A. (1993). Selective serotonin reuptake inhibitors: meta-analysis of efficacy and acceptability. *Br. Med. J.* **306**: 683–7.

9. Arjas, E., Liu, L. (1996). Non-parametric approach to hazard regression: a case study with a large number of missing covariate values. *Stat. Med.* **15**: 1771–8.

10. Best, N.G., Spiegelhalter, D.J., Thomas, A., Brayne, C.E.G. (1996). Bayesian analysis of realistically complex models. *J. Roy. Stat. Soc. A* **159**: 323–42.

11. Kong, A., Liu, J.S., Wong, W.H. (1994). Sequential imputations and Bayesian missing data problems. *J. Am. Stat. Assoc.* **89**: 278–88.

12. Lambert, P.C., Abrams, K.R., Sanso, B. *et al.* (1997). Synthesis of incomplete data using Bayesian hierarchical models: an illustration based on data describing survival from neuroblastoma. Department of Epidemiology and Public Health, University of Leicester, England. Technical Report 97–03.

13. Abrams, K.R., Lambert, P.C., Sanso, B. *et al.* (2000). In: Meta-analysis of heterogeneously reported study results – a Bayesian approach. Stangl, D.K., Berry, D.A. editors. *Meta-Analysis in Medicine and Health Policy*. New York: Marcel Dekker.

14. Cooper, H., Hedges, L.V. (editors). (1994). Potentials and limitations of research synthesis. *The Handbook of Research Synthesis*. New York: Russell Sage Foundation. 521–30.

CHAPTER 14

Meta-analysis of Different Types of Data

14.1   Introduction

This section outlines methods for combining several specific types of data. Part A of this book outlined methods for combining results which were generalizable to many situations; however, this was not all encompassing, so this and the following chapters aim to cover the remainder of the methods, many specific to certain situations. In this chapter, meta-analysis of ordinal data, rates and rare outcomes are all considered briefly. Scale issues pertinent to meta-analysis, including transformation of scale when studies report results in different formats, are then discussed. Meta-analysis of diagnostic test results require different methods from those already considered, as do those for surrogate markers. Methods of meta-analysis using vote counting and combining p-values are described. These may be necessary if effect sizes are not reported for the primary studies. A section considering further novel applications of meta-analysis concludes this chapter.

14.2   Combining ordinal data

Two situations which arise when combining ordinal data from different studies are usefully distinguished: (a) When the response variable is the same in each study; and (b) When different response variables are used in different studies. A unified, but restrictive, framework is presented below that can incorporate both possibilities. This approach reduces the outcome to a binary response by combining categories in order to produce two categories [1]. Log odds ratios can then be calculated for each study and combined using methods previously described. Define the treatment effect for the ith study as:

$$\theta_{ji} = \log\left\{\frac{Q_{jCi}(1 - Q_{jTi})}{Q_{jTi}(1 - Q_{jCi})}\right\},\qquad(14.1)$$

where $Q_{jTi} = p_{1Ti} + \ldots + p_{jTi}$, $Q_{jCi} = p_{1Ci} + \ldots + p_{jCi}, j = 1, \ldots, m - k(k < m)$, and p_{jTi} and p_{jCi} are the probability of a patient in the ith trial being in the jth outcome category. So, the outcome is partitioned with 1 to $m - k$ outcomes in one category and k to m outcomes forming the other. The proportional odds model, which assumes the value for θ_{ji} would stay constant if the partitioning of outcomes, had been at some other value (e.g. $m - 2$, $m - 5$, etc.) is adopted. Thus, θ_{ji} can be considered as the log odds of success on the experimental treatment relative to control for the ith study, irrespective of how the ordered categories might be divided into success or failure [2].

This analysis has the following advantages: (a) no studies need to be omitted from the meta-analysis because of differences between the category definitions/ordinal scoring systems used because they can all be dichotomized; and (b) the data can be used in their original form. For computational details, along with an example, see Whitehead and Jones [2]. The proportional odds assumption within each study may be investigated by calculating the estimates of the log-odds-ratios, θ_{ji}, from the various binary splits. These estimates should be similar. Currently, only fixed effect estimates can be calculated using this approach.

14.3   Issues concerning scales of measurement when combining data

Often, primary studies will present their results in differing forms, and even use different effect measures. Present meta-analysis methodology dictates that it is necessary to combine binary outcome data on the same scale (the standardized mean difference can be used to combine different continuous scales). Before these estimates can be combined it is often necessary to transform, at least some of them, to a common scale (if this is possible). With limited data from published reports, this may be a non-trivial task. In extreme cases, it may be necessary to change the type of data (e.g. from continuous to categorical) for some of the studies before data is combined. Other factors needing to be taken into account before any decision on which scale to transform the data to is taken are (i) will the choice of scale used affect the final result, and (ii) if a choice of scales is possible for the analysis, which one is most desirable? Perhaps surprisingly, the answer to the first question is occasionally 'yes' [3]. Critiques of the relative merits of the different binary scales are given elsewhere [4, 5]. Methods of transforming data between scales are given below.

14.3.1   Transforming scales, maintaining same data type

Different scales may have been used to present effect size estimates in different studies. It is common practice to transform to a standardized effect size for continuous variables (see Section 2.4.2), and methods for combining ordinal data reported on different scales have been discussed above. Methods for transforming results reported on different binary outcome scales are discussed below. Additionally, a method of combining continuous with binary outcomes through changing data type is also discussed.

14.3.2   Binary outcome data reported on different scales

If 2×2 tables are available for all studies to be combined, then any of the measures based on such tables (OR, RR, risk difference, etc.) can be calculated. The choice of scale used to combine, in this instance, should take into account the properties of each scale, and the design of the studies being combined. It would often seem sensible to convert data to the (log) odds ratio scale, because of its statistical advantages [4], although this is not universally agreed [5]. Another factor to consider is that using different scales can give different values for the test for heterogeneity (for an example of this, see [4, p.258]). For this reason, an investigation of the sensitivity of overall results to combining on different scales is recommended [6].

If the full 2×2 table is not given for a trial, it may be possible to reconstruct it from other data sources within the paper (such as p-values), or through making contact with the original investigators. However, if this data cannot be obtained, the transformations which are possible may be restricted, and thus dictate to a greater or lesser degree the scale on which studies may be combined.

There are many permutations of possible transformations of scales, and with the amount of information available varying greatly between studies it is impossible to suggest a standard set of procedures for data transformation. However, a useful transformation when the 2×2 table is not known, but the risk of the event in the control group is known, involves changing the Relative Risk (RR) from the Odds Ratio (OR) (or vice versa) [5], using

$$RR = \frac{OR}{1 + Risk_c(OR - 1)} \qquad (14.2)$$

where $Risk_c$ = risk of the event in the control group. This may be useful in the context of combining observational studies, possibly with different designs (see Chapter 16 which also outlines ways in which binary outcome estimates can be reconstructed or estimated from other information *in this context*).

14.3.3   Combining studies whose outcomes are reported using different data types

In some situations, primary studies may not only have used different scales, but even different data types as outcome measures. For example, if change in blood pressure was of interest, some studies may report the absolute change on a continuous scale, while other studies report the proportion of patients who achieve a prespecified level of change. First, a methodology for combining continuous and binary outcomes is considered. This is followed by a (non-parametric) method that can be used to combine many different outcome comparisons.

14.3.4   Combining summaries of binary outcomes with those of continuous outcomes

A method for combining trials, some of which report continuous outcome measures, and others binary outcomes created by dichotomizing a continuous measurement, has been based upon estimating the probability of an event when the outcome variable is continuous [7, 8]. The method is based on the assumption of a normal distribution and a maximum likelihood theory approach; it has been shown to be more efficient than the binary approach applied to dichotomised data. However, when studies' results are measured on different scales, the ideal solution would be to obtain individual patient data and dichotomise the continuous outcome at the patient level, and then to investigate through a sensitivity analysis whether choice of the cut-point influences the overall result.

14.3.5   Non-parametric methods of combining different data type effect measures

The Mann–Whitney statistic can be used to combine treatment effects when different scales/measures are used for the primary studies [9]. For an RCT, the Mann–Whitney statistic can be used to estimate the probability that a randomly selected patient will perform better given the new treatment, than a randomly selected patient given the standard treatment. Denoting this as $P(I > S)$, it can be estimated from the proportions of treatment failures

$$0.5 + 0.5(pS - pI) \tag{14.3}$$

where p_S and p_I are proportions of treatment failures on the standard and on the new treatment, respectively. The Mann–Whitney statistic can be calculated for many different statistical measures, for example, proportion surviving, mean change in blood pressure, and frequency of side effects and this allows combination of results of statistics using different data types. For more details, see Colditz *et al.* [9].

14.4 Meta-analysis of diagnostic test accuracy

A 'diagnostic test' may most generally be defined as 'any measurement aimed at identifying individuals who could potentially benefit from intervention' [10]. Though randomised trials of screening may be used to assess the effectiveness of a test regarding patient outcome, the conduct of such trials is infeasible for assessment of every new diagnostic test. Hence, often the focus of investigation is on the test's ability to accurately detect conditions for which randomised trials show effective intervention [10].

A number of comprehensive review articles on meta-analysis of diagnostic test accuracy have been written [11–16]. The Cochrane Methods Working Group on Systematic Review of Screening and Diagnostic Tests [10] has summarized the objectives of systematic reviews of diagnostic test accuracy, and also give practical guidelines (checklists) for literature retrieval, quality appraisal, data presentation and analysis. The various sources of bias in the assessment of diagnostic tests necessitate efforts and methods to adjust for them, some of them distinct from those for RCTs.

Diagnostic tests may be differentiated by the type of their outcomes – binary, ordered categorical, or continuous. For each outcome type specific meta-analytic procedures have been proposed. The following exposition borrows from Irwig *et al.* [13].

14.4.1 Combining binary test results

Suppose, $n = n_1 + n_2$ subjects undergo both the index and the reference test. The binary test results may be condensed in a 2×2-table (see Figure 14.1). The table corresponds to a certain threshold (cut-off, positively criterion) CO such as

$$\text{index test} = \begin{cases} + : & \text{(latent) test variable} \geq CO \\ - : & \text{(latent) test variable} < CO \end{cases} \qquad (14.4)$$

where the '(latent) test variable' is either observable or non-observable.

Widely used indices of test accuracy are

		gold standard (reference test)	
		+	−
index test	+	a	b
	−	c	d
		n_1	n_2

Figure 14.1 2×2-table for a binary test.

$$
\begin{aligned}
TPR &= a/n1 \quad \text{(True Positive Rate, sensitivity)} \\
FPR &= b/n2 \quad \text{(False Positive Rate, 1–specificity)} \\
FNR &= c/n1 \quad \text{(False Negative Rate)} \\
TNR &= d/n2 \quad \text{(True Negative Rate, specificity)}
\end{aligned}
\qquad (14.5)
$$

which are monotone in CO. The Odds Ratio (OR)

$$
OR = \left(\frac{TPR}{1 - TPR} \right) \Big/ \left(\frac{FPR}{1 - FPR} \right) \qquad (14.6)
$$

measures the discriminatory power of the index test (the odds of a positive test result among diseased persons relative to the odds of a positive test result among non-diseased persons), and may also vary with the chosen threshold CO.

Suppose k studies each reported as a 2×2-table are to be summarized. Midgette *et al.* [17] suggested weighted averages of and TPR_i and $FPR_i (i = 1, 2, \ldots, k)$, provided these are not positively correlated (which can be assessed using Spearman's test) but homogeneous. Heterogeneous data should not be combined, except maybe within subgroup analysis. If the TPR and FPR values are positively correlated, rendered by different thresholds, Midgette *et al.* [17] recommend estimation of a summary ROC curve (see below).

The (smooth) Receiver Operating Characteristic (ROC) curve of the test (using specific patient covariates) is obtained by plotting sensitivity versus 1-specificity for arbitrary cut-off values θ (Figure 14.4, described later, displays a Summary ROC, which is of the same format as a normal ROC). The area under the ROC curve may easily be interpreted as the probability of correctly ranking a randomly chosen pair consisting of a diseased and a non-diseased subject. It is the most important summary index of the test's performance [18, 19].

The estimation of a Summary Receiver Operating Characteristic (SROC) curve [15, 20–24] uses a linear model of the form

$$
D = \alpha + \beta S \qquad (14.7)
$$

where $D = \text{logit(TPR)} - \text{logit(FPR)} = \text{log(odds(TPR)} / \text{odds(FPR))} = \text{log(OR)}$, $S = \text{logit(TPR)} + \text{logit(FPR)}$, α is the intercept, and β is the regression coefficient of S. This allows the combination of TPRs, FPRs and ORs with varying corresponding thresholds. Hence, α is an odds ratio, and β examines the extent to which the odds ratio is dependent on the threshold value used. Model (14.7) can be analysed in an unweighted, weighted or robust manner. If the regression coefficient β is near zero ($\beta \approx 0$) the accuracy for each primary study can be summarized by a common odds ratio given by the intercept, α. In this special case, other fixed or random effects approaches may also be appropriate [14, 25]. Different diagnostic tests may be compared by examination of regression residuals (e.g. with a *t*-test) or introducing 'type of test' as covariate in model (14.7). Inclusion of appropriate covariates in model (14.7) also permits consideration of whether study quality or patient characteristics affect test characteristics, and allows for their adjustment (as confounders) in a comparison between tests. If data about two or more thresholds are available the use of Generalised Estimating Equations (GEE) [26] may be useful for estimation of the (unweighted) SROC [13]. A SROC, together with the individual study estimates from which it is derived, is presented in Figure 14.4. This is described in detail in the example which follows.

Example: Meta-analysis combining binary diagnostic test data

A dataset of studies comparing the accuracy of duplex doppler ultrasound with angiography as the standard for stroke prevention, described previously [14] is used here to illustrate the methods of combining binary diagnostic test data. Table 14.1 provides the data from each of the 14 primary studies, where data corresponding to a single 2×2 table of the form of Figure 14.1 were available (summarized in the table).

Independently pooled estimates of sensitivity and specificity, using either fixed or random effects could easily calculated for these studies [17]. However, these frequently used methods have come under strong criticism because they do not take into account the fact that different studies may have used different test thresholds [13]. For example, some studies may require considerable abnormality to be present before they declare a test to be positive, whereas others may require only a hint of abnormality before they declare a test to be positive. There is inevitably a trade-off between sensitivity and specificity, so if high levels of abnormality are

Table 14.1 Studies of duplex doppler ultrasound using angiography as the standard (modified from Hasselblad and Hedges [14]).

Study id	True positive	False positive	False negative	True negative
1	26	2	4	83
2	11	2	1	5
3	68	8	3	34
4	74	0	12	111
5	84	13	20	99
6	40	7	3	41
7	16	9	1	109
8	96	15	20	206
9	11	2	2	57
10	91	5	5	57
11	46	3	9	42
12	15	2	1	93
13	58	16	10	121
14	26	1	4	74

required, sensitivity will be low and specificity high, and vice versa. Failure to take this into account can lead to serious bias, usually resulting in underestimation of the performance of the test [13]. The SROC approach (see above), which takes this into account, is applied here.

First, the sensitivity (True Positive Rate (TPR)) and specificity (1–False Positive Rate (1–FPR)) are calculated for each study using equation (14.5), as shown it Table 14.2. These are plotted against each other in Figure 14.2.

It can be seen that TPR and FPR vary only moderately between studies. To use the regression equation (14.7), D and S need calculating. These are the difference between and sum of the logits of TPR and FPR, respectively. The values for the 14 studies are provided in Table 14.2. Linear regression is then performed using these two variables. The choice between using unweighted or weighted regression for such a model is not obvious. Irwig et al. [13] note that while an unweighted analysis does not emphasise the larger studies, a weighted analysis may bias the estimate. This can be seen by considering two studies of equal sample size which are both measuring the same underlying test accuracy. If one reports a poorer accuracy than the other, then it will be given more weight. For this reason it is sensible to report both the weighted and unweighted result. Lines of best fit for both analyses are displayed in Figure 14.3.

The intercept terms are similar for both models (4.61 for the unweighted and 4.26 for the weighted). Both slope coefficients are non-negligable (-0.5113 and

Table 14.2 Quantities required for SROC curve analysis.

Study id	TPR	FPR	Logit (TPR)	Logit (FPR)	Sum $= S$	Diff $= D =$ ln(OR)	Var	1/Var $= Wt$
1	0.87	0.02	1.87	−3.73	−1.85	5.60	0.59	1.68
2	0.92	0.29	2.40	−0.92	1.48	3.31	1.20	0.83
3	0.96	0.19	3.12	−1.45	1.67	4.57	0.41	2.43
4	0.86	0.00	1.79	−5.41	−3.62	7.19	1.11	0.90
5	0.81	0.12	1.44	−2.03	−0.60	3.47	0.14	7.04
6	0.93	0.15	2.59	−1.77	0.82	4.36	0.43	2.31
7	0.94	0.08	2.77	−2.49	0.28	5.27	0.70	1.43
8	0.83	0.07	1.57	−2.62	−1.05	4.19	0.13	7.93
9	0.85	0.03	1.70	−3.35	−1.65	5.05	0.81	1.24
10	0.95	0.08	2.90	−2.43	0.47	5.34	0.37	2.73
11	0.84	0.07	1.63	−2.64	−1.01	4.27	0.40	2.48
12	0.94	0.02	2.71	−3.84	−1.13	6.55	0.95	1.06
13	0.85	0.12	1.76	−2.02	−0.27	3.78	0.18	5.65
14	0.87	0.01	1.87	−4.30	−2.43	6.18	0.77	1.31

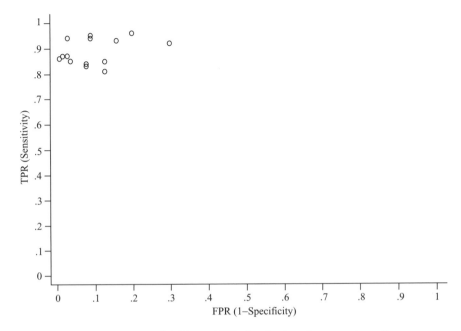

Figure 14.2   Scattergram of TPR vs. FPR for duplex ultrasound studies.

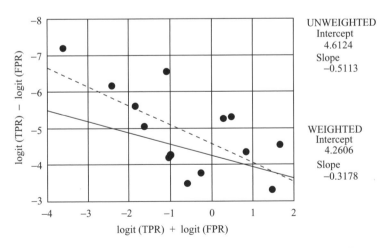

Figure 14.3   Regression analysis of D (= logit(TPR) – logit(FPR)) on S (= logit(TPR) + logit(FPR)).

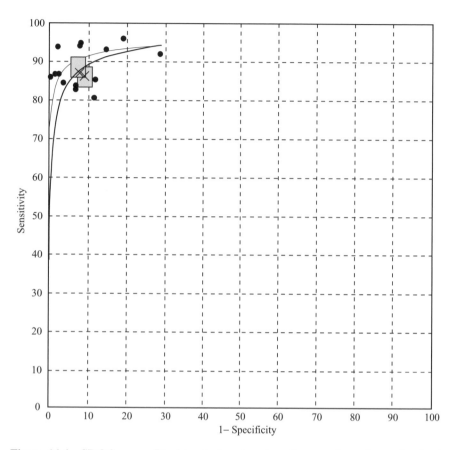

Figure 14.4 SROC curve fitted to duplex doppler ultrasound studies, using both weighted and unweighted regression.

−0.3178 for the unweighted and weighted analyses, respectively); this suggests that the odds ratio is not constant across different thresholds, and hence cannot be ignored. The results of these analyses are transformed onto sensitivity vs. 1–specificity axes to produce a SROC, displayed in Figure 14.4.

The thinner line corresponds to the unweighted analysis and the thicker line to the weighted analysis. Each circle represents a study. The two shaded regions represent the zone of 95% CI of the pooled sensitivity and 95% CI of pooled specificity, for the analysis combining sensitivity and specificity *independently* for fixed and random effect models (the lighter shading corresponds to the random effect analysis). The crosses in these zones represent the point estimates. Note that both these point estimates lie below the SROC curves, indicating a slight underestimation of performance. Unlike standard meta-analysis models which produce one answer, the performance of the diagnostic performance changes with changes in sensitivity and specificity, which is reflected in the curves on Figure 14.4.

14.4.2 Combining ordered categorical test results

Suppose the test result Y can fall into one of J categories ('ratings'). The probability of Y falling in a given category j or below can be modelled as a non-linear function of explanatory variables by means of ordinal regression equations (27–29).

Meta-analysis can either make use of this model either directly to combine data from appropriate studies [27–30], or pool the areas under the corresponding ROC curves using fixed or random effects models [31, 32]. Hellmich *et al.* [33] extend the work of Zhou to a fully Bayesian hierarchical model in a similar manner to (11.8).

14.4.3 Combining continuous test results

For continuous test outcomes, and in particular, if they are normally distributed with equal variances, Hasselblad and Hedges [14] advocate the use of the standardized difference of empirical means as a measure of discrimination or effectiveness (see Section 4.2.2). Alternatively, the area under the ROC curve can be estimated either parametrically or non-parametrically [18, 19]. For either measure, fixed or random effects approaches can be used to combine information from a collection of studies. Furthermore, the likelihood ratio, i.e. 'the ratio of the probability that a given level of a test result occurs in people with the disease to the probability of that test result in people without the disease' [34] can be modelled by means of the linear model [35, 36]. This likelihood ratio provides information on how informative a test is (i.e. how much extra information is gained compared to not doing the test, and only having the prior estimate (usually the population prevalence) of the probability of the condition).

In practice, methods for ordered categorical data are often used for categorized continuous test results, because costs-benefit arguments rule out collecting data for arbitrarily many threshold values.

14.5   Meta-analysis using surrogate markers

A surrogate marker may be used in the place of the real outcome, when the latter is difficult to measure [37]. An example would be use of CD4 count as a marker for time until the onset of AIDS in an HIV trial [38], or cholesterol reduction in trials of the effect of lipid lowering drugs on mortality, allowing the evaluation of an intervention to be completed more quickly (provided the surrogate marker is valid). However, there is the danger that interventions which perform well when evaluated using surrogates may perform less well when using the main outcomes of interest [39].

A'Hern *et al.* [40] investigated the relationship between response rate (a surrogate outcome) and median survival (the clinical outcome) in advanced breast cancer. The authors used the results of RCTs, and investigated the association between the odds ratio that summarised the difference in response rates in pairs of arms within the same study and the corresponding ratio of median survivals. Several limitations of the method they used for doing this have been highlighted. Daniels and Hughes [38] and Torri *et al.* [41] both comment that, although the precision in estimating the treatment difference on the clinical outcome is allowed for, the method did not take into account the level of precision in estimating the treatment difference on the response variable. This problem is addressed in two approaches described below.

Tori *et al.* [41] calculate a correlation coefficient (Kendall's tau) between antitumour response and survival in patients with advanced ovarian cancer in each study using values for median response, and then combining them using standard methods.

Daniels and Hughes [38] present a method to model the association between the treatment difference on a potential surrogate marker and the treatment difference on the clinical outcome of interest. This model is then used to determine the surrogate's reliability for predicting the treatment difference on clinical outcome, given an observed difference on the surrogate marker. The approach taken to modelling is a Bayesian one, and this model is a development of that of DuMouchel [42], used for more traditional meta-analysis applications (see Chapter 11). They illustrate the model with an example in AIDS medicine, using CD4 count as a surrogate for the time to AIDS onset [38]. It is interesting to note that the 15 trials included used *different* interventions. This is permissible, since it is not treatment effect which is of interest here, but the relationship between CRD4 and AIDS.

14.6 Combining a number of cross-over trials using the patient preference outcome

An outcome special to cross-over trials (in which patients receive both interventions sequentially, but in random order) is the patient preference outcome, and a simple methodology has been developed to combine trials using this scale. In a simple two-period cross-over design, patients receive both treatments sequentially, and are asked which one they prefer.

The proportions P_A and P_B preferring treatments A and B respectively – ignoring patients who did not express a preference–are calculated for each trial. The difference $P_A - P_B$ is then calculated, and is used as a (continuous) outcome measure. The variance of this outcome can be calculated using the formula below:

$$\text{Var}(P_A - P_B) = (P_A + P_B - [P_A - P_B]^2)/n \qquad (14.10)$$

where n = total number of patients in the study analysis. These estimates can then be combined using the standard methods [43].

14.7 Vote-counting methods

Light and Smith [44] were among the first to describe formally the 'taking a vote' procedure, the attraction of which is its apparent simplicity and wide applicability. Put simply, each study is assigned to one of three categories: a positive relationship, a negative relationship, or no relationship in either direction, depending upon the estimated effect size. The category with the highest count is assumed to give the best estimate of the direction of the true relationship. Clearly, this method could not be simpler. It has been criticized, however, for the following reasons:

1. The sample size, and therefore the precision of each estimate from the primary studies is not incorporated into the vote [44].
2. It does not provide an effect size estimate [45].
3. It has very low power for medium to small sample sizes and effect sizes [46].
4. The power of this test can decrease as the number of studies increases [46].

For these reasons, vote counting procedures have been described as naive, with no statistical rationale. They can lead to erroneous conclusions [46, 47]. A more detailed explanation of the limitations is given elsewhere [48]. Hence they are only recommended as a last resort, when effect magnitudes and significance levels of the primary studies are not available. Alternatively, they may be used to complement the more sophisticated methods described in this book. For example, if treatment estimates are only available for a proportion of the studies, effect sizes could be combined on that subset, but a vote-counting procedure could be carried out on all studies [49].

A related procedure is based on the (non-parametric) sign test [49]. The rationale behind this test is that if there is no treatment effect then the chance of a study showing a positive effect is 0.5. Hence, the null hypothesis is

$$H_0 : \text{probability of a positive result } (p) = 0.5,$$

and the alternative

$$H_A : p > 0.5.$$

Let U = the number of positive results in k independent studies being considered. Then an estimator of p is $\hat{p} = U/k$. Tables for the binomial distribution are then consulted to calculate how extreme $\hat{p}$ is, and hence whether to reject the null hypothesis. Again, this test also has its disadvantages, namely it does not incorporate sample size into the vote, and it does not produce an effect estimate.

A vote-counting procedure has been devised which *does* give an estimate of the treatment effect, but it assumes all studies to be combined have the same sample sizes, and that the numbers in both arms of each study are also the same [49]. Clearly, this is very restrictive and unlikely to be appropriate in most instances. Hedges and Olkin [46] recommend treating studies as if they have the same sample size, if in reality they do not vary much [48]. This method produces a confidence interval for the treatment effect of interest; from this, inferences can be made about the effectiveness of the intervention. A clear account, with several examples, using different scales has been presented elsewhere [49]. Because unequal sample sizes are the rule in research synthesis rather than the exception, the counting estimators are likely to be most useful for providing quick approximate estimates, rather than as the analytic tool for the final analysis [48]. An extension of this methodology to handle unequal sample sizes for the primary studies has been developed [48, 49]. The method involves maximum likelihood calculation, and is considerably more complex than that for equal sample sizes. The interested reader is referred to either of the previously mentioned texts for more details. It should also be noted that accurate estimates from both these methods rely on a reasonably large sample of studies [48].

In addition to the theoretical drawbacks of vote counting procedures, there are also certain practical problems. First, if all the results are in the same direction, the method of maximum likelihood cannot be used to obtain an estimate for p, although a Bayesian estimate is available [48, p.300, 49, p.211]. Secondly, it is difficult to know how to define a positive result: it could be defined as one that is statistically significant ($p < 0.05$, or some other value), or just positive, $p < 0.5$. Hedges and Olkin [48] advocate that 0.05 is a good practical choice, as a paper may state that the result reached statistical significance, even if it does not give any other details. Taking a positive result to be $p < 0.5$ allows synthesis in situations where the data is so sparse that only the direction of the result needs to be known.

See Section 13.4.2 for methods which have been developed to combine vote counting procedures for studies where effect size estimates are not known, with effect size estimates form studies where they are reported.

14.8   Combining *p*-values/significance levels

Methods of combining probability (*p*-)values from independent significance tests have a long history [50]. Several of these methods are closely related to the vote counting techniques [51] outlined above. Like vote counting procedures, these methods do not produce an effect size estimate, and hence analyses based on effect magnitude measures are usually preferable [51]. Hasselblad suggests two situations when combining *p*-values may be appropriate [52]: (i) when some studies do not report any effect measures but do report *p*-values; (ii) when the study designs or treatment levels are so different that combining effect measures would be inappropriate. So again, this method could be used to supplement combination of treatment effects when these are not available for every study.

The formal null and alternative hypotheses for combining *p*-values, where T_i represents the effect of interest in the *i*th study, are

$$H_0 : T_i = 0, \quad \text{for } i = 1, \ldots, k,$$

(i.e. for the joint null hypothesis to be true, all the individual null hypotheses must be true).

A possible alternative hypothesis is

$$H_A : T_i \neq 0.$$

Under this alternative, the population parameters are not required to have the same sign. This is very general, and is uninformative about the specific structure of variability [51]. Another possible alternative hypothesis is

$$H_A : T_i \geq 0, \quad \text{for } i = 1, \ldots, k, \text{ with}$$
$$T_j > 0, \text{for at least one j·}$$

This is used if the effect such as a correlation can not be negative, or if negative values are not of interest *per se*, such as in a variance test where negative values are evidence of zero variance.

Methods for combining *p*-values are described below; all are non-parametric [48]. This summary draws heavily on a review by Becker [51], which should be consulted if more details are required.

14.8.1 Minimum *p* method

This method was proposed by Tippett in 1931 [53]. The null hypothesis is rejected if any of the *p*-values (from the *k* studies) is less than α, where α is computed as $1 - (1 - \alpha^*)^{1/k}$, and α^* is the significance level for the combined significance test (e.g. traditionally set α^* to $0.05 = 5\%$). Put formally, one rejects H_0 if

$$\text{Min}(p_1, \ldots, p_k) = p_{[1]} < \alpha = 1 - (1 - \alpha^*)^{1/k}. \tag{14.11}$$

A generalization suggested by Wilkinson [54] is to use the *r*th smallest *p*-value [55]: H_0 is rejected if $p_{[r]} < C_{\alpha, k, r}$, where $C_{\alpha, k, r}$ is a constant that can be obtained from the Beta distribution. (These are tabulated in Hedges and Olkin [48, p.37].) The advantage of this method is that it does not rely on the most extreme result, and therefore is more resistant to outliers in the data than Tippett's original method.

14.8.2   Sum of z's method

This method, first described by Stouffer *et al.* [56], is based on the sum of the areas (sometimes known as z-scores) corresponding to the p-value associated with every study being combined ($z(p_i)$). The test statistic is

$$\frac{1}{\sqrt{k}} \sum_{i=1}^{k} (p_i). \tag{14.12}$$

This is compared with critical values of a standard Normal distribution, as the sum of p-values is assumed approximately normally distributed [51].

14.8.3   Sum of logs method

First described by Fisher in 1932 [50], the test statistic is

$$-2 \sum_{i=1}^{k} \log(p_i). \tag{14.13}$$

This is compared with the $100(1 - \alpha^*)\%$ critical value of the chi-square distribution with $2k$ degrees of freedom.

14.8.4   Logit method

This method was proposed by George in 1977 [57], and uses a test statistic defined by

$$\frac{-\sum_{i=1}^{k} \log(p_i/1 - p_i)}{[k\pi^2(5k + 2)/3(5k + 4)]^{1/2}}. \tag{14.14}$$

This test statistic is approximately distributed t with $5k + 4$ degrees of freedom.

14.8.5   Other methods of combining significance levels

There are many variants on the methods discussed above. In addition to extensions given below, see Becker [51] who gives a classified table of 16 test statistics that can be used for this purpose, and the following references [48, 58] for further information. Additionally, the whole concept of p-values is at odds with the Bayesian philosophy, but Berger and Mortera [59] investigate the interpretation of a p-value from a Bayesian perspective.

14.8.6   Appraisal of the methods

Many studies have investigated the power of the various tests available [51]. No one test is the most powerful in all situations. However, as Elston observes [60], Fisher's method (Section 14.8.3) is asymptotically optimal among essentially all methods of combining independent tests. Hence, Hedges and Olkin [48] advise that Fisher's test is perhaps the best one to use if there is no indication of particular alternatives.

There is confusion in the literature as to whether combining studies via p-values, weights the studies according to their power to detect a treatment effect. Although p-values do contain information relating to sample size and variability, the extent to which this is true in any specific situation will depend upon a number of factors, including the type of test used. Hence, no weighting scheme analogous to the inverse-variance weights used in a fixed effect meta-analysis of treatment effects is available for combining p-values.

14.8.7   Example of combining p-values

Consider the cholesterol-lowering trials dataset. For illustration, the p-values corresponding to each effect size estimate are combined. P-values can be obtained by looking up each study's z-score ($\ln(OR)/se(\ln(OR))$) on normal distribution tables, so for example, the $\ln(OR)$ for study 2 is -0.072, and $se(\ln(OR))$ is 0.232 (extracted from Table 3.1). Hence, The z-score for this study is $-0.072/0.258 = -0.310$; comparing this score to a standard normal distribution produces a p-value of 0.62. p-values for each study are displayed in Table 14.3.

Combining the p-values by the method of Fisher [50] using equation (14.13) proceeds as follows:

$$-2\sum_{i=1}^{k} \log(p_i) = -2[0 + (-0.48) + \ldots + (-0.98) + (-0.49)]$$

$$= 58.62.$$

This test statistic is compared to a χ^2_{68} distribution, which produces a p-value of 0.78. Recall that pooling these studies using a random effect model produced a result that was marginally non-significant ($p = 0.09$). The much higher p-value produced here highlights the low power of this type of method, and illustrates the limited value of such an analysis. (Note: although the p-values combined here were derived from effect sizes, combining p-values should normally only be required when some effect sizes are missing from reports and only the levels of significance are given.)

Table 14.3 *p*-values corresponding to the comparative treatment effect estimate for 34 trials of cholesterol lowering interventions.

Study	p-value	$\ln(p\text{-value})$
1	1.00	0.00
2	0.62	−0.48
3	0.96	−0.04
4	0.96	−0.04
5	0.90	−0.11
6	0.98	−0.02
7	0.95	−0.05
8	0.69	−0.38
9	0.57	−0.57
10	0.51	−0.68
11	0.99	−0.01
12	0.14	−1.97
13	0.24	−1.42
14	0.85	−0.16
15	1.00	0.00
16	0.63	−0.46
17	0.70	−0.36
18	0.67	−0.40
19	0.64	−0.45
20	0.90	−0.10
21	0.04	−3.21
22	0.11	−2.22
23	0.71	−0.34
24	0.19	−1.67
25	0.91	−0.09
26	0.75	−0.29
27	0.09	−2.36
28	0.61	−0.50
29	0.48	−0.74
30	0.06	−2.86
31	0.00	−5.57
32	0.75	−0.29
33	0.38	−0.98
34	0.61	−0.49

14.9 Novel applications of meta-analysis using non-standard methods or data

Meta-analysis methods have been developed and applied for non-typical problems in a variety of contexts. Li [61] presents a multiplicative [48, p.315] random effects model for meta-analysis of admixture component in genetic epidemiology. Freedman [62] describes the meta-analysis of animal experiments using (fixed effect) logistic regression models designed to investigate the effects of dietary fat intake upon mammary tumour development. Keller *et al.* [63] present a comparison of four methods for meta-analysis to standardise the different results of studies in pharmacokinetics, using an example based on renal insufficiency data.

Several environmental risk models combining information from exposure to man-made environmental agents, diet and cigarette smoking [55] have been developed (see Section 17.6.6) [55, 64–68]. Finally, meta-analysis has been employed in forecasting the future size of the Acquired ImmunoDeficiency Syndrome (AIDS) epidemic [55, p.165]. Taylor [69] presents a solution to the forecasting problem which involves formulating stochastic models. More recently, Cooley *et al.* [70] conducted a meta-analysis of estimates of the AIDS incubation distribution.

14.10 Summary/Discussion

This chapter considers various issues related to the type of data used in a meta-analysis. The methods in this chapter have been developed to deal with data which is somewhat 'non-standard', and cannot be combined directly using the outcome measures described in Chapter 2. Ways of transforming data so that standard methods can be used are discussed first. The combination of data from diagnostic test studies and studies using surrogate endpoints is considered next. Vote counting and combining *p*-values, approaches which can be used when measures of effect are not available, are presented.

References

1. Whitehead, A., Whitehead, J. (1991). A general parametric approach to the meta-analysis of randomised clinical trials. *Stat. Med.* **10**: 1665–77.

2. Whitehead, A., Jones, N.M.B. (1994). A meta-analysis of clinical trials involving different classifications of response into ordered categories. *Stat. Med.* **13**: 2503–15.

3. Deeks, J.J., Altman, D.G., Dooley, G., Sackett, D.L.S. (1997). Choosing an appropriate dichotomous effect measure for meta-analysis: empirical evidence of

the appropriateness of the odds ratio and relative risk. *Controlled Clin. Trials* **18**: 84s–5s.

4. Fleiss, J.L. (1994). Measures of effect size for categorical data. In: Cooper, H., Hedges, L.V., editors. *The Handbook of Research Synthesis*. New York: Russell Sage Foundation. 245–60.

5. Sinclair, J.C., Bracken, M.B. (1994). Clinically useful measures of effect in binary analyses of randomized trials. *J. Clin. Epidemiol* **47**: 881–9.

6. Huque, M.F., Dubey, S.D. (1994). A metaanalysis methodology for utilizing study-level covariate-information from clinical-trials. *Comm. in Statistics–Theory and Methods* **23**: 377–94.

7. Whitehead, A., Bailey, A.J., Elbourne, D. (1999). Combining summaries of binary outcomes with those of continuous outcomes in a meta-analysis. *J. Biopharmaceutical Stat.* **9**: 1–16.

8. Suissa, S. (1991). Binary methods for continuous outcomes: a parametric alternative. *J. Clin. Epidemiol* **44**: 241–8.

9. Colditz, G.A., Miller, J.N., Mosteller, F. (1989). How study design affects outcomes in comparisons of therapy. I: Medical. *Stat. Med.* **8**: 441–54.

10. Cochrane Methods Working Group on Systematic Review of Screening and Diagnostic Tests. Recommended methods, updated 6 June 1996. Available at http://som.flinders .edu.au/FUSA/COCHRANE/cochrane/sadt.htm.

11. Irwig, L., Tosteson, A.N., Gatsonis, C., Lau, J., Colditz, G., Chalmers, T.C., Mosteller, F. (1994). Guidelines for meta-analyses evaluating diagnostic tests. *Ann. Intern. Med.* **120**: 667–76.

12. Irwig, L., Tosteson, A.N.A., Gatsonis, C. (1994). Metaanalyses evaluating diagnostic-tests – response. *Ann. Intern. Med.* **121**: 817–8.

13. Irwig, L., Macaskill, P., Glasziou, P., Fahey, M. (1995). Meta-analytic methods for diagnostic test accuracy. *J. Clin. Epidemiol* **48**: 119–30.

14. Hasselblad, V., Hedges, L.V. (1995). Meta-analysis of screening and diagnostic tests. *Psychol. Bull.* **117**: 167–78.

15. Shapiro, D.E. (1995). Issues in combining independent estimates of the sensitivity and specificity of a diagnostic test. *Acad Radiol* **2**: S37–47.

16. Ohlsson, A. (1994). Systematic reviews – theory and practice. *Scand. J. Clin. & Laboratory Investigation* **54**: 25–32.

17. Midgette, A.S., Stukel, T.A., Littenberg, B. (1993). A meta-analytic method for summarizing diagnostic test performances: receiver-operating-characteristic-summary point estimates. *Med. Decis Making* **13**: 253–7.

18. Bamber, B. (1975). The area above the ordinal dominance graph and the area below the reciever operating characteristic graph. *J. Math. Psychol.* **12**: 387–415.

19. Hanley, J.A., McNeil, B.J. (1982). The meaning and use of the area under the receiver operating characteristic (ROC) curve. *Radiology* **143**: 29–36.

20. Kardaun, J.W.P.F., Kardaun, O.W.J.F. (1990). Comparative diagnostic performance of three radiological procedures for the detection of lumbar disk herniation. *Meth. Inform. Med.* **29**: 12–22.

21. Moses, L.E., Shapiro, D., Littenberg, B. (1993). Combining independent studies of a diagnostic test into a summary ROC curve: Data-analytic approaches and some additional considerations. *Stat. Med.* **12**: 1293–316.

22. Littenberg, B., Moses, L.E. (1993). Estimating diagnostic accuracy from multiple conflicting reports: a new meta-analytic method. *Med. Decis. Making.* **13**: 313–21.

23. Rutter, C.M., Gatsonis, C.A. (1996). Regression methods for meta-analysis of diagnostic test data. *Acad. Radiol* **2**: S48–56.

24. Devries, S.O., Hunink, M.G.M., Polak, J.F. (1996). Summary receiver operating characteristic curves as a technique for metaanalysis of the diagnostic performance of duplex ultrasonography in peripheral arterial-disease. *Acad. Radiol.* **3**: 361–9.

25. Klassen, T.P., Rowe, P.C. (1992). Selecting diagnostic tests to identify febrile infants less than 3 months of age as being at low risk for serious bacterial infection: A scientific overview. *J. Pediatrics* **121**: 671–6.

26. Zeger, S., Liang, K.Y. (1986). Longitudinal data analysis for discrete and continuous outcomes. *Biometrics* **42**: 121–30.

27. Tosteson, A.N., Begg, C.B. (1988). A general regression methodology for ROC curve estimation. *Med. Decis. Making* **8**: 204–15.

28. Tosteson, A.N., Weinstein, M.C., Wittenberg, J., *et al.* (1994). ROC curve regression analysis: the use of ordinal regression models for diagnostic test assessment. *Environmental Health Perspectives* **102**: 73–8.

29. Peng, F., Hall, W.J. (1996). Bayesian analysis of ROC curves using Markov-chain Monte Carlo methods. *Med. Decis. Making* **16**: 404–11.

30. Mossman, D., Somoza, E. (1989). Maximizing diagnostic information from the dexamethasone suppression test: an approach to criterion selection using receiver operating characteristic analysis. *Arch General Psychiatry* **46**: 653–60.

31. McClish, D.K. (1992). Combining and comparing area estimates across studies or strata. *Med. Decis. Making* **12**: 274–9.

32. Zhou, X.H. (1996). Empirical Bayes combination of estimated areas under ROC curves using estimating equations. *Med. Decis. Making* **16**: 24–8.

33. Hellmich, M., Abrams, K.R., Sutton, A.J. (1999). Classical and Bayesian approaches to meta-analysis of ROC curves: a comparative review. *Med. Decis. Making* **19**: 252–64.

34. Sackett, D.L., Haynes, R.B., Guyatt, G.H. *et al.* (1991). *Clinical Epidemiology: a Basic Science for Clinical Medicine. 2nd ed.* Boston: Little Brown.

35. Albert, A. (1982). On the use and computation of likelihood ratios in clinical chemistry. *Clin. Chem.* **28**: 1113–9.

36. Irwig, L. (1992). Modelling result-specific likelihood ratios. *J. Clin. Epidemiol* **45**: 1335–8.

37. Wittes, J., Lakatos, E., Probstfield, J. (1989). Surrogate endpoints in clinical trials: cardiovascular diseases. *Stat. Med.* **8**: 415–25.

38. Daniels, M.J., Hughes, M.D. (1997). Meta-analysis for the evaluation of potential surrogate markers. *Stat. Med.* **16**: 1965–82.

39. Gotzsche, P.C., Liberati, A., Torri, V., Rossetti, L. (1996). Beware of surrogate outcome measures. *Int. J. Technol. Assessment in Health Care* **12**: 238–46.

40. A'Hern, R.P., Ebbs, S.R., Baum, M.B. (1988). Does chemotherapy improve survival in advanced breast cancer? A statistical overview. *Br. J. Cancer.* **57**: 615–8.

41. Tori, V., Simon, R., Russek-Cohen, E., Midthune, D., Friedman, M. (1992). Statistical model to determine the relationship of response and survival in patients with advanced ovarian cancer treated with chemotherapy. *J. Nat Cancer Inst* **84**: 407–14.

42. DuMouchel, W. (1994). Hierarchical Bayes linear models for meta-analysis. Technical Report #27, National Institute of Statistical Sciences, PO Box 14162, Research Triangle Park, NC 27709.

43. Gotzsche, P.C. (1989). Patients' preference in indomethacin trials: An overview. *Lancet* **1**: 88–91.

44. Light, R.J., Smith, P.V. (1971). Accumulating evidence: procedures for resolving contradictions among different research studies. *Harvard Educ. Rev.* **41**: 429–71.

45. Glass, G.V., McGraw, B., Smith, M.L. (1981). *Meta-analysis in Social Research.* Newbury Park, CA: Sage.

46. Hedges, L.V., Olkin, I. (1980). Vote-counting methods in research synthesis. *Psychol Bull* **88**: 359–69.

47. Greenland, S. (1987). Quantitative methods in the review of epidemiological literature. *Epidemiol Rev.* **9**: 1–30.

48. Hedges, L.V., Olkin, I. (1985). *Statistical Methods for Meta-Analysis.* London: Academic Press.

49. Bushman, B.J. (1994). Vote-counting procedures in meta-analysis. In: Cooper, H., Hedges, L.V., editors. *The Handbook of Research Synthesis*. New York: Russell Sage Foundation. 193–214.

50. Fisher, R.A. (1932). *Statistical Methods for Research Workers. 4th ed.* London: Oliver and Boyd.

51. Becker, B.J. (1994). Combining significance levels. In: Cooper, H., Hedges, L.V., editors. *The Handbook of Research Synthesis*. New York: Russell Sage Foundation. 215–30.

52. Hasselblad, V. (1995). Meta-analysis of environmental health data. *Sci. of the Total Environment* 160–161: 545–58.

53. Tippett, L.H.C. (1931). *The Methods of Statistics. 1st ed.* London: Williams & Norgate.

54. Wilkinson, B.A. (1951). A statistical consideration in psychological research. *Psychol Bull* **48**: 156–8.

55. National Research Council. (1992). Combining Information: Statistical Issues and Opportunities for Research. Washington, DC.: National Academy Press.

56. Stouffer, S.A., Suchman, E.A., De Vinney, L.C. *et al.* (1949). The American soldier: Adjustment during army life (Vol.1). Princeton, NJ: Princeton University Press.

57. George, E.O. (1977). Combining independent one-sided an two-sided statistical tests – Some theory and applications. University of Rochester.

58. Mosteller, F., Bush, R.R. (1954). Selected quantitative techniques. In: Lindzey, G., (editor). *Handbook of Social Psychology*. Cambridge, MA: Addison-Wesley. 289–334.

59. Berger, J.O., Mortera, J. (1991). Interpreting the stars in precise hypothesis-testing. *Int. Stat. Rev.* **59**: 337–53.

60. Elston, R.C. (1991). On Fisher's method of combining p-values. *Biometrical J.* **33**: 339–45.

61. Li, Z. (1995). A multiplicative random effects model for meta-analysis with application to estimation of admixture component. *Biometrics* **51**: 864–73.

62. Freedman, L.S. (1994). Meta-analysis of animal experiments on dietary fat intake and mammary tumours. *Stat. Med.* **13**: 709–18.

63. Keller, F., Erdmann, K., Giehl, M., Buettner, P. (1993). Nonparametric meta-analysis of published data on kidney-function dependence of pharmacokinetic parameters for the aminoglycoside netilmicin. *Clin. Pharmacokinetics* **25**: 71–9.

64. Carroll, R.J., Simpson, D.G., Zhou, H. *et al.* (1994). Stratified ordinal regression: A tool for combining information from disparate toxicological studies. Technical Report #26, National Institute of Statistical Sciences, PO Box 14162, Research Triangle Park, NC 27709.

65. Cox, L.H., Piegorsch, W.W. (1994). Combining environmental information: Environmetric research in ecological monitoring, epidemiology, toxicology, and environmental data reporting. Technical Report #12, National Institute of Statistical Sciences, PO Box 14162, Research Triangle Park, NC 27709.

66. Cox, LH., Piegorsch, W.W. (1996). Combining environmental information I: Environmental monitoring, measurement and assessment. *Environmetrics* **7**: 299–308.

67. Mathew, T., Sinha, B.K., Zhou, L. (1993). Some statistical procedures for combining independent tests. *J. Am. Stat. Assoc* **88**: 912–9.

68. Piegorsch, W.W., Cox, L.H. (1996). Combining environmental information 2: Environmental epidemiology and toxicology. *Environmetrics* **7**: 309–24.

69. Taylor, J.M. (1989). Models for the HIV infection and AIDS epidemic in the United States. *Stat. Med.* **8**: 450–8.

70. Cooley, P.C., Myers, L.E., Hamill, D.N. (1996). A meta-analysis of estimates of the AIDS incubation distribution. *Euro. J. Epidemiology* **12**: 229–35.

CHAPTER 15

Meta-analysis of Multiple and Correlated Outcome Measures

15.1   Introduction

This chapter describes the methodology used to combine results when several outcomes have potentially been reported in each study. It is assumed that each study consists of two groups of patients. When this occurs, one could simply conduct separate analyses for each outcome measure, or discard all but one outcome [1], and combine the results using standard methods. However, this latter approach may not always be desirable because a single treatment may have different effects on different outcomes; it may be misleading to average such effects for each study, or simply to choose a single outcome for analysis and ignore the others [2]. Hence, questions such as 'Does a treatment have larger effects on some outcomes than on others? Does the duration of treatment affect different outcomes differently?' are hard to answer.

Further, the analysis of individual outcomes is not optimally efficient as it does not use statistical information about the errors of estimation contained in the other estimated effect sizes (i.e. information on how the outcomes are related) [3]. An alternative approach would be to combine all outcome measures within each study, and then combine these across studies. This approach too has its problems, including the fact that because each study in a series of related studies typically uses a different set of outcome variables, hence a standard multivariate, linear-model approach, requiring the same set of outcomes for each unit, cannot be used without either discarding data or imputing missing values [2].

A further problem is that it cannot be assumed *a priori* that the relationship between study features and effect sizes are the same for each outcome. Thus, it has been suggested that a method of analysis should allow different predictors (covariates) of effect magnitude for different outcomes [2].

229

Various approaches to combine studies reporting multiple outcome measures, retaining as much information as possible, are discussed below. Many of these were developed by researchers in education, but similar situations do arise in health research.

Different effect-size estimates calculated for any one sample are typically correlated so the statistical methods that assume independence between outcomes may be inappropriate [2]. Gleser and Olkin [3] provide a discussion on when one can ignore correlations among estimated effect sizes, at the cost of being conservative, and use univariate approaches, and when such univariate approaches are not advisable. They conclude that univariate approaches can be used for certain across-study inferences on individual effect sizes, but that multivariate methods are needed for most within-study inferences on effect sizes.

Sections 15.2, 15.3 and early parts of Section 15.4 consider relatively early approaches to the synthesis of multiple outcome measures. More recent and sophisticated modelling approaches are described in more detail in the latter parts of Section 15.4. In most cases, more recent models have superseded previous ones, and for this reason, in most instances, the final 'basic' model (Section 15.4.4) should be used, and hence is considered in most detail here.

A further situation exists where multiple results (distinct from outcomes) may be available from each study. If all, or a proportion, of the primary studies were designed with more than two arms, then different comparisons between the arms produce multiple estimates (but using the same outcome measure). Methods for dealing with such data are considered in Section 17.4.

15.2   Combining multiple *p*-values

Strube [4] describes a method for combining significance levels (*p*-values) when the outcomes are not independent. The motivating example is from psychology, where two trials are considered in which both the patient and the therapist evaluated the treatment. If the four results were combined in the standard way (using Stouffers's formula—see Section 14.8.2), this would assume independence, which is clearly disputable with pairs of results coming from the same experiments, and this will inflate the Type I error rate. To avoid this, terms for the covariance of the within-study results are included in the denominator of Stouffers's formula.

This would require the correlation to be given in the report or access to individual patient data. If neither of these are available then it may be possible to estimate it from other results in the paper [4, 5], or adopt an alternative, more accurate, procedure [5] that can be used when knowledge of the degrees of freedom used is available; this is described in the next section.

15.3 Method for reducing multiple outcomes to a single measure for each study

This method was developed for dealing with the situation where several subscales are to be combined into a single study estimate [5]. Its aim is to:

(i) derive a single summary statistic incorporating the information from all the effect measures of a single study (appropriate when outcomes are parallel measures of a single construct [2]). This statistic could then be combined with, and compared to, the results from other studies using standard meta-analytic procedures;

(ii) test specific hypotheses about the relative magnitudes of effects on different covariates, and estimate the magnitude of these contrasts.

This methodology is developed for combining either significance levels or measures of effect magnitude. To use this method, and most of those which follow, the correlations of the within study results ideally should be known. If a proportion of them are not, two alternatives exist: (1) estimate the correlation using other studies where the value is known; or (2) carry out two analyses with upper and lower bounds for the correlation used [5]. Technical details on how to calculate this are given in Rosenthal and Rubin [5].

Hedges and Olkin [1, p210] present an alternative rendering of the above. A more generalized form of this method is presented by Gleser and Olkin [3] using a generalized least squares regression approach.

15.4 Development of a multivariate model

15.4.1 Model of Raudenbush *et al.*

The model presented by Raudenbush *et al.* [2] is more general than the methods outlined above, and it has formed the base on which further work (described below) has developed. Their model uses a generalized least squares regression approach, which allows different outcomes (and different numbers of outcomes) to be measured across studies, and also different covariates to be used in regression models to explain the variation in effect sizes for each outcome. Essentially, this can be viewed as an extension of the fixed effect meta-regression methods of Chapter 6. A clear and thorough explanation of the model is given by Raudenbush *et al.* [2]. As for the methods presented earlier, it is necessary to estimate the correlations between outcomes. Several limitations of the method are noted [2], including the danger of model mis-specification, and that a mixed modelling approach is not developed.

15.4.2   Model of Gleser and Olkin

An alternative formulation of essentially the same type of generalized least squares regression model is given by Gleser and Olkin [3]. In this presentation, the authors point out a mistake in a previously developed method described by Hedges and Olkin, [1] which combined a *vector* of effect sizes from each study (a limitation of this early model is that it assumes the same outcomes are available on all studies to be combined, and hence has not been considered here). This mistake was carried over into the model of Raudenbush *et al.* [2]. The error is taking the large-sample correlation between two estimated effect sizes in a given multiple-endpoint study as equal to the observed correlation of the outcomes. This model is also capable of incorporating studies with more than two arms, and this aspect of the model is considered further in Section 17.5.1.

15.4.3   Multiple outcome model for clinical trials

Berkey *et al.* [6] consider multiple-outcome meta-analysis in a clinical trials setting, where the outcomes are measured on a continuous scale. This method handles data on multiple outcomes, where each study reports all or a subset of these outcomes. Study level covariates can also be included in the model. It also allows computation of effect sizes when more than two treatment types are considered, and no single treatment or control group appears in every study to serve as the common group [6]. (This issue is considered separately in Section 17.4.) Essentially, the problem being addressed is one where multiple outcomes and multiple treatments are considered simultaneously in a single meta-analysis. The model can incorporate more than two treatment groups from multi-arm trials, and single arms from randomized trials that include only one of the treatments; in doing so, it allows the meta-analysis to use more of the available data. The authors acknowledge the model by Dear [7] (see Section 18.6) for meta-analysis of survival data, and use the model of Raudenbush *et al.* [2] as a starting point.

If common 'treatment arms' are available for all studies, then within trial comparisons can be directly analysed. If this is not the case, an analysis can still proceed using simply the outcomes of each arm. Adjustment by study level and treatment-group level covariates when evaluating treatment effectiveness are both possible. Again, the correlation between outcomes is required, but if necessary, values can be assumed and their impact assessed by a sensitivity analysis [6, 8].

15.4.4   Random effect multiple outcome regression model

A random effect multiple outcome regression model has been developed [9]. This is appropriate when two or more continuous outcomes are combined from comparative two-arm studies. It also allows the inclusion of study level covariates, and hence it can be viewed as an amalgamation of the model described in Section 15.4.3 and the mixed models of Chapter 6. This is the most sophisticated model of this kind developed, and its use is recommend. The general form of the model is

$$\mathbf{y}_i = X_i\boldsymbol{\beta} + \boldsymbol{\delta}_i + \mathbf{e}_i, \tag{15.1}$$

where $\mathbf{y}_i$ is a vector of p outcomes reported by trial i; X_i is a matrix containing the observed trial-level covariates for trial i; $\boldsymbol{\beta}$ is the vector of regression coefficients to be estimated; $\boldsymbol{\delta}_i$ is a vector of p random effects associated with trial i. The $\text{cov}(\boldsymbol{\delta}_i) = D$ needs to be estimated, and it is assumed that the $\boldsymbol{\delta}_i$ arise from a multivariate Normal distribution (MVN $(\mathbf{0}, D)$); and $\mathbf{e}_i$ is the vector of random sampling errors within trial i, having $p \times p$ covariance matrix S_i, which is assumed known (but is usually estimated/reported by the individual trials). If each n_i is sufficiently large, then the vector $\mathbf{e}_i$ is approximately MVN $(\mathbf{0}, S_i)$. This leads to:

$$cov\,(y_i) = D + S_i \quad \text{and} \quad y_i \sim MVN(X\beta, D + S_i).$$

Solutions are possible using both generalized least squares and Multivariate Maximum Likelihood (MML) methods, if all outcomes and covariates are available for each study, and MML methods only if some are missing.

15.4.5   DuMouchel's extended model for multiple outcomes

DuMouchel [10] has developed a very general mixed model for combining studies reporting multiple outcomes. The model differs from those discussed above because binary as well as continuous outcomes can be modelled (although a common scale is required for all outcomes in any particular model), and correlations are estimated by the model, and hence not required as data. This model also allows several other extensions, and is considered further in Section 17.4.4.

15.4.6   Illustration of the use of multiple outcome models

A dataset originally used by Berkey et al. [9] to illustrate their models provides an excellent illustration of the use of multiple outcome models. The dataset, reproduced in Table 15.1, consists of results from five RCTs comparing the effects of surgical and non-surgical treatments for medium-severity periodontal (gum) disease, one year after treatment. Two outcomes are considered – improvement in Probing Depth (PD) and improvement in Attachment Level (AL), both measured in millimetres. The final column provides the within-trial covariance matrix for these outcomes (obtained for the original study reports). Additionally, publication year is included in the modelling as a covariate.

Table 15.1   Results from five trials comparing surgical and non-surgical treatments for medium-severity periodontal disease, one year after treatment (modified with permission from Berkey *et al.* [9] © John Wiley & Sons, Ltd.).

Trial	Publication year	Total number of patients	Improvement in		S_i	
			Probing depth	Attachment level	PD	AL
1	1983	14	+0.47	−0.32	$\begin{bmatrix} 0.0075 & 0.0030 \\ 0.0030 & 0.0077 \end{bmatrix}$	
2	1982	15	+0.20	−0.60	$\begin{bmatrix} 0.0057 & 0.0009 \\ 0.0009 & 0.0008 \end{bmatrix}$	
3	1979	78	+0.40	−0.12	$\begin{bmatrix} 0.0021 & 0.0007 \\ 0.0007 & 0.0014 \end{bmatrix}$	
4	1987	89	+0.26	−0.31	$\begin{bmatrix} 0.0029 & 0.0009 \\ 0.0009 & 0.0015 \end{bmatrix}$	
5	1988	16	+0.56	−0.39	$\begin{bmatrix} 0.0148 & 0.0072 \\ 0.0072 & 0.0304 \end{bmatrix}$	

Four models are fitted to this data:

   (i)   separate fixed effects models for each outcome (using a fixed effect regression model of Section 6.3.1);
   (ii)   separate random effects models for each outcome (using mixed models of Section 6.3.3);
   (iii)   the fixed-effects multiple outcome model described in Section 15.4.3; and
   (iv)   random-effects multiple outcome model described in Section 15.4.4 (by multivariate maximum likelihood methods).

The results from these analyses are presented in Table 15.2. In each instance the covariate year was included. The first row of output displays the between-study covariance matrices, D. No between study heterogeneity is assumed for the fixed effect models (models 1 and 3), and in these each element of D is restricted to 0. For model 2, separate models are fitted to each outcome so the random effect covariance term (represented by both off diagonal elements of D) is fixed at 0.

Table 15.2 Results of four different regression models used to combine the periodontal trial results (modified with permission from Berkey *et al.* [9] © John Wiley & Sons, Ltd).

	Model 1 Separate outcomes – Fixed effects	Model 2 Separate outcomes – Random effects	Model 3 Multiple outcomes – Fixed effect	Model 4 Multiple outcomes – Random effect
D matrix				
$PD \begin{bmatrix} 0 & 0 \\ AL & 0 & 0 \end{bmatrix}$		$\begin{bmatrix} 0.020 & 0 \\ 0 & 0.036 \end{bmatrix}$	$\begin{bmatrix} 0 & 0 \\ 0 & 0 \end{bmatrix}$	$\begin{bmatrix} 0.008 & 0.009 \\ 0.009 & 0.025 \end{bmatrix}$
Outcome PD models				
$Y = out_1 + \beta_1 x$ $(SE\,(out_1))\ (SE(\beta_1))$				
$PD = 0.345 - 0.008x$ (0.029) (0.008)		$0.363 + 0.005x$ (0.073) (0.022)	$0.304 + 0.005x$ (0.029) (0.008)	$0.348 + 0.001x$ (0.052) (0.015)
Outcome AL models				
$Y = out_2 + \beta_2 x$ $(SE(out_2))\ SE(\beta_2))$				
$AL = -0.394 - 0.012x$ (0.019) (0.007)		$-0.340 - 0.014x$ (0.092) (0.028)	$-0.394 - 0.009x$ (0.019) (0.007)	$-0.335 - 0.011x$ (0.079) (0.024)

The regression parameters for all four models are quite similar; it is the standard errors of the model parameters (from which confidence intervals, for the treatment effect can be constructed, using equations such as (4.4)) where the greatest differences are seen. Recall that fixed effect models generally produce smaller standard errors, and hence narrower confidence intervals, than random effect models as between study heterogeneity is not accounted for. This is observed here by comparing model 1 to model 2, and model 3 to model 4. However, further insight is gained from comparing the estimated standard errors of model 2 with those of model 4. Here, for both PD and AL the multiple outcome model (model 4) estimates smaller standard errors than the separate outcome model (model 2). The reason for this is the efficiency gained from combining both outcomes in the same model, and is a direct result of incorporating a covariance term (the off diagonal elements of S_i) for each outcome. In this example, the standard errors for the random effect multiple outcome model (model 4) were in-between those estimated by the fixed and random effect separate outcome models (models 1 and 2). Generally, the decision to use a fixed or random effect model will have more impact than that of modelling outcomes separately or simultaneously, but there are moderate benefits to be gained from using multiple outcome models [9].

15.5   Summary/Discussion

This chapter has considered extensions to fixed, random and regression meta-analysis models to allow more than one outcome measure to be modelled simultaneously. Although it is always possible to combine data from all trials relating to each outcome in a separate analysis using standard methods, this is not optimally efficient because the relationship between outcomes has not been taken into account. Hence, simultaneous modelling may produce estimates with narrower confidence intervals. In order to make use of these models, however, the correlations between outcomes are required, or have to be estimated. The models considered in this chapter (with the exception of that in Section 15.4.5) are not appropriate for binary outcomes.

References

1. Hedges, L.V., Olkin, I. (1985). *Statistical Methods for Meta-Analysis*. London: Academic Press.

2. Raudenbush, S.W., Becker, B.J., Kalaian, H. (1988). Modeling multivariate effect sizes. *Psychol. Bull.* **103**: 111–20.

3. Gleser, L.J., Olkin, I. (1994). Stochastically dependent effect sizes. In: Cooper, H., Hedges, L.V., editors. *The Handbook of Research Synthesis*. New York: Russell Sage Foundation. 339–56.

4. Strube, M.J. (1985). Combining and comparing significance levels from nonindependent hypothesis tests. *Psychol Bull* **97**: 334–41.

5. Rosenthal, R., Rubin, D.B. (1986). Meta-analytic procedures for combining studies with multiple effect sizes. *Psychol. Bull.* **99**: 400–6.

6. Berkey, C.S., Anderson, J.J., Hoaglin, D.C. (1996). Multiple-outcome meta-analysis of clinical trials. *Stat. Med.* **15**: 537–57.

7. Dear, K.B.G. (1994). Iterative generalized least squares for meta-analysis of survival data at multiple times. *Biometrics* **50**: 989–1002.

8. Berkey, C.S., Antczak-Bouckoms, A., Hoaglin, D.C., Mosteller, F., Pihlstrom, B.L. (1995). Multiple-outcomes meta-analysis of treatments for periodontal disease. *J. Dent. Res.* **74**: 1030–9.

9. Berkey, C.S., Hoaglin, D.C., Antczak-Bouckoms, A., Mosteller, F., Colditz, G.A. (1998). Meta-analysis of multiple outcomes by regression with random effects. *Stat. Med.* **17**: 2537–50.

10. DuMouchel, W. (1998). Repeated measures meta-analysis. *Bull. Int. Stat. Inst.* Tome LVII, Book 1(Session 51):285–8.

CHAPTER 16

Meta-analysis of Epidemiological and Other Observational Studies

16.1 Introduction

Many of the methods and controversies that have been presented with respect to RCTs are also pertinent to non-experimental and observational studies, but there are additional concerns [1]. It is often difficult to confirm a relationship between exposure and disease, because of small prevalence or incidences, moderate effect sizes, and long latency periods in epidemiological studies [2]. It has been noted that in such situations, meta-analysis could be a powerful tool for combining the results of several studies to reach an overall estimate [3]. Indeed, with increased numbers in a pooled analysis, rare exposures can be more easily studied [4], and this can help an epidemiologist decide if a particular association does or does not exist, and if so provides an indication of the quantitative relationship between them [3].

The techniques used for meta-analysis of epidemiological studies are often similar to those used for RCTs; guidelines for meta-analysis of observational data should at the very least follow those for clinical trials [5]. However, additional factors need consideration, and specific methodology exists [6–8]. Two publications of particular importance are (1) Blair *et al.* [7], a report from an expert working group for the application of meta-analysis in environmental epidemiology, which aims to develop a consensus guidelines for meta-analysis of environmental health issues; and (2) Greenland [6], a detailed overview which lays out much of the methodology outlined here.

It is commonly accepted that observational studies are prone to a greater degree of bias than RCTs. For this reason, Spitzer [1] questions whether meta-analytic techniques can be applied to epidemiological studies at all, but considers the answer to be 'a guarded yes'. For example, one has to assess if proper control or adjustment has been made for biases which frequently occur [9]. There has also been the suggestion that the potential for publication bias is greater for observational studies [9], though this has not been systematically investigated [10]. Even if observational studies are published, authors may only report the outcomes which gained statistical significance, which may be a subset of all the outcomes examined and tested [10].

A further issue concerns meta-analysis of ecologic (cross-sectional studies) as opposed to planned prospective studies. Ecologic studies are often carried out because someone thinks there is a raised incidence of a disease in a specific geographical location (e.g. near an incinerator). Therefore, the studies tend to be carried out when there is an association. Rarely do they occur when no one thinks there is a problem. This raises a big problem of bias since the 'universe' of studies is biased. This is not a publication bias issue but a research initiation bias question. Ideally one wants only studies which are planned and not initiated because of the signs in the data already. Clearly, this is not a problem for prospective epidemiological/experimental studies.

For these reasons, caution should be exercised when combining and reporting a meta-analysis of observational studies. Sensitivity analysis can help tackle some of these shortcomings, and its importance cannot be overstated.

Whether observational, non-randomized studies and RCTs both investigating the same or similar questions can be combined together, remains controversial. Methodology for combining studies of different designs is dealt with in Chapter 17, and the section below deals just with combining several observational studies.

A meta-analysis of the use of Oral Contraceptive (OC) use and breast cancer risk [11] is examined to illustrate several of the methods described in this chapter. It also provides a good illustration of the problems inherent in carrying out a meta-analysis of observational studies. The meta-analysis included 27 epidemiological studies, of cohort and case-control designs, assessing the association between oral contraceptive use and the risk of breast cancer published between 1980 and 1989. Aspects of this meta-analysis are considered throughout the chapter.

16.2   Extraction and derivation of study estimates

Chêne and Thompson [12] consider the problem of (large) differences between the (style of) presentation of results of epidemiological studies. Their illustrative example is a meta-analysis of nine studies investigating the relation between serum albumin and subsequent mortality. The paper highlights major differences in the reporting of studies, and highlights many of the issues facing a meta-analysis of observational studies. These include:

1. presentation of crude numbers of deaths, mortality rates, or relative risks in groups defined according to serum albumin concentration;
2. different studies used between three and six groups;
3. most also expressed the risk relation either as a logistic regression coefficient (or equivalently an odds ratio for a given increment in serum albumin), or as the mean difference in serum albumin concentrations between those subjects who died and those who survived;
4. different confounding factors were used in different studies in the logistic regression analyses; and
5. standard errors for the logistic regression coefficients or the mean differences were not always available.

Similar problems were present in the meta-analysis of OC use and breast cancer. First, it was necessary to derive a single common set of definitions for the outcome, explanatory and confounding variables, compatible with those used in the primary studies. *A priori* it was decided to investigate variations with age at diagnosis, parity and total duration of oral contraceptive via subgroup analyses (see Chapter 6). To do this, relative risks of breast cancer in OC users in the age, parity and duration subgroups where required from each of the 27 primary studies, where possible.

In reporting the results of the primary studies, 27 different sets of age-at-diagnosis categories, 11 parity categories, and 42 duration of use categories were employed. The common definitions of age at diagnosis categories adopted for the meta-analysis were: (1) < 25, 25–29, 30–34, 35–39, 40–44, 45–49, 50–54, 55+; and (2) < 45, 45 and over. The parity categorization used was dichotomous (nulliparous or parous), and the total duration of OC use was categorized as < 2, 2–4, 4–8, 8+ years. The majority of the categorizations in the primary studies fell into these categorizations.

For some studies, it was necessary to estimate key items of data not reported in the published account, such as numbers of cases and controls (or exposed and unexposed subjects) in subgroups of the study samples. For example, in some analyses these were estimated in proportion to marginal distributions, and the impact of these assumptions subjected to sensitivity analysis.

A further issue is that different studies matched for different factors when selecting study samples. Table 16.1 illustrates the main matching variables in the 27 studies. Covariates encoded indicating whether year of interview, geographical area or hospital were used in the matching or not, and recording the total number of matching factors were used in a regression analysis to examine whether they had an influence on outcome, and hence explained between study heterogeneity. Similarly, the outcomes from the primary studies were adjusted by combinations of a wide range of adjustment factors (full details given in the original report [11]). The effect of the most commonly used of these (age at diagnosis, age at first full term pregnancy, parity, age at menarche, family history of breast cancer, history of benign breast disease) were encoded, as were the total number of adjustment factors

Table 16.1 Matching factors in selection of study samples for the 27 studies investigation the association between OC and breast cancer (modified from Rushton and Jones [11]).

Study No.	Age	Marital Status	Hospital/geographical	Year of interview or year of diagnosis	Ethnic group	Parity	Age at end of education
Case-control studies							
1	*			*	*		
2	*		*		*		
3	*		*	*	*		
4	*		*		*		
5	*	*	*			*	
6	*		*		*		
7	*						
8	*						
9	*		*	*			
10	*						
11	*						
12	*		*	*			
13	*		*				
14	*						
15	*	*	*	*			
16	*		*	*			
17	*		*	*			
18	*		*	*			
19	*		*	*	*		
20	*						
21	*						
Cohort studies							
22	*						
23	*	*					
24	*		*	*	*	*	*
25	*	*	*				
26	*	*					
27	*	*			*		

used, and the method of standardization of study results (external/internal standard).

For synthesis to proceed, it is necessary to express the results in a consistent and comparable manner. The section below outlines many of the shortcomings of published reports, where the necessary information (treatment/exposure effect estimate and its variance) is often either missing or disguised, and presents ways of deriving/estimating these values on a relative risk scale. Most of these methods were originally outlined by Greenland [6].

16.2.1 Scales of measurement used to report and combine observational studies

Greenland notes that if the outcome under study is rare in all populations and subgroups under review, one can generally ignore the distinctions among the various measures of relative risk (e.g. odds ratios, rate ratios, and risk ratios). The distinctions can, however, be important when considering common outcomes, especially in case-control design and analysis [6]. The majority of the methods in this chapter deal with estimates on the relative risk scale. Such estimates may come from coefficients of a logistic model as well as simply from 2×2 tables previously presented; for example, Piegorsch and Cox [13] found in a meta-analysis of studies examining the effects of passive smoking that the case-control studies estimated the relative risk via the odds ratio, while the cohort studies used more complex risk ratio estimators. Their combined analysis therefore included a mix of different estimators for the RR endpoint.

16.2.2 Data manipulation for data extraction

1. *Calculating the standard error of an effect estimate as a relative risk from a confidence interval* If a 95% (or any other specified level) Confidence Interval [CI] is given instead of a standard error, i.e. from equation – 95%CI for $\ln(RR) = (\ln(RR) \pm 1.96 \times SE)$, then the standard error is given by [6]:

$$SE = (UCI(\ln(RR)) - LCI(\ln(RR))/3.92, \qquad (16.1)$$

where $UCI(\ln(RR))$ and $LCI(\ln(RR))$ are the upper and lower confidence interval bounds, and 3.92 ($2 \times Z_{(1-\alpha/2)}$) the corresponding critical value of the standard normal distribution.

2. *Calculating the standard error of an effect estimate from a p-value* If the p-value is given accurately enough (to at least two significant digits if p is over 0.1 and one digit if p is under 0.1), one can compute a 'test-based' standard error estimate:

$$SE = (\ln(RR))/Z_p, \qquad (16.2)$$

where Z_p is the value of a standardized normal test statistic corresponding to the p-value [6]. (See also Rosenthal [14] for methods for deriving continuous outcomes from levels of significance.)

16.2.3   Methods for transforming and adjusting reported results

Different studies may have adjusted for different confounding variables, and have different patient inclusion criteria. In addition, stratified results may have been presented for different levels of exposure. Biases may be known to exist in some studies, but it may be possible to adjust for these also for the purposes of a meta-analysis.

1. *Qualitative/Categorical exposure variables – Adjustments using external estimates of confounding*   If a number of the studies to be combined have not adjusted for suspected important confounders (in the original study analysis), then it may be possible to estimate and adjust for the degree of confounding present using data from *other* studies which report the same outcome provided data on the effects of the putative confounder have also been reported [6].

2. *Adjusting an unadjusted relative risk: A method using confounder-exposure information*   If a cohort study gives only an unadjusted estimate (RR_u), of exposure effect, but provides the joint distribution of exposure and the putative confounder in the total cohort, then these data can be combined with an external estimate of the confounder's effect on risk within levels of exposure. In this way, one can obtain an externally adjusted estimate of exposure effect that is potentially more accurate than the type given above [6].

3. *Adjusting for subject selection bias*   If the data are available, it may be possible to make the studies more similar by applying more strict inclusion/exclusion criteria to the subjects, and then re-analysing the study using only the subjects meeting the new criteria [6]. However if the parameters determining bias vary across studies, external correction could actually increase bias, although these estimates could be used as a starting point for a sensitivity analysis. Other methods are discussed below and in Section 16.7.1.

4. *Adjusting for misclassification*   There are no simple methods to allow estimation of a correction factor for misclassification bias. Corrections for misclassification should be based on reconstruction of the correctly classified data (described elsewhere [15–17]). If this cannot be done, informal sensitivity analysis should be carried out using hypothesized values of its magnitude.

5. *Calculating exposure coefficients from stratified results*   Ordered exposure variables often lead to presentations in terms of exposure-specific rates or

ratios, and these ratios are usually computed without taking into account the ordering of exposure levels. An estimate of an exposure coefficient from such presentations can be achieved using a weighted least squares regression model if either the standard errors or confidence intervals for the estimate in each stratum are reported. If they are not, but the size of the denominator for the rate in each exposure group is reported, then *ad hoc* approximate standard errors may be computed [6].

6. *Standardized morbidity ratios derived using an external reference population* An outcome related to exposure-specific ratios is the Standardized Morbidity Ratio (*SMR*). This is often constructed by computing the expected values based on an external reference population. When these external reference rates are assumed known without error, an estimate of the exposure coefficient in a regression may be obtained by a weighted linear regression of log(*SMR*) on exposure. If the confidence intervals, or standard errors, are reported for each *SMR* then a simple calculation yields the standard error for log(*SMR*) [6].

7. *Ratios derived using an internal reference group* When a report presents results in terms of relative risk estimates that are computed by using a single internal exposure group as the reference category, it is possible perform a weighted linear regression of the log relative risk on exposure. Since the log relative risk for the reference level is necessarily zero, the computations employ only the non-reference exposure groups, and the fitted line must be forced to pass through zero. Because the numbers in the reference group are subject to *statistical* error and are employed in all the log relative risk estimates, the estimates will have non-zero covariances [6].

8. *Repeated estimates which use only broad exposure categories* Greenland [6] states that many reports treat continuous exposures in a categorical fashion, computing relative risks for broad categories of exposure. In such cases, it is necessary to assign numeric values to the categories before estimating coefficients. When the categories are broad, results will be sensitive to the method of assignment. A common method is to assign midpoint values to categories. This has no general justification, and gives no answer for open ended categories. If, however, no frequency distribution for exposure is available, it may be the only choice, along with arbitrary assignments to open-ended categories. If the frequency distribution is available from the study, or from another study with a similar exposure distribution, each broad category may be assigned a numeric value corresponding to another measure of the centre of the category such as the mean.

9. *Estimation of coefficients from reports presenting only means* Pre-1980 studies sometimes present results for continuous exposures in terms of mean exposure levels among cases and non-cases, rather than in terms of relative risk estimates. If such a report supplies a cross-classification of the data by

exposure levels and outcome status, crude relative risk and coefficient estimates may be computed. If this is not the case, but standard errors for the means are given, crude logistic coefficient estimates can be constructed by the linear discriminant function method [6].

10. *Summarizing the risk associations of quantitative variables in epidemiologic studies in a consistent form*   Chêne and Thompson [12] present an approach to re-expressing results in a uniform manner. They convert results given in quartile groups or as logistic regression coefficients as a mean difference between those subjects who died and those who survived (along with the standard deviation). The appropriateness of the methods used depends upon the approximate normality of the continuous variable.

16.3   Analysis of summary data

16.3.1   Heterogeneity of observational studies

One of the particular advantages of meta-analysis of observational studies is that it allows us to assess whether the associations between an exposure and a disease (or health state) observed in a single study depend upon the composition of the population under study, the level of exposure in the study population, the definition of disease employed in the study, or any of a number of measures of the methodological quality of the study [18]. The homogeneity assumption is less likely to be satisfied with observational studies than it is with RCTs [6, 7].

Methods outlined in Chapter 3 can be used to assess heterogeneity of observational studies. In addition, the homogeneity assumption may also be assessed by categorizing the studies in terms of characteristics likely to be associated with heterogeneity [6].

1. *Dealing with heterogeneity when it is present*   Heterogeneity can be controlled by stratification or regression, in a similar way to that for RCTs [7]. Where unacceptable or excessive heterogeneity exists, and its source(s) cannot be identified and taken into account, combining disparate studies is not recommended [7]. Blair *et al.* [7] stress that the decision as to whether estimated differences are large enough to preclude formal combination should depend upon the scientific context, not just statistical significance, for example, a 25% difference among relative risks may be considered unimportant in a study of a very rare cancer, but important in a study of a more prevalent disease.

16.3.2 Fixed or random effects?

With considerable potential for between-study heterogeneity, the appropriateness of the 'fixed effect' assumption needs to be carefully considered in the epidemiological context [19]. Hence, it will often be appropriate to use random effect models, but as with RCTs it is more desirable to control and/or explain any heterogeneity where possible, by using either sub-group analyses or meta-regression [7].

16.3.3 Weighting of observational studies

Alternative weighting schemes taking into account study quality were considered in Section 8.3. Poor observational studies can be downweighted in this manner. Additionally, downweighting studies can sometimes be justified when the uncertainty of the result is not entirely reflected by the computed standard error estimate. For example, after external adjustment, it can be argued that the weight of the corrected coefficient should be less than that computed from the standard error of the original unadjusted estimate [6]. However, quantifying the extra uncertainty is difficult. A sensitivity analysis can go some way to alleviate this problem, although Colditz *et al.* [20] have commented that the random effects weighting probably puts too much weight on large observational studies (even though it weights them less than the fixed effect method), and they call for more empirical investigation. Note that the significant variations in quality which are often found between observational studies also are likely to make assessment of publication bias using funnel plots and related tests unreliable if quality and study size are associated.

16.3.4 Methods for combining estimates of observational studies

The techniques for combining estimates discussed in previous sections of this chapter can be used, where appropriate, for observational studies. Problems may arise when outcome measures from small studies are combined that do not approximate normality. However, the studies have to be very small, and a large proportion of the studies to be combined have to be small before this becomes a serious problem. In such cases, one may be able to use variants of large sample regression theory to derive heterogeneity and regression statistics, or choose to simply focus the meta-analysis on tabulations and graphic plots [6]. For a meta-analysis of a continuous exposure Dyer [21] proposed a Z-score approach. This method can be suitable for testing for effects across small studies if some suitable normalizing transformation of the exposure can be found. However, it does not provide a measure of exposure effect on risk, and so is unsuitable for quantifying strength of heterogeneity of effects [6].

16.3.5   Dealing with heterogeneity and combining the OC and breast cancer studies

Analysis consisted of examining the three subgroup sets separately (age at diagnosis, parity, and total duration of OC use). Considerable heterogeneity existed in the results of these subgroups between studies. Meta-regression (see Chapter 6) was used to assess the influence of the degree of matching, and adjustment variables, as defined above. Additionally, further study design characteristics were examined, including type of study (case-control/cohort), place of study (UK/USA/other), and source of study subjects (hospital/neighbourhood, or friend/resident list/GP list/ family planning clinic/screening system/nurse occupational group), to assess their impact on the results.

16.4   Reporting the results of meta-analysis of observational studies

Reporting the results of a meta-analysis was the topic of Chapter 10, where guidelines of how observational meta-analyses should be reported were referred to [7]. Specific points to note are that (a) in addition to basic information about the studies, a review should present a table of the results of the study re-analyses, showing at least the point estimate, net correction, and standard error (or confidence interval) from each study [6]; and (b) where stratification of study results by factors such as exposure metric, study design, and health outcome is conducted, an array of results both in a table and in the text is useful [7]. Weighted histograms of subgroup analyses results have also been suggested [6].

16.5   Use of sensitivity and influence analysis

It has been noted that the use of sensitivity analysis or influence analysis has been relatively instructive for environmental epidemiological studies [7]. Chapter 9 discussed several aspects which could be addressed in a sensitivity analysis – many of which are relevant to observational studies. For example, examining the influence of each individual study on the pooled result, as done in Section 9.2, is recommended for meta-analyses of observational studies. Further, meta-analysis specific explorations may also be useful in a synthesis of observational studies; for example, Greenland provides a specific sensitivity analysis from a meta-analysis of the relationship between myocardial infarction and coffee consumption:

> ... one may have externally controlled for cigarette smoking in all studies that failed to control for smoking by subtracting a bias correction factor

from the unadjusted coefficients in those studies. The sensitivity of inferences to the assumptions about the bias produced by failure to control for smoking can be checked by repeating the meta-analysis using other plausible values of the bias, or by varying the correction across studies. If such re-analyses produces little change in the inference, one can be more confident that the inferences appear deceptively precise relative to the variation that can be produced by varying assumptions, and thus choose to base the meta-analysis only on those studies that present results adjusted for smoking [6, p.23].

Sensitivity analyses was performed in the OC and breast cancer meta-analysis, where data encoding was problematical. The influence of individual studies was also checked. Additionally, the careful and comprehensive assessment of hetero-geneity using many exploratory variables further adds to the overall assessment of robustness of the results, although mixed modelling (with random effects) would now be regarded as preferable by many researchers (including the original authors!).

16.6   Study quality considerations for observational studies

It was noted in Chapter 8 that adjusting a meta-analysis by use of quality assess-ments was controversial. The assessment of the quality of observational studies is even more difficult (though probably more important) than that of RCTs. Since no one scale is available relevant to all study designs [4], the *weighting* of studies using a quality score is problematic if more than one study type is being combined. Criteria to be considered when evaluating the quality of case-control [22, 23] and cohort studies [24] are available elsewhere. Friedenreich [4] considers that too few observa-tional studies have used quality scores for a full assessment to be made on their usefulness, and regards it as a challenge for epidemiologic studies to identify the factors that represent the quality of the study most adequately, recognizing that these factors may differ across different exposure-disease relations [4]. Her paper includes a table of characteristics common to all observational studies, and those for specifically case-control and cohort studies that could be included in a quality assessment [4]. Dickersin and Berlin review the evidence for the effect of study quality on outcome. Their findings suggest that quality is sometimes shown to have an association with study outcome, but that this is not consistent [18].

Section 8.3.3 provides an example of the use of regression methods for examin-ing the effect of adjusting for study quality on the outcome of a set of observational studies.

Blair *et al.* [7] report that their expert working group on meta-analysis of environmental epidemiological studies failed to reach a consensus on the use of quality scores. They commented that some group members rejected any use of such

scores in favour of quality-component analysis, i.e. investigating study characteristics believed to be associated with study quality separately. They suggest that
sensitivity analysis and influence analysis provide alternatives to quality scoring [7].

No quality assessment was carried out for the OC and breast cancer meta-
analysis because of the difficulty of assessment of quality across epidemiological
studies with different designs (at the time of the original analysis, less empirical
research and fewer instruments to assess quality were available than is now the
case). However, the exploratory regression analysis, described previously, did investigate the effect of individual components of study quality.

16.7   Other issues concerning meta-analysis of observational studies

16.7.1   Analysing individual patient data from observational studies

Meta-analysis of individual patient data was discussed in Chapter 12, but a number
of aspects which are related particularly to meta-analysis of observational studies
are considered below.

There are several advantages of individual patient data [ipd] meta-analysis of
observational studies. Confounding and interactions between established and suspected risk factors can be more easily examined, allowing more valid and precise
conclusions regarding a particular exposure-disease relation than are possible with a
meta-analysis of summary data [4]. Additionally, an ipd meta-analysis may reveal
previously unrecognized errors in the data and associations or dose-response effects
that were either previously unknown or only suggested [4]. To do an individual
patient data meta-analysis simply, all the studies being combined need to be of the
same type. If both matched and unmatched case-control studies are to be combined,
a method has been developed to make this possible (see Section 17.3).

Friedenreich [4] discusses methodology for pooling observational studies at the
patient level. She considers the question of whether differences in the populations
and methods used in the original studies influence the results obtained from the
pooled analysis to be of central importance, and not yet addressed. She comments
on several drawbacks of current methodology; these include: (a) little or no consideration of how study sample, design and data collection characteristics influence
the results obtained, (b) pooled analyses of epidemiologic studies have not combined qualitative with quantitative assessments; (c) no pooled epidemiologic studies
(at the individual patient level) have included a sensitivity analysis. Her paper [4]
presents eight steps to follow and methodologic issues to consider when conducting
such a pooled analysis, including methods for examining heterogeneity, influence of

study design and data allocation methods on the pooled results, assessment of study quality and integration of qualitative assessments in the analysis.

16.7.2   Combining dose-response data

There are two reasons for assessing dose-response relationships in epidemiology, namely [25]: establishing such a relationship is one of several standard criteria for causality of the exposure (if increased risk occurs with increased dose, it is a strong step in proving that a causal association exists); and when an association has been established between an agent and a disease, the dose response relationship is of crucial use in establishing the levels of risk or benefit for individuals at different levels of exposure.

Information about multiple levels of exposure may be available from within studies, as well as among different studies [18]. This can be used in meta-analyses to great advantage, particularly if standard categories have been developed. Evidence of a broadly-defined dose-response pattern can be provided by a separate combination of relative risks defined by comparing a reference group with the each of the others in turn [26].

If it is possible to quantify exposure levels more precisely, then more sophisticated methods are available as illustrated in the meta-analysis of the association between alcohol consumption and breast cancer [27]. Methods available when effect estimates for several precise exposure categories are available from each study are discussed below. Data reporting results of a dose-response relationship usually appear in two forms. If the data have been modelled using a continuous variable for exposure, then a single regression coefficient (along with its standard error) will usually be available and can be combined using standard weighted methods of Chapters 4 and 5. If, on the other hand, a series of risk estimates derived for defined exposure levels have been presented (the more common reporting procedure) [28] alternative methods are required. Methods for each of these situations are discussed, together with a method for combining both of the above types of outcome.

Two general notes of caution relating to the combination of such data are [25]: (a) inclusion of the unexposed group may have an important confounding effect: an observed dose-response relationship may be in fact simply evidence of overall association but not of increasing (or decreasing) risk with increasing dose; and (b) the inconsistency of dose measurement may influence not only within-study regressions but also across-study equivalence of regression parameters.

Combining continuous exposure parameter estimates from several studies

From each study an estimate of one coefficient in a regression model, say β, that represents the change in the natural logarithm of the relative risk (or odds ratio, etc.) per unit of exposure is obtained. Then, provided its variance or standard error is given, standard methods can be used to combine the estimates of this coefficient.

Smith *et al.* [29] and DuMouchel [30] discuss dose-response models for this situation and note three issues: (i) a method of weighting the studies that gives greater influence to those whose dose-response slopes conform closest to a linear relationship between the relative risk and duration (which can lead to large differences in calculated weights as a function of non-linearity); (ii) the nature of the intercept of the slope in the dose-response model for each individual study. There are two alternatives: a model with zero intercept, and one estimating the intercept on the basis of the data (variable intercept). Selecting a model with a variable intercept implies that the risk between the two groups may differ before initial dose [6], and a model with a zero intercept implies that the risk among subjects taking very low doses is the same as the risk among untreated subjects; and (iii) the use of fixed and random effect models.

Combining risk/odds estimates for several exposure levels

This section deals with combining dose response data when the results are presented in the same format as those for the meta-analysis of breast cancer according to duration of oral contraceptive used in Table 16.2.

1. Number in parenthesis is assigned value in a dose-response regression model to control for other covariates.
2. The standard error required can be worked out from the confidence interval (see above) so either could be reported.

It may be possible to obtain a regression slope from a report by pooling estimates for responses at different levels of exposure (or treatment) [31]. However, standard methods for pooling estimates assume independence of the estimates, an assumption that is never true in this meta-analysis context because the estimates for separate exposure levels depend on the same reference (unexposed) group [31].

Greenland presents two methods of pooling responses at different levels that take account of the correlation between estimates. The first approach [28, 31] is based on constructing an approximate covariance estimate for the adjusted log odds ratios from a fitted table that conforms to the adjusted log odds ratios. The

Table 16.2 Odds ratios for breast cancer according to duration of oral contraceptive use (reproduced from Berlin *et al.* [28]).

Duration of contraception use in years	Cases	Controls	Odds ratio	95% Confidence Interval
Never [0]	96	156	1.0	
0–3 (1.5)	156	205	1.1	0.8–1.6
4–7 (5.5)	80	93	1.2	0.8–1.9
8–11 (9.5)	51	50	1.4	0.8–2.3
≥ 12 (14.4)	39	23	2.2	1.2–4.0

objective of the above method is to approximate the logistic coefficient that would have been obtained had either more complete study data, or the estimated logistic coefficient, been reported, and to provide a less biased variance than was previously available [31]. The derived estimates from this method can then be combined using standard methods, which allows both types of data (single coefficient and several exposure levels) to be combined.

A second, more flexible, method which involves pooling of study data *before* trend analysis is the 'pool-first' method [31]. This method is algebraically equivalent to the method of pooling the corrected coefficient estimates from each study. The advantage of the 'pool-first' method is that it is easily extended to fitting and testing non-linear logistic models. A limitation of this method is that it cannot incorporate studies that report only a slope estimate.

Tweedie and Mengersen [25] present a technique for dose response meta-analysis, for calculating dose response estimates for each individual study based on these approaches. These are:

(a) A non-parametric test for equality of response across dose levels (sometimes called the Armitage test for equality).

(b) Assumption of an exponential model (that is, a linear trend in the logarithms of the response) and test of significance of the regression parameter. This is a weighted regression model with zero intercept – essentially the same as the method of Greenland and Longnecker [31].

(c) Assumption of a direct linear trend in rates of occurrence and test of significance of the regression parameter. This can be used for studies which provide numbers of cases and controls in each exposure category, and thus the analysis of actual rates of occurrence of cases is possible (in this situation (a) and (b) are also possible).

The relative merits of these approaches are discussed by the authors. One important word of caution is that, under the linear model, care is needed to ensure that rates of the same magnitude are being combined. In case-control studies, if the number of controls for each case varies between studies, then the slope will also vary. Also, if rates for case-control and cohort studies are being compared, then one would typically obtain different orders of magnitude for the slope parameters.

16.7.3 Meta-analysis of single case research

A substantial number of papers have been written on the subject of meta-analysis of studies with single case designs [32–37] (and this list, we suspect, is not exhaustive). Single case designs are often used in psychology, but their use in medicine and related areas is somewhat rarer. There is considerable debate in the literature over appropriate measures of effect size. As well as effect differences, percentage non-overlapping data and other more complex measures have been used as outcome variables. For a good review and introduction to the area see Allison and Gorman [32].

16.8   Unresolved issues concerning the meta-analysis of observational studies

Whilst many of the issues considered above require further development, there are a number of other issues for which there are currently no satisfactory answers [1]. Several of the issues highlighted below concern the synthesis of data from studies with different designs; methods for combining such studies are considered in the next chapter.

Issues concerning the pooling of observational studies with different design features

(i) Is it permissible to integrate exposed patients sampled from hospitals with those from primary care settings?

(ii) When is it permissible to combine different types of cohort? For instance, for both exposed cohorts and comparison cohorts should one integrate data from a fixed cohort with an open one?

(iii) For reference cohorts, *not exposed* to an intervention or risk factor, other questions arise. For example,

    (a) Is a comparison cohort from Sweden combinable with one from Italy or Japan?

    (b) Are cohorts taken from occupational sampling frames sufficiently similar to those from the corresponding general population (or another geographically-defined one) to synthesize them?

    (c) How separate in time must the accrual or demarcation of unexposed cohorts become to be ineligible for aggregation? (The question is also pertinent for exposed cohorts.)

(iv) Is it admissible to merge hospital-based with population-based case groups? Or in Miettinen's terms, can two or more case series be combined if they are not representative of the same type of base experience?

(v) Are data provided by proxy informants similar enough to data from respondents to be considered equivalent?

(vi) Should one include case-control or cohort studies in which interviewers/ researchers were unblinded with blinded studies in one meta-analysis?

Other issues

(i) Operationally, what are the conditions under which both case-control studies and cohort studies may be included in one single meta-analysis? Should such analyses ever be done without access to the raw data of the component studies?

Specifically for case control studies:

(ii) Conceptually, and in execution, is a nested case-control study similar enough to a conventional case-control study for both to be included in the same meta-analysis?

(iii) When there are two or more control groups in a case-control study does one merge all the control groups? If not, what criteria must one use to exclude any control group from the meta-analysis. There is no parallel between multiple arms defined by exposure in a randomized controlled trial and multiple reference samples demarcated by outcome in a case-control study.

(iv) Should control groups assembled by matching be combined with independent samples of referenced populations?

(v) What constitutes 'proper control or adjustment for the biases that frequently occur in epidemiological studies'?

(vi) How homogeneous must the outcome be? For instance, can one pool data from a study that ascertained 'all cancers of the lung', with one that did so only for 'oat cell cancer', or only 'adenocarcinoma'?

16.9   Summary/Discussion

Most of the major considerations when combining observational studies are the same as those outlined in the rest of the book (where the examples are mainly focused on RCTs). One major new issue that needs addressing is [9], 'Has proper control or adjustment been made for the biases that frequently occur in epidemiological (observational) studies, such as sociodemographic or clinical differences between study populations, misclassification of subjects with regard to case-control status and to levels of exposure, factors other than the level of exposure that may affect whether a subject is a case or a control (i.e. confounding variables)?' Methods to deal with such issues have been considered and the use of sensitivity analysis to deal with the above problems is emphasized.

References

1. Spitzer, W.O. (1991). Meta-meta-analysis: unanswered questions about aggregating data. *J. Clin. Epidemiol.* **44**: 103–7.

2. Herbold, M. (1993). Meta-analysis of environmental and occupational epidemiological studies: a method demonstrated using the carcinogenicity of PCBs as an example. *Soz Praventivmed* **38**: 185–9.

3. Doll, R. (1994). The use of meta-analysis in epidemiology: Diet and cancers of the breast and colon. *Nutr. Rev.* **52**: 233–7.

4. Friedenreich, C.M. (1993). Methods for pooled analyses of epidemiologic studies. *Epidemiology* **4**: 295–302.

5. Cook, D.J., Sackett, D.L., Spitzer, W.O. (1995). Methodologic guidelines for systematic reviews of randomized control trials in health care from the Potsdam Consultation on Meta-Analysis. *J. Clin. Epidemiol.* **48**: 167–71.

6. Greenland, S. (1987). Quantitative methods in the review of epidemiological literature. *Epidemiol. Rev.* **9**: 1–30.

7. Blair, A., Burg, J., Foran, J., Gibb, H., Greenland, S., Morris, R., Raabe, G., Savitz, D., Teta, J., Wartenberg, D. *et al.* (1995). *Guidelines for application of meta-analysis in environmental epidemiology.* ISLI Risk Science Institute. Regul Toxicol Pharmacol **22**: 189–97.

8. Egger, M., Schneider, M., Davey Smith, G. (1998). Meta-analysis: Spurious precision? Meta-analysis of observational studies. *Br. Med. J.* **316**: 140–4.

9. Fleiss, J.L., Gross, A.J. (1991). Meta-analysis in epidemiology, with special reference to studies of the association between exposure to environmental tobacco smoke and lung cancer: a critique. *J. Clin. Epidemiol.* **44**: 127–39.

10. Morris, R.D. (1994). Meta-analysis in cancer epidemiology. *Environmental Health Perspectives* 102 Suppl **8**: 61–6.

11. Rushton, L., Jones, D.R. (1992). Oral contraceptive use and breast cancer risk: a meta-analysis of variations with age at diagnosis, parity and total duration of oral contraceptive use. *Br. J. Obstet. Gynaecol.* **99**: 239–46.

12. Chêne, G., Thompson, S.G. (1996). Methods for summarizing the risk associations of quantitative variables in epidemiologic studies in a consistent form. *Am. J. Epidemiol.* **144**: 610–21.

13. Piegorsch, W.W., Cox, L.H. (1996). Combining environmental information 2. environmental epidemiology and toxicology. *Environmetrics* **7**: 309–24.

14. Rosenthal, R. (1994). Parametric measures of effect size. In: Cooper, H., Hedges, L.V., editors. *The Handbook of Research Synthesis.* New York: Russell Sage Foundation. 231–44.

15. Copeland, K.T., Checkoway, H., McMichael, A.J. (1977). Bias due to misclassification in the estimaton of relative risk. *Am. J. Epidemiol.* **105**: 488–95.

16. Barron, B.A. (1997). The effects of misclassification on the estimation of relative risk. *Biometrics.* **33**: 414–8.

17. Greenland, S., Kleinbaum, D.G. (1983). Correcting for misclassification in two-way tables and matched-pair studies. *Int. J. Epidemiol.* **12**: 93–7.

18. Dickersin, K., Berlin, J.A. (1992). Meta-analysis: state-of-the-science. *Epidemiol. Rev.* **14**: 154–76.

19. Jones, D.R. (1992). Meta-analysis of observational epidemiological studies: A review. *J. Roy. Soc. Med.* **85**: 165–8.

20. Colditz, G.A., Burdick, E., Mosteller, F. (1995). Heterogeneity in meta-analysis of data from epidemiologic studies: Commentary. *Am J. Epidemiol.* **142**: 371–82.

21. Dyer, A.R. (1986). A method for combining results from several prospective epidemiological studies. *Stat. Med.* **5**: 307–17.

22. Lichtenstein, M.J., Mulrow, C.D., Elwood, P.C. (1987). Guidelines for reviewing case-control studies. *J. Chron. Dis.* **40**: 893–903.

23. Horwitz, R.I., Feinstein, A.R. (1979). Methodologic standards and contradictory results in case-control research. *Am. J. Med.* **66**: 550–64.

24. Feinstein, A.R. (1985). Twenty scientific principles for trohoc research. In: *Clinical Epidemiology: The Architecture of Clinical Research.* Philadelphia: WB Saunders. 543–7.

25. Tweedie, R.L., Mengersen, K.L. (1995). Meta-analytic approaches to dose-response relationships, with application in studies of lung cancer and exposure to environmental tobacco smoke. *Stat. Med.* **14**: 545–69.

26. Berlin, J.A., Colditz, G.A. (1990). A meta-analysis of physical activity in the prevention of coronary heart disease. *Am. J. Epidemiol.* **132**: 612–28.

27. Longnecker, M.P., Berlin, J.A., Orza, M.J., Chalmers, T.C. (1988). A meta-analysis of alcohol consumption in relation to risk of breast cancer. *J. Am. Med. Assoc.* **260**: 652–6.

28. Berlin, J.A., Longnecker, M.P., Greenland, S. (1993). Meta-analysis of epidemiologic dose-response data. *Epidemiology* **4**: 218–28.

29. Smith, S.J., Caudill, S.P., Steinberg, K.K., Thacker, S.B. (1995). On combining dose-response data from epidemiological studies by meta-analysis. *Stat. Med.* **14**: 531–44.

30. DuMouchel, W. (1995). Meta-analysis for dose-response models. *Stat. Med.* **14**: 679–85.

31. Greenland, S., Longnecker, M.P. (1992). Methods for trend estimation from summarized dose-response data, with applications to meta-analysis. *Am. J. Epidemiol.* **135**: 1301–9.

32. Allison, D.B., Gorman, B.S. (1993). Calculating effect sizes for meta-analysis: the case of the single case. *Behaviour Research & Therapy* **31**: 621–31.

33. Scruggs, T.E., Mastropieri, M.A., Casto, G. (1987). The quantitative synthesis of single subject research: Methodology and validation. *Remedial and Special Education* **8**: 24–33.

34. White, O.R. (1987). Some comments concerning 'The quantitative synthesis of single-subject research'. *Remedial and Special Education* **8**: 34–9.

35. Corcoran, K.J. (1985). Aggregating the idiographic data of single subject research. *Social Work Research and Abstracts* **21**: 9–12.

36. White, D.M., Rusch, F.R., Kazdin, A.E., Hartmann, D.P. (1989). Applications of meta analysis inindividual subject research. *Behavioral Assessment* **11**: 281–96.

37. Jayaratne, S., Tripodi, T., Talsma, E. (1988). The comparative analysis and aggregation of single case data. *J. App. Behavioral Sci.* **24**: 119–28.

CHAPTER 17

Generalized Synthesis of Evidence – Combining Different Sources of Evidence

17.1   Introduction

Thus far the methodology reviewed has been concerned principally with combining evidence from studies of a single type. Consideration to methods required for RCTs has been given throughout the book, and Chapter 16 dealt with issues of particular relevance to observational studies. This chapter describes methodology for combining results from different sources, including different study designs.

Methodology is presented for combining single arm studies, i.e. studies which use historical in a meta-analysis with other RCTs. Then methods to combine studies comparing different combinations of treatments are considered. This is followed by two general methods of combining evidence from different sources (or different groups of studies), namely the confidence-profile method and cross-design synthesis. Much of this work uses hierarchical models, and reference should be made to Chapter 11 on Bayesian methods in meta-analysis, and especially Section 11.7.4. Finally, methods for combining information from disparate toxicological studies are briefly considered.

17.2   Incorporating single-arm studies: models for incorporating historical controls

Begg and Pilote [1] present a model to estimate an overall treatment effect when some comparative studies are to be combined with non-comparative, historical control studies (or other studies with a single arm design). This potentially allows extra information to be combined in the synthesis. The model differs from the standard random effects model for two reasons: an estimate for each treatment (arm) is calculated (rather than a treatment effect difference or a ratio); and the treatment effects are fixed, but a random effect baseline term is included. Hence, the model has the potential to combine extra information from the non-controlled (and non-comparative) studies.

An outline of this model is as follows. Consider two treatments to be compared, where there are n comparative trials, yielding data summaries (x_i, y_i), $i = 1, \ldots, n$, where x_i is the observed effect of treatment A (possibly an estimate of the odds for binary data) and y_i is the observed effect of treatment B. Additionally, there are k uncontrolled studies of treatment A with observed effects u_i, and m uncontrolled studies of treatment B with observed effects v_i. These are all assumed to be normally distributed with known variances. Hence, algebraically:

$$
\begin{aligned}
x_i &\sim N\left(\theta_i, s_i(x)^2\right) & i &= 1, \ldots, n; \\
y_i &\sim N\left(\theta_i + \delta, s_i(y)^2\right) & i &= 1, \ldots, n; \\
u_i &\sim N\left(\theta_i, s_i(u)^2\right) & i &= (n+1), \ldots, (n+k); \\
v_i &\sim N\left(\theta_i + \delta, s_i(v)^2\right) & i &= (n+k+1), \ldots, n+k+m,
\end{aligned}
\tag{17.1}
$$

where θ_i is the true effect of treatment A in study i, and δ is the additional benefit of treatment B over treatment A (assumed constant across studies). Further, it is assumed that it is conditional on (i) x_i and y_i being uncorrelated, and (ii) the sampling variances $s_i(x)^2$, $s_i(y)^2$, $s_i(u)^2$ and $s_i(v)^2$ being known. Additionally, Begg and Polite [1] assume that the random effects are normally distributed, i.e. $\theta_i \sim N(\mu, \sigma^2)$, and use MLE theory to obtain solutions for model. Li and Begg [2] present a generalized least squares approach which removes the normality assumptions of the baseline random effects and the summary estimates of the effects in each study. An empirical Bayes estimator of the between-study variance is applied.

Extensions to this model have been suggested [1]: first, a preliminary test for systematic bias; and secondly, a methodological extension to include a random effects term for the treatment effect as well as the baseline effect. The authors stress that caution should be used at the interpretation stage when using this method.

17.2.1   Example

To illustrate how single arm studies can be incorporated into a meta-analysis, randomized and non-randomized studies of adjuvant chemotherapy in the treatment of childhood medulloblastoma are considered. Six randomized trials comparing treatment by radiotherapy alone with radiotherapy and adjuvant chemotherapy have been carried out. Additionally, seven single arm studies reporting on patients give radiotherapy and 10 single arm studies reporting on patient given radiotherapy and chemotherapy were identified. The proportion of patients surviving five years, and a standard error for this estimate are given in Table 17.1 for each arm of each study.

For this example, the model defined above is implemented using Bayesian estimation methods and the software WinBUGS [3] (see Appendix I), rather than the maximum likelihood methods used originally by Begg and Pilote [1]. Using this model, the difference in the proportion of patients surviving between the two treatment at five years is 0.131 (−0.257 to 0.519), indicating a non-statistically significant benefit of radiotherapy and chemotherapy over radiotherapy alone. If a standard random effects meta-analysis were performed (as described in Chapter

Table 17.1 Five year survival proportion estimates for patients receiving radiotherapy (Rtx) or Radiotherapy and Chemotherapy (Chemo) for the treatment of childhood medulloblastoma.

Study id	RTx + Chemo		RTx	
	S_5	$SE(S_5)$	S_5	$SE(S_5)$
1	0.55	0.026	0.42	0.020
2	0.58	0.058	0.60	0.054
3	0.74	0.083	0.56	0.099
4	0.59	0.060	0.50	0.065
5	0.17	0.217	0.63	0.341
6	0.46	0.114	0.30	0.118
7	0.83	0.030	—	—
8	0.82	0.120	—	—
9	0.96	0.039	—	—
10	0.82	0.384	—	—
11	0.55	0.188	—	—
12	0.64	0.170	—	—
13	0.26	0.196	—	—
14	0.60	0.097	—	—
15	0.36	0.170	—	—
16	0.93	0.120	—	—
17	—	—	0.71	0.184
18	—	—	0.48	0.223
19	—	—	0.41	0.087
20	—	—	0.32	0.057
21	—	—	0.34	0.080
22	—	—	0.71	0.068
23	—	—	0.33	0.071

11), combining *only* the two arm RCTs a difference of 0.076 (-0.575 to 0.450) is obtained. Note how the point estimate differs slightly between models, and also how the 95% credibility interval for the estimate from the model incorporating the single arm studies is narrower than that from the model just including the RCTs, due to the fact that it allows more information to be synthesized.

17.3 Combining matched and unmatched data

Duffy *et al.* [4] present a method for combining matched and unmatched data from RCT's, though the same methodology is directly applicable to case-control studies, including those with several controls per case. The example comes from combining RCTs in which each patient had one of their eyes (selected at random) treated while the other one remained untreated. Figure 17.1 illustrates the data structure and notation used for describing this methodology.

The method derived is an extension of the Mantel–Haenszel procedure (see Section 4.3.1). This method is used to combine the results from the m unmatched studies. For clarity, the formula (4.5) is re-expressed using the notation from Figure 17.1:

	Event (deterioration)	No event (no deterioration)	Total
Unmatched studies: ith study of m studies			
Treated	a_i	b_i	$a_i + b_i$
Not treated	c_i	d_i	$c_i + d_i$
	$a_i + c_i$	$b_i + d_i$	N_i
Matched studies: jth matched pair of such pairs in the kth of n studies			
Treated	e_{kj}	f_{kj}	1
Not treated	g_{kj}	h_{kj}	1
	$e_{kj} + g_{kj}$	$f_{kj} + h_{kj}$	2
Matched studies: kth study of n studies[a]		Not treated	
Treated: Event (deterioration)	x_k	y_k	
Treated: No event (no deterioration)	z_k	w_k	

[a] $x_k = \sum_{j=1}^{n_k} e_{kj}g_{kj}; y_k = \sum_{j=1}^{n_k} e_{kj}h_{kj}; z_k = \sum_{j=1}^{n_k} f_{kj}g_{kj}; w_k = \sum_{j=1}^{n_k} f_{kj}h_{kj}$

Figure 17.1 Structure and notation for combining data from matched and unmatched studies. (Reproduced from Duffy *et al.* [4]).

$$\hat{\varphi}_1 = \frac{\sum_{i=1}^{m} \frac{a_i d_i}{N_i}}{\sum_{i=1}^{m} \frac{b_i c_i}{N_i}}. \tag{17.2}$$

To combine the results from the matched studies, each matched pair within a study is treated as a stratum. By doing this, stratification by study is performed automatically. The formula for combining the n matched studies is:

$$\hat{\varphi}_2 = \frac{\sum_{k=1}^{n} \sum_{j=1}^{n_k} \frac{e_{kj} h_{kj}}{2}}{\sum_{k=1}^{n} \sum_{j=1}^{n_k} \frac{f_{kj} g_{kj}}{2}} = \frac{\sum_{k=1}^{n} y_k}{\sum_{k=1}^{n} z_k}. \tag{17.3}$$

A pooled estimate of both the matched and the unmatched studies can then be obtained using

$$\hat{\varphi}_3 = \frac{\sum_{i=1}^{m} \frac{a_i d_i}{N_i} + 1/2 \sum_{k=1}^{n} y_k}{\sum_{i=1}^{m} \frac{b_i c_i}{N_i} + 1/2 \sum_{k=1}^{n} z_k}. \tag{17.4}$$

A test statistic (based on Mantel–Haenszel's chi squared statistic) can be derived and confidence intervals calculated. [4]

Moreno *et al.* [5] also consider the synthesis of matched and unmatched studies of case-control design using logistic regression of individual patient data. The logistic regression model proposed combines the conditional logistic regression likelihood function for the matched cases and controls, and an unconditional logistic regression likelihood function for the unmatched study. The likelihood expression that is derived can be implemented in specialist software; however, if there is only one case in each matched set (which is common) then the model can be simplified, so that standard logistic regression programs can be used. The method also allows the estimation of the ratio of odds ratios adjusted by potential confounding factors.

17.4 Approaches for combining studies containing multiple and/or different treatment arms

Chapter 15 discussed methods for combining multiple outcome measures simultaneously. This sections considers the combination of RCTs using a single outcome measure, but where different treatments have been allocated to the various arms in each study, so different comparisons have been made in the different studies. For example, suppose two new treatments (Treatment A and Treatment B) are developed at similar times. Initially, trials comparing each treatment to placebo (Treatment C) are carried out. Hence, trials of Treatment A vs. Treatment C, and Treatment B vs. Treatment C exist. Later, in an attempt to establish which of the new treatments is superior, trials of Treatment A vs. Treatment B are carried out. If one is carrying out meta-analyses of Treatment A, for example, then both Treatment A vs. Treatment C and Treatment A vs. Treatment B trials provide information, but it would not be sensible to combine the two types of trial without taking account of the different comparisons being made. Further, trials with three arms (Treatment A, Treatment B and Treatment C) may also exist.

17.4.1   Approach of Gleser and Olkin

The modelling approach of Gleser and Olkin [6] was first considered in Section 15.4.2, as a method for dealing with multiple outcomes. Their approach is also capable of combining studies in which more than one type of treatment has been compared to a control group. Hence, it could combine studies comparing Treatment A vs. Treatment C and Treatment B vs. Treatment C, but not Treatment A vs. Treatment B, because no control group is included in such studies.

17.4.2   Models of Berkey *et al.*

The fixed effect, and random effect multiple outcome models of Berkey *et al.* [7, 8] outlined in Sections 15.4.3 and 15.4.4 are capable of combining studies comparing multiple treatments, where any study may consider only a subset of treatments.

17.4.3   Method of Higgins

Higgins [9] has also developed a Bayesian method to combine studies which have arms in common, but make different comparisons.

17.4.4   Mixed model of DuMouchel

DuMouchel [10] describes a model which allows for many extensions of the standard random effect model of Chapter 5. It is included here because it allows studies of heterogeneous designs to be combined. It can also be considered a generalization of the mixed models of Chapter 6. First, it has the capacity to model multiple outcomes from trials, provided they are all reported on the same scale (e.g. they are all binary events). It does not treat these as truly multivariate and no correlations between outcomes are included, hence unlike the models of Chapter 15, no correlations between outcomes within studies are required. This enables binary as well as continuous outcomes to be modelled (however, not in any one model).

It is assumed that each study reports results for two or more of the treatments, however no common treatment arm is required across all studies (a requirement of Gleser and Olkin's model of Section 15.4.2). Different study designs are accounted for because different treatment groups can consist of results from separate (sub)-groups of subjects, or from groups that cross-over and are subject to multiple treatments.

A key feature which distinguishes this model from the random effect models of Chapter 5 is that each group of patients is modelled separately. So, for example, if one is considering a binary outcome, the log odds in each group (see Section 2.2.1) could be modelled, rather than a comparative effect such as the log odds ratio; this is conceptually very similar to the modelling approach of Section 17.2 which allows the inclusion of single arm trial in a meta-analysis. By constructing contrasts of parameters in the model comparative effects can be calculated after parameters in the model have been estimated. The most general form of the model is

$$y_{ijktm} = \mu_m + \beta_k + \gamma_{mk} + a_i + b_{im} + c_{ik} + d_{j(i)} + e_{t(i)} + \varepsilon_{ijktm}, \qquad (17.5)$$

where i indexes the N studies in the analysis, m indexes the M different outcomes considered, j indexes the J_i cohorts of patients in the ith study, k indexes the K treatments being compared in the analysis, and t indexes the T_i time periods or cross-over states in the ith study. A variance component model is used to model the expected correlations among reported results from the same study and cohorts of subjects. Outcome and treatment are considered fixed, while study, cohort and period are considered random effects. Hence, μ, β and γ denote fixed effects due to outcome, treatment and their interaction, and a, b, c, d and e are random effects due to study, study* outcome, study* treatment, cohort within study, and period within study, respectively, and ε denotes pure error.

For details of how this model can be implemented, see DuMouchel's original paper [10]. This model was applied effectively to a meta-analysis of 16 RCTs to evaluate computer-based clinical reminder systems for preventative care in the ambulatory setting [11]. It should be stressed that equation (17.5) is the most general form of the model, and it can be simplified as required, depending on the features represented among the studies being combined.

17.5   The confidence profile method

The confidence profile method was first presented by Eddy [12] in 1989. It has been advocated as a Bayesian method for assessing health technologies [12], although it can be formulated in classical terms where maximum likelihood estimates and covariances for the model parameters are derived [13]. The method has been described as:

...a set of quantitative techniques for interpreting and displaying the results of individual experiments; exploring the effects of biases that affect the internal validity of experiments; adjusting experiments for factors that affect their comparability or applicability to specific questions (external validity); and combining evidence from multiple sources [13].

Multiple pieces of evidence, different experimental designs, different types of outcomes, different measures of effect, biases to internal validity, biases to comparability and external validity, indirect evidence, mixed comparisons, and gaps in experimental evidence can all be dealt with using this approach [13]. Hence, it provides a very general method for combining virtually any kind of evidence about various parameters, so long as those parameters can be described in the model. An application of this model provides one of the first examples of evidence from randomized and non-randomized studies being combined together [14]. Several accounts describe the details of this method [12, 13, 15, 16], and there is also a computer program to implement it available [17, 18].

Since its conception in the early 1990s, the uptake of the use the method has been relatively poor, quite possibly due to its radically different conceptual approach to meta-analysis. However, there are close similarities between this approach and the graphical modelling used by BUGS [19]. Indeed, a fully Bayesian implementation of the modelling ideas of the confidence profile method has been demonstrated recently using this software [20]. A potential drawback of the method is that it is necessary to model biases in individual study estimates explicitly. Identification and quantification of such biases will usually be very difficult in practice.

17.6   Cross-design synthesis

This section discusses the idea of combining different types of studies further, and specifically the issue of synthesizing data from randomized and non-randomized studies, such as cohort studies, case-control studies, animal experiments or database analyses, etc. This issue is contentious, because the inclusion of non-randomized evidence into a meta-analysis of effectiveness may introduce bias, and hence have a detrimental effect [21–23].

Although RCTs may be the 'gold standard' against which the value of evidence from other types of study is judged, it is probably too difficult, costly and time-consuming to perform a (large) randomized study to answer every question of clinical interest [24–26]. Further, in some clinical areas, it can be difficult to perform RCTs, for ethical and other reasons [27], while in other areas, the low participation rate of patients in RCTs limits the generalizability of the results obtained in them. If the ideal RCT could be conducted, there would be no need for subjective interpretations or for systemic methods of supplementing RCT results [28], however in these, and other, situations, evidence from studies other than RCTs is valuable. The factors motivated the methodology outlined below.

17.6.1 Beginnings

The initial work on cross-design synthesis was carried out by the Program Evaluation and Methodology Division of the U.S. General Accounting Office (GAO) [29], although the previous work of Rubin [30] (see Section 20.2.1) and Eddy [13] (see Section 17.5) is acknowledged.

The idea was to create a methodology that captures the strengths of multiple-study designs while minimising their weaknesses [31]. For example, there are difficulties in applying results of RCTs to clinical practice: patients seldom conform to the characteristics of participants in RCT's are recognized, and meta-analysis can make it harder to move from judging whether a treatment is, in principle, efficacious, to deciding how to manage a particular patient, while subgroup analyses of (individual) RCTs lead to *post hoc* verdicts of questionable reliability [32]. In contrast, observational data cannot provide definite answers to questions about intervention efficacy because of susceptibility to bias.

Although it is recognized that different study designs have different strengths, it cannot be assumed that, in combining their study results, their strengths will be preserved while their weaknesses are eliminated. To address this issue, a three-stage strategy for minimizing weaknesses of study design was developed:

(a) a focused assessment of the study biases that may derive from characteristic design weaknesses;

(b) individual adjustment of each study's results to 'correct for' identified biases;

(c) development of a synthesis framework and an appropriate model for combining results (within and across designs) in the light of all assessment information.

A limitation of this method is that it is necessary to rely on investigator judgement for many decisions. Until refinements of this strategy are developed, the GAO believes it is best applied by those knowledgeable about both a specific medical treatment and evaluation methods in general [28]. Indeed, it has not yet been widely adopted, not least due to the fact that the method was described [29] in conceptual rather than practical terms, and there is controversy about its use [28, 32].

17.6.2 Bayesian hierarchical models

Prevost *et al.* [33] have developed a model to include studies with disparate designs into a single synthesis. The authors point out that whilst it may be appropriate to consider randomised studies alone when assessing the efficacy of an intervention, when considering the effectiveness of such an intervention within a more general population evidence from non-randomised studies may be relevant as well. They also point out that in certain situations the randomized evidence may be less than adequate due to economic, organizational or ethical considerations. Although this work follows in the cross design synthesis spirit of Droitcour *et al.* [29], the methodology used is somewhat different and is more specifically operationalized. More generally, this approach presents a framework for sensitivity analysis to include evidence from different study types.

A Bayesian hierarchical model approach is used (see Section 11.7.4). The hierarchical nature of the model specifically allows for the quantitative within and between sources heterogeneity, whilst the Bayesian approach can accommodate *a priori* beliefs regarding qualitative differences between the various sources of evidence. These prior distributions may represent subjective beliefs elicited from experts, or other data-based evidence, which though pertinent to the issue in question is not of a form that can be directly incorporated, such as data from animal experiments [34]. This model can be viewed as an extension of the standard random effects model of Chapter 5, but with an extra level of variation to allow for variability in effect sizes between different sources of evidence.

The method is illustrated in the context of screening for breast cancer, where evidence is available from both randomized clinical trials and non-randomized studies. Figure 17.2 outlines the three parameter levels: (i) the overall population effect of screening μ, (ii) type-of-study parameters $\theta_i (i = 1, 2, \ldots$ where $i = 1$ denotes the effect associated with randomized studies, $i = 2$ denotes the effect associated with case-control studies (if cohort studies were available they could be denoted by $i = 3$, and so on)); and (iii) $\varphi_{ji} (i = 1, 2, 3, ; j = 1, \ldots, n_i)$ study-specific parameters, there being n_1 RCTs.

Algebraically, such a model would be an extension of equation (11.5), and could be formulated as

$$
\begin{aligned}
&T_{ij} \sim N[\theta_{ij}, \sigma_{ij}^2/n_{ij}] \quad \sigma_{ij}^2 \sim [-,-] \quad i = 1, \ldots, I_j \quad j = 1, \ldots, J \\
&\theta_{ij} \sim N[V_j, \omega_j] \quad \omega_j \sim [-,-] \\
&v_j \sim N[\mu, \tau^2] \\
&\mu \sim [-,-] \quad \tau^2 \sim [-,-],
\end{aligned}
\tag{17.6}
$$

where subscript ij refers to the ith study in the jth category, and thus v_j refers to the overall pooled effect in the jth category, and ω_j is a measure of the between study heterogeneity in the jth category, whilst μ is the overall population effect and τ^2 is a measure of the heterogeneity in this effect between the J categories.

Extensions including incorporation of prior constraints, prior beliefs and study level covariates are also feasible. Additionally, beliefs about the relative merits of individual studies or types of study can be incorporated in the model. For example, beliefs about the relative value of RCTs, cohort-study and case-control study results may be modelled explicitly, and the dependence of the conclusions of the review on these beliefs investigated. The model is fitted in BUGS [35].

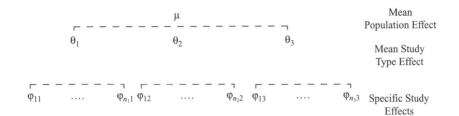

Figure 17.2   Hierarchical model for breast-cancer-screening synthesis.

Example: Combining evidence from different study types using a hierarchical model

This model has been applied to the evidence regarding the benefits of screening via a mammography for breast cancer in terms of reduction in mortality [33]. Evidence relating to the effect in women over the age of 50 is considered here. Results from RCTs and non-randomized studies (consisting of geographic (cross-sectional) and case-control studies) were pooled using their three level hierarchical model with two study types at the second level. Figure 17.3 displays the synthesis results, in terms of relative risks. Several estimates are displayed on this plot. First, for each study, the observed and the shrunken estimates (see Section 11.6) are displayed; the shrunken estimates having solid lines representing 95% credibility intervals. The pooled result for the three level model is shown at the bottom of the plot (0.65 (0.46 to 0.86)), again with solid lines representing credibility intervals. As well as an overall pooled result, pooled results for each study type produced by the model are also displayed (0.70 (0.58 to 0.88) for the randomized studies, and 0.48 (0.27 to 0.76) for the non-randomized studies). These can be compared with the results obtained by pooling the study types separately using a standard Bayesian meta-analysis model (see Chapter 11), which are plotted with dashed credibility intervals (0.68 (0.56 to 0.82) for the randomized studies, and 0.62 (0.42 to 0.81) for the non-randomized studies). Finally, the pooled result combining all studies using a standard model, and hence ignoring study type is plotted with a dashed credibility interval adjacent to the three level model overall result (0.65 (0.45 to 0.80)) (refer to the legend of Figure 17.3 for clarification of the different estimates described above).

Shrinkage at two levels can be observed here. First, individual study estimates are shrunken towards their study type estimate, and these study type estimates are themselves shrunken towards the overall pooled estimate. Although the overall estimated relative risks are almost identical for the three level model, compared to a standard meta-analysis ignoring study type, the credibility interval is wider for the three level model because heterogeneity between study types has been accounted for. Generally, using this model, the greater the difference between pooled individual study type estimates, the greater the between study type heterogeneity, and

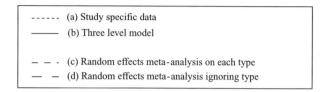

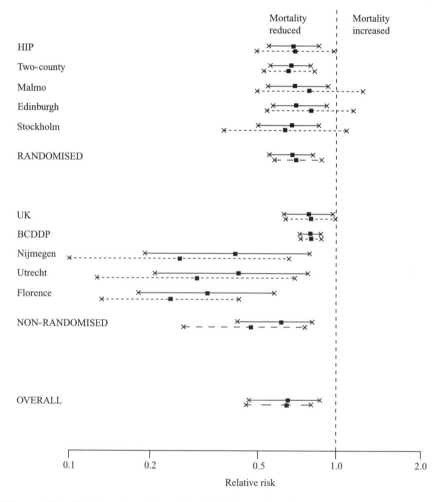

Figure 17.3 Estimated relative risk (95% intervals) of breast cancer mortality in studies of breast cancer screening for women aged 50 and over (55 in Malmo, 45 in UK), obtained from three level Bayesian hierarchical model and random effects meta-analysis. (Reproduced from Prevost *et al.* [33].)

the greater the inflation to the credibility interval for the overall pooled result will be.

17.6.3 Grouped random effects models of Larose and Dey

Larose and Dey [36] present another Bayesian hierarchical model for more appropriately combining results from dissimilar studies. Their illustrative example considers 15 comparative studies of progabide (a new anti-epileptic drug), all of which use a cross-over design. A distinction is made between double-blinded (closed) studies and those in which the investigator had knowledge of the treatment regime (open studies). A random effects model is used which calculates an overall mean plus a group-specific random effect, in the example, for the open and closed studies, though these group-specific effects could be defined in terms of any other design feature. This model allows each group's effect size estimate to have its own separate variance term, and is implemented via Markov Chain Monte Carlo methods [37]. It is assumed that there is exchangeability (see Chapter 11) between studies within each group. The authors' main purpose was not to produce a single overall pooled estimate, but to demonstrate that in the example if no groupings were made (i.e. using a traditional fixed or random effect model), a conclusion of no treatment effect would be made, but that using the grouped random effects model demonstrates that the open studies support the efficacy of progabide while the closed studies do not. The model could easily incorporate covariate information through the use of linear structure in the overall mean [38].

17.6.4 Synthesizing studies with disparate designs to assess the exposure effects on the incidence of a rare adverse event

A further method for synthesizing studies with disparate designs is provided by Brumback *et al.* [39], who describe a model for combining case-control and cohort studies examining exposure effects on the incidence of a rare adverse event. Fixed effect logistic models are used to model the case-control studies, while random effect Poisson models for the modelling of the cohort studies. This model has similarities with those of Section 17.2, because again, comparative studies (here, case control studies rather than RCTs) are combined with non-comparative cohort studies. A 'hybrid' Bayesian approach is taken to implement the method.

17.6.5   Combining the results of cancer studies in humans and other species

DuMouchel and Harris [40] propose a class of Bayesian statistical methods for interspecies extrapolation of dose-response functions. The motivation for considering information from non-human species is that there was an abundance of precise data available from these concerning the assessment of cancer risks from environmental agents, but little accurate information on direct effects in humans. A formal distinction is made between conventional measurement error within each dose-response experiment and a novel error of uncertain relevance *between* experiments. Dose-response data from many substances and species is used to estimate the intra-experimental error. From the data, the estimated error of interspecies extrapolation, and prior biological information on the relations between species or between substances, posterior densities of human dose-response are calculated.

17.6.6   Combining biochemical and epidemiological evidence

Tweedie and Mengersen [41] investigate the relationship between lung cancer and passive smoking. Previously, two approaches had been taken for investigating this: the biochemical approach, using cotinine in the main as a marker; and the epidemiological approach. The paper uses both sorts of studies in one meta-analysis.

17.6.7   Combining information from disparate toxicological studies using stratified ordinal regression

Cox and Piegorsch [42, 43] discuss the development of methods for combining studies on acute inhalation assessment. Their work involves combination of data from studies of inhalation damage from various airborne toxins in order to estimate human health risk. The primary studies vary greatly in their endpoints: short- and long-term exposures in laboratory animals, acute exposures to humans in chemical and/or community accidents, chronic exposure studies in urban areas, etc. Their goal was to develop a methodology for data combination that incorporates the range of endpoint severity, exposure concentrations, and exposure durations. Particular emphasis is directed at acute exposures, since these are thought to be more common than chronic, long-term exposures in many human situations. The method is based on severity modelling, wherein concentration, duration and response are integrated to determine potential risks to humans after acute inhalation exposure to some environmental toxin. The method groups the response data into ranked severity categories, and assumes that duration and concentration are independent explanatory variables for predicting response. This is essentially an ordinal regression, using a logistic or another discrete-data regression model for the concentration-duration response.

17.7 Summary/Discussion

This chapter considers methods for combining studies of different designs. Many of the techniques utilize modern statistical models, including Bayesian methods, and hence many of the methods are extensions of those of Chapter 11. Their implementation is, however, facilitated by recent advances in computer software (see Appendix I). When such analyses are appropriate is still open to debate, as there is concern that including studies with poorer designs will weaken the analysis through the introduction of biases. However, the methods do provide a framework for carrying out a sensitivity analysis, including a broader range of evidence than traditional methods allow.

References

1. Begg, C.B., Pilote, L. (1991). A model for incorporating historical controls into a meta-analysis. *Biometrics* **47**: 899–906.

2. Li, Z.H., Begg, C.B. (1994). Random effects models for combining results from controlled and uncontrolled studies in a metaanalysis. *J. Am. Stat. Assoc.* **89**: 1523–7.

3. Stevenson, J. (1998). Meta-analysis of trials and studies of adjuvant chemotherapy in the treatment of childhood medulloblastoma. MSc Thesis, Department of Epidemiology and Public Health, Leicester University, UK.

4. Duffy, S.W., Rohan, T.E., Altman, D.G. (1989). A method for combining matched and unmatched binary data. Application to randomized, controlled trials of photocoagulation in the treatment of diabetic retinopathy. *Am. J. Epidemiol* **130**: 371–8.

5. Moreno, V., Martin, M.L., Bosch, F.X., De Sanjose, S., Torres, F., Munoz, N., Desanjose, S. (1996). Combined analysis of matched and unmatched case-control studies: Comparison of risk estimates from different studies. *Am. J. Epidemiol* **143**: 293–300.

6. Gleser, L.J. Olkin, I. (1994). Stochastically dependent effect sizes. In: Cooper, H., Hedges, L.V. (editors). *The Handbook of Research Synthesis*. New York: Russell Sage Foundation. 339–56.

7. Berkey, C.S., Anderson, J.J., Hoaglin, D.C. (1996). Multiple-outcome meta-analysis of clinical trials. *Stat. Med.* **15**: 537–57.

8. Berkey, C.S., Hoaglin, D.C., Antczak-Bouckoms, A., Mosteller, F., Colditz, G.A. (1998). Meta-analysis of multiple outcomes by regression with random effects. *Stat. Med.* **17**: 2537–50.

9. Higgins, J.P.T., Whitehead A. (1996). Borrowing strength from external trials in a metaanalysis. *Stat. Med.* **15**: 2733–49.

10. DuMouchel, W. (1998). Repeated measures meta-analysis. *Bull. Int. Stat. Inst.* Tome LVII, Book 1(Session 51):285–8.

11. Shea, S., DuMouchel, W., Bahamonde, L. (1996). A meta-analysis of 16 randomised controlled trials to evaluate computer-based clinical reminder systems for preventive care in the ambulatory setting. *J. Am. Med. Informatics Assoc.* **3**: 399–409.

12. Eddy, D.M. (1989). The confidence profile method: A Bayesian method for assessing health technologies. *Operat. Res.* **37**: 210–28.

13. Eddy, D.M., Hasselblad, V., Shachter, R. (1992). *Meta-analysis by the Confidence Profile Method.* San Diego: Academic Press.

14. Eddy, D.M., Hasselblad, V., McGivney, W., Hendee, W. (1988). The value of mammography screening in women under the age of 50 years. *Am. Med. Assoc.* **259**: 1512–9.

15. Eddy, D.M., Hasselblad, V., Shachter, R. (1990). An introduction to a Bayesian method for meta-analysis: The confidence profile method. *Med. Dec. Making* **10**: 15–23.

16. Eddy, D.M., Hasselblad, V., Shachter, R. (1990). A Bayesian method for synthesizing evidence: The confidence profile method. *Int. J. Technol. Assess. in Health Care* **6**: 31–55.

17. Behar, D. (1992). FastPro – software for metaanalysis by the confidence profile method. *Am. Med. Assoc.* **268**: 2109.

18. Eddy, D.M., Hasselblad, V. (1992). FastPro: Software for MetaAnalysis by the Confidence Profile Method [computer program]. San Diego, California: Academic Press, 3.5–inch disk. IBM-PC.

19. Spiegelhalter, D., Thomas, A., Gilks, W. BUGS (1994). Examples 0.30.1. Cambridge: MRC Biostatistics Unit.

20. Spiegelhalter, D.J., Miles, J., Jones, D.R., Abrams, K.R. (2000). A systematic review of Bayesian methods in HTA. *Health Technol. Assess.* (to appear).

21. Byar, D.P. (1980). Why data bases should not replace randomized clinical trials. *Biometrics* **36**: 337–42.

22. Mantel, N. (1983). Cautions on the use of medical databases. *Stat. Med.* **2**: 355–62.

23. Green, S.B. (1982). Patient heterogeneity and the need for randomized clinical trials. *Controlled Clin Trials* **3**: 189–98.

24. Hlatky, M.A., Lee, K.L., Harrel, F.E.J., Califf, R.M., Pryor, D.B., Mark, D.B., Rosati, R.A. (1984). Tying clinical research to patient care by use of an observational database. *Stat. Med.* **3**: 375–84.

25. Hlatky, M.A., Califf, R.M., Harrel, F.E.J., Lee, K.L., Mark, D.B., Muhlbaier, L.M., Pryor, D.B. (1990). Clinical judgment and therapeutic decision making. *J. Am. College of Cardiology* **15**: 1–14.

26. Feinstein, A.R. (1984). Current problems and future challanges in randomized clinical trials. *Circulation* **70**: 767–74.

27. Abrams, K.R., Jones, D.R. (1995). Meta-analysis and the synthesis of evidence. *IMA J. Math. Appl. Med. Biol.* **12**: 297–313.

28. Chelimsky, E., Silberman, G., Droitcour, J. (1993). Cross design synthesis [letter]. *Lancet* **341**: 498.

29. General Accounting Office. (1992). *Cross design synthesis: a new strategy for medical effectiveness research.* Washington, DC: General Accounting Office.

30. Rubin, D. (1992). A new perspective. In: Wachter, K.W., Straf, M.L., eds. *The Future of Meta-analysis.* New York: Russell Sage Foundation, 155–65.

31. Droitcour, J., Silberman, G., Chelimsky, E. (1993). Cross-design synthesis: A new form of meta-analysis for combining results from randomized clinical trials and medical-practice databases. *Int. J. Technol. Assess. in Health Care* **9**: 440–9.

32. Anonymous. (1992). Cross design synthesis: a new strategy for studying medical outcomes? *Lancet* **340**: 944–6.

33. Prevost T.C., Abrams, K.R., and Jones, D.R. Using hierarchical models in generalised synthesis of evidence: an example based on studies of breast cancer screening. *Stat. Med.* (forthcoming).

34. Abrams, K.R., Hellmich, M., and Jones, D.R. (1997). Bayesian approach to health care evidence. Department of Epidemiology and Public Health. Technical Report 97–01. University of Leicester.

35. Thomas, A., Spiegelhalter, D.J., Gilks, W.R. (1992). A program to perform Bayesian inference using Gibbs sampling. In: Bernardo, J., Berger, J., Dawid, A., Smith, A., eds. *Bayesian Statistics 4.* Oxford: Oxford University Press.

36. Larose, D.T., Dey, D.K. (1997). Grouped random effects models for Bayesian meta-analysis. *Stat. Med.* **16**: 1817–29.

37. Gilks, W.R., Richardson, S., Spiegelhalter, D.J. (1996). *Markov Chain Monte Carlo in Practice.* London: Chapman & Hall.

38. Larose, D.T., Dey, D.K. (1997). Modeling dependent covariate subclass effects in Bayesian meta-analysis. Technical report #96–22, Department of Statistics, University of Connecticut.

39. Brumback, B.A., Holmes, L.B., Ryan, L.M. (1999). Adverse effects of chorionic villus sampling: a meta-analysis. *Stat. Med.* **18**: 2163–75.

40. DuMouchel, W.H., Harris, J.E. (1983). Bayes methods for combining the results of cancer studies in humans and other species (with comment). *J. Am. Statist Assoc.* **78**: 293–308.

41. Tweedie, R.L., Mengersen, K.L. (1992). Lung cancer and passive smoking: Reconciling the biochemical and epidemiological approaches. *Br. J. Cancer* **66**: 700–5.

42. Cox, L.H., Piegorsch, W.W. (1994). Combining environmental information: Environmetric research in ecological monitoring, epidemiology, toxicology, and environmental data reporting. Technical Report #12, National Institute of Statistical Sciences, PO Box 14162, Research Triangle Park, NC 27709.

43. Carroll, R.J., Simpson, D.G., Zhou, H. *et al.* (1994). Stratified ordinal regression: A tool for combining information from disparate toxicological studies. Technical Report #26, National Institute of Statistical Sciences, PO Box 14162, Research Triangle Park, NC 27709.

CHAPTER 18

Meta-analysis of Survival Data

18.1   Introduction

Special methods are needed if one wishes to combine studies whose outcome of interest is time to an event, and data may be censored. The censoring makes the analysis of this type of data unique, and often complex [1], and much has been written on the analysis of survival data from single studies [2, 3].

Several different approaches to combining survival data are reported in this chapter. Which method is most suited to a given situation is largely dictated by the type of data available from each of the individual studies. Obtaining accurate data is often a problem; for example, it is sometimes necessary to resort to the extraction of information from Kaplan–Meier curves presented in papers. These may be small, inaccurate, and difficult to read, all of which will reduce the accuracy of the data. Techniques available for different situations include weighting and combining survival-rate differences at a fixed point(s) in time, calculating and combining a summary parameter describing the survival curves, combining 'log-rank' odds ratios, and combining data on individual patients from different studies.

Methods available, but not included in this section, include a confidence profile survival model [4] (the general confidence-profiling approach to meta-analysis has been considered in Section 17.5), and meta-analysis of surrogate measures of survival (see Section 14.5). All the methods considered below (with the exception of Section 18.9.1) assume comparative trial data is being combined.

18.2   Inferring/estimating and combining (log) hazard ratios

The simplest way to carry out a meta-analysis of survival data would be to summarize each contributing trial by a single number, along with its standard error, and use standard methods of meta-analysis to combine them as presented in Chapters 4 and 5 [5].

The most sensible summary estimate is the (log) hazard ratio because it is the only summary statistic which allows for both censoring and time to an event, and is a measure of the difference between two Kaplan–Meier curves [6]. The hazard ratio measures the reduction in (instantaneous) risk of death (or other event) on treatment compared to control over the period of follow-up in a trial. Hence, to combine studies only the (log) hazard ratio and its variance are required from each trial. If these are available, for each trial, then the standard fixed and random effect models can be used [6, 7]. Unfortunately, published reports seldom report these statistics, which means the researcher often has to estimate them using other information that is provided. Parmar *et al.* [6] consider in detail the various ways of estimating the hazard ratio and its variance depending on available information. The log hazard ratio and its variance can be estimated directly if the observed number of events and the logrank expected number of events are available for each treatment group. Additionally, it can be estimated indirectly if the *p*-value for the logrank, Mantel–Haenszel, or chi-squared test is reported. Finally, if none of these are possible the statistics can be estimated from survival curves. Parmar *et al.* [6] is essential reading for the researcher who is faced with estimating hazard ratios.

18.3   Calculation of the overall 'log-rank' odds ratio

This method combines standard two-arm RCTs by generating an overall index of relative effectiveness, expressed in terms of an odds ratio [8–11]. Hence, unlike the hazard ratio, which measures the instantaneous risk at a given point in time on a survival curve, a comparison across the whole length of the survival curve is made here. A test for heterogeneity between trials, as well as interaction and trend tests for indirect comparisons of trial subgroups has been developed [11]. An outline of the method is as follows:

1. Split time into *j*-1 consecutive intervals, identical across all trials, but whose duration need not be constant.
2. Work out the observed minus expected number of deaths (O-E) for each arm in each period using standard methods.

3. Sum the values of O-E and its variance over all k studies separately for each time period.

4. Divide these summed O-E values by the square root of the summed O-E variances to obtain test statistics (compared against the standardized normal distribution) for the difference in survival in each respective time interval.

18.4 Calculation of pooled survival rates

This method, described by Coplen *et al.* [10] pools the individual survival rates from each study at specified time points using

$$P_t = \frac{\sum_{j=1}^{k}(S_{tj}W_{tj})}{\sum_{j=1}^{k}W_{tj}}, \tag{18.1}$$

where P_t is the pooled survival rate at time t (i.e. the proportion of patients surviving at time t estimated by the meta-analysis), S_{tj} is the survival rate at time t in the jth study (S_{tj} can be either the Kaplan–Meier estimate or an actuarial estimate (estimated from Kaplan–Meier curves, or life tables), W_{tj} is the inverse of the variance of S_t, and k is the number of trials included in the meta-analysis. The variance of S_t for each of the j trials can be calculated by Greenwood's formula:

$$\mathrm{Var}(S_t) = \sum_{i=1}^{h}\frac{D_t}{N_t(N_t - D_t)}, \tag{18.2}$$

where h is the number of time intervals into which the follow-up from time 0 to time t has been divided in to survival analysis (e.g. intervals may be one year each), $D_t =$ the number of deaths during an individual time interval, and $N_t =$ the number of patients at risk during the same interval.

Once the proportion surviving in each trial at a given time point has been estimated, standard fixed or random effect models can be used to combine studies (see Chapters 4 and 5), using an outcome measure such as the risk difference or the odds ratio (see Chapter 2).

18.5 Method of Hunink and Wong

Hunink and Wong [12] present a method for combining failure-time data from various sources, adjusting for differences in case-mix among studies (which causes heterogeneity between them) by the use of covariates. The approach uses a proportional-hazards model and the actuarial life-table approach. The life-table approach is used because the alternative Kaplan–Meier approach would need individual patient data. This model implies survival follows an exponential distribution within each interval.

The model is capable of combining results from non-controlled cohort studies as well as controlled studies. The method is limited because it adopts a fixed effects approach and data must be available to construct life-tables for each study; this may not always be possible from study reports. A brief outline of the method follows:

1. Summarize the available data in the form of a life-table for each study.
2. Estimate the hazard-rate ratio for each covariate, and calculate the hazard-rate ratio for each stratum, defined by combinations of the covariates compared with the reference stratum (calculated across all studies).
3. Combine the data to estimate the hazard-rate of the reference stratum for every interval.
4. Calculate the hazard-rate and survival curve for every stratum.

Once this is carried out, a sensitivity analysis is recommended using techniques such as:

1. Varying the censoring rates among studies and subgroups.
2. Determining the contribution of each study using a jack-knife type of sensitivity analysis.
3. Stratifying the studies by level of detail presented, and including studies with decreasing levels of detail at each step.

Monte Carlo analysis may be performed to derive empirical estimates of the standard errors and confidence intervals.

18.6   Iterative generalized least squares for meta-analysis of survival data at multiple times

The need for this model in addition to the method of combining hazard ratios (see Section 18.2) is two-fold. First, the difficulty of finding the necessary data of sufficient quality in trial reports to combine hazard ratios; and secondly, combining hazard ratios does not allow inclusion of single-arm studies in the analysis [5]. Dear [5] presents a method for the analysis of survival proportions reported at multiple times (e.g. yearly intervals), which can include single arm studies. Generalized least squares are used to fit a linear model, including between and within trial covariates. This is an extension of the method of Raudenbush et al. [13] for combining multiple outcome data (see Section 15.5.1). Here, the multiple outcomes are treated as the same outcome reported repeatedly. Additionally, this method allows the incorporation of multi-arm studies and non-randomized historical controls, and thus also combines the methodological advances of Begg and Polite [14] (see Section 17.2). A limitation of this approach is that it cannot incorporate a random effects baseline term. Dear [5] suggests the possibility of an alternative formulation of this problem using a logistic model through the use of generalized estimating equations.

18.6.1   Application of the model

Dear [5] applied the model to a meta-analysis containing 14 studies on patients with myelogenous leukaemia. Six of the studies compared two treatments, allogeneic bone-marrow transplantation (BMT) and chemotherapy alone. Two included BMT, and the remaining six included only chemotherapy. Table 18.1 gives the empirical probabilities of disease-free survival at five one year intervals after the start of the treatment (where available).

Each line shows results from one clinical trial on patients with acute myelogenous leukaemia. The first six trials compared bone-marrow transplantation (BMT) with chemotherapy alone; the other eight trials tested only one of these alternative therapies.

Previously these data were analysed and results were summarized for the first four years of follow up separately; this permits a joint analysis. Two advantages of this are that: the treatment effect can be estimated with greater precision by combining information between years. Indeed, in this example additive year

Table 18.1   Percent disease-free survival (standard error in parentheses) by year (1 to 5). (Reproduced from Dear [5]).

Trial	BMT					Chemotherapy alone				
	1	2	3	4	5	1	2	3	4	5
1	49(12)	46(12)	42(12)	40(12)	40(12)	54(8)	25(8)	23(7)	23(7)	23(7)
2	55(10)	50(10)	36(9)			40(8)	23(7)	23(7)	23(7)	
3	54(10)	47(13)	40(13)	40(13)		54(9)	42(8)	28(8)	28(8)	
4	70(23)	70(23)	70(23)	70(23)		48(17)	48(17)	17(13)		
5	54(4)	46(5)	42(6)			40(5)	21(4)	16(4)	16(4)	
6	54(2)	43(3)	40(3)	39(3)		50(4)	32(4)	24(4)	18(4)	
7	59(8)	49(9)	47(9)	47(9)	47(9)					
8	61(8)	53(8)	53(8)	53(8)	53(8)					
9						60(9)	48(9)	32(9)	32(9)	32(9)
10						44(5)	26(4)	17(5)	16(4)	
11						50(3)	33(3)	26(3)	22(3)	19(3)
12						62(3)	38(3)	29(3)	24(3)	22(3)
13						50(10)	24(8)	16(7)	12(6)	
14						76(7)	53(8)	53(8)	50(8)	50(8)

Table 18.2 Pooled results of Chemotherapy vs. BMT example – percent disease-free survival (standard error in parentheses). (Adapted from Dear [5].)

Year	BMT	Chemotherapy	Hazard ratio	Survival difference
1	59.0(2.6)	53.2(2.1)	1.1	5.7(3.1)
2	49.6(2.9)	33.6(2.0)	2.3	16.0(3.3)
3	45.9(3.0)	26.1(2.0)	3.0	19.8(3.4)
4	45.1(3.0)	22.8(2.0)	7.3	22.3(3.4)
5	45.1(3.0)	20.8(2.1)	∞	24.3(3.4)

effects were found to significantly improve the fit of the model; and it is possible to test the hypothesis for an interaction between treatments (incorporating all the studies). The results obtained for this example are displayed in Table 18.2

18.7 Identifying prognostic factors using a log (relative risk) measure

To pool studies, Voest *et al.* [15] derived a parameter, the log (relative risk), which summarizes the whole survival curve in a single figure. This involves computing an average survival curve for each treatment regimen, from those obtained from the individual studies, and then finding the difference between each regimen curve and the mean curve to yield a log (relative risk) index for each regimen. The authors use this method (and derived a measure) to detect prognostic factors rather than compare treatments.

18.8 Combining quality of life adjusted survival data

Cole *et al.* [16] present a methodology for meta-analysis to compare treatments in terms of quality-of-life-adjusted survival that does not require individual patient-level data. It allows one to investigate the trade-off between treatment toxicity and improved survival. In the motivating example, a Q-TWiST (quality adjusted time without symptoms or toxicity) analysis was carried out on each trial. This measure allows one to make treatment comparisons that incorporate differences in quality of life associated with various health states, as well as length of survival. The method combines outcomes using regression models for recurrence-free survival and overall survival (separately).The model adjusts for the differing lengths of follow-up, while incorporating the estimated covariance of the restricted means for each trial. Gelber *et al.* [17] apply this method to a meta-analysis of adjuvant chemotherapy plus tamoxifen versus tamoxifen alone for postmenopausal breast cancer.

18.9   Meta-analysis of survival data using individual patient data

All the methods discussed so far for pooling survival data combine data aggregated in some way, whether as a single parameter estimator for the study or periodic summaries, such as those obtained by life tables. However, the ideal way to carry out a meta-analysis of survival data is to use individual patient data [12]. Indeed, some believe that this is the only reliable way [18]. The methodology required is similar to that for survival analysis of a single study, or especially a multi-centre trial. A test for balanced follow-up [19] has been developed. A covariate for study effect can be included, this could be fixed, but often it would be appropriate to make it random to allow for variation between studies. Chapter 12 gives a general explanation of meta-analysis of individual patient data.

18.9.1   Pooling independent samples of survival data to form an estimator of the common survival function

Srinivasan and Zhou [20] consider how to combine possibly censored IPD from different studies efficiently to obtain an estimate of a common (single) survival function. Two parametric settings are discussed. The first considers an optimal weighted average (inversely proportional to the dispersion matrices of the individual estimators) of the two estimators from the two independent studies; and the second pools data form the joint likelihood function and finds the maximum likelihood estimate of the data from the joint likelihood. This method may be difficult to implement in practice because data for dispersion matrices is required.

18.9.2   Is obtaining and using survival data necessary?

It has been suggested that the only way to carry out a survival meta-analysis is to use IPD [18], largely for practical reasons, however. It has yet to be established when the superiority of IPD analysis outweighs the costs it incurs.

Buyse and Ryan [21] compare the Asymptotic Relative Efficiency (ARE) of the Mantel–Haenszel (M–H) test with respect to the stratified logrank test. The M-H test is for difference in proportions between arms across trials ignoring time to event information, while the logrank test takes this into account. When only summary data is available the M–H method of comparing groups has to be used. It was concluded that the M–H test has high efficiency relative to the stratified log-rank test if the M–H test has high efficiency relative to the logrank test for each individual trial. In general, the loss of efficiency for using the M–H test may not be excessively large. However, full survival data should be used whenever possible.

18.10   Summary/Discussion

Survival analysis data requires specialist meta-analysis techniques (as well as specialist statistical methods in general) because of data censoring. If this censoring is ignored this may bias the overall estimates. Methods such as finding summary measures for survival data (such as the hazard ratio), and then combining them are possible. However, some authors consider using individual patient data is the only reliable way to carry such an analysis out.

References

1. Abrams, K.R. (1997). Regression models for survival data. In: Everitt, B., Dunn, G., (editors). *Recent Advances in the Statistical Analysis of Medical Data*. London: Arnold.

2. Parmar, M.K.B., Machin, D. (1995). *Survival Analysis: A Practical approach*. Chichester: John Wiley & Sons.

3. Collett, D. (1994). *Modelling Survival Data in Medical Research*. London: Chapman & Hall.

4. Eddy, D.M. 'Hasselblad, V. 'Shachter, R. (1992). *Meta-analysis by the Confidence Profile Method*. San Diego: Academic Press.

5. Dear, K.B.G. (1994). Iterative generalized least squares for meta-analysis of survival data at multiple times. *Biometrics* **50**: 989–1002.

6. Parmar, M.K.B., Torri, V., Stewart, L. (1998). Extracting summary statistics to perform meta-analysis of the published literature for survival endpoints. *Stat. Med.* **17**: 2815–34.

7. Whitehead, A., Whitehead, J. (1991). A general parametric approach to the meta-analysis of randomised clinical trials. *Stat. Med.* **10**: 1665–77.

8. Pignon, J.P., Arriagada, R., Ihde, D.C., Johnson, D.H., Perry, M.C., Souhami, R.L., Brodin, O., Joss, R.A., Kies, M.S., Lebeau, B. *et al.* (1992). A meta-analysis of thoracic radiotherapy for small-cell lung cancer. *New Engl. J. Med.* **327**: 1618–24.

9. Early Breast Cancer Trialists' Collaborative Group. (1990). *Treatment of Early Breast Cancer. Volume 1: Worldwide Evidence 1985–1990*. Oxford: Oxford University Press.

10. Coplen, S.E., Antman, E.M., Berlin, J.A., Hewitt, P., Chalmers, T.C. (1990). Efficacy and safety of quinidine therapy for maintenance of sinus rhythm after cardioversion: A meta-analysis of randomized control trials. *Circulation* **82**: 1106–16.

11. Messori, A., Rampazzo, R. (1993). Metaanalysis of clinical-trials based on censored end-points – simplified theory and implementation of the statistical algorithms on a microcomputer. *Computer Methods and Programs in Biomed.* **40**: 261–7.

12. Hunink, M.G.M., Wong, J.B. (1994). Meta-analysis of failure-time data with adjustment for covariates. *Med. Decis. Making* **14**: 59–70.

13. Raudenbush, S.W., Becker, B.J., Kalaian, H. (1988). Modeling multivariate effect sizes. *Psychol. Bull.* **103**: 111–20.

14. Begg, C.B., Pilote, L. (1991). A model for incorporating historical controls into a meta-analysis. *Biometrics* **47**: 899–906.

15. Voest, E.E., Van Houwelingen, J.C., Neijt, J.P. (1989). A meta-analysis of prognostic factors in advanced ovarian cancer with median survival and overall survival (measured with the log (relative risk)) as main objectives. *Euro. J. Cancer and Clinical Oncology* **25**: 711–20.

16. Cole, B.F., Gelber, R.D., Goldhirsch, A. (1995). A quality-adjusted survival meta-analysis of adjuvant chemotherapy for premenopausal breast cancer. International Breast Cancer Study Group. *Stat. Med.* **14**: 1771–84.

17. Gelber, R.D., Cole, B.F., Goldhirsch, A., Rose, C., Fisher, B., Osborne, C.K., Boccardo, F., Gray, R., Gordon, N.H., Bengtsson, N.O. *et al.* (1996). Adjuvant chemotherapy plus tamoxifen compared with tamoxifen alone for postmenopausal breast cancer: meta-analysis of quality-adjusted survival. *Lancet* **347**: 1066–71.

18. Clarke, M.J., Stewart, L.A. (1994). Systematic reviews – obtaining data from randomized controlled trials – how much do we need for reliable and informative meta-analyses. *Br. Med. J.* **309**: 1007–10.

19. Stewart, L.A., Clarke, M.J. (1995). Practical methodology of meta-analyses (overviews) using updated individual patient data. Cochrane Working Group. *Stat. Med.* **14**: 2057–79.

20. Srinivasan, C., Zhou, M. (1993). A note on pooling Kaplan–Meier estimators. *Biometrics* **49**: 861–4.

21. Buyse, M., Ryan, L.M. (1987). Issues of efficiency in combining proportions of deaths from several clinical-trials. *Stat. Med.* **6**: 565–76.

CHAPTER 19

Cumulative Meta-analysis

19.1   Introduction

Cumulative meta-analysis of RCTs has been defined as the process of performing a new or updated meta-analysis prospectively every time a new trial is published [1]. Updating meta-analysis on a regular basis in this way ensures that overview estimates are up to date [2]. Meta-analyses, such as those found within the Cochrane Database of Systematic Reviews [3] are good examples of ones which are continually updated as new evidence becomes available. Lau *et al.* [4] demonstrated that if a cumulative meta-analysis of the RCTs of streptokinase for the treatment of myocardial infarction had been carried out concurrently with the publication of trial results, then it could have been shown conclusively to be life-saving several years before it was adopted as routine practice.

However, this definition has been broadened, as cumulative meta-analysis can also be performed retrospectively as part of the process of a single meta-analysis, by adding trials sequentially to the analysis, and re-analysing after every addition to the set of trials. In this way, the contribution of individual studies to the cumulatively pooled result can be ascertained. Hence, it can be used as an exploratory tool, as well as a mechanism of updating a meta-analysis. For example, trials may be ordered and added sequentially according to their year of completion or publication, but the impact of any continuous exploratory variable on the analysis can be explored in this way.

No new statistical techniques are required to combine studies sequentially; standard fixed or random effect models (see Chapters 4 and 5) can be used to carry out the analysis, which is repeated each time a new study is added. This can also be carried out to test the robustness of subgroup analyses (or meta-regression) over time. There is, however, an issue of interpretability of repeated analyses (see Sections 19.4 and 19.5).

19.2   Example: Ordering by date of publication

The year of publication for the 34 cholesterol lowering trials (described originally in Chapter 2) is provided in Table 19.1.

Table 19.1   Year of publication of cholesterol lowering RCTs.

Study number	Year of publication	Cumulative estimate (95% CI)	P-value
12	1961	1.61 (0.68 to 3.81)	0.279
2	1962	1.05 (0.70 to 1.57)	0.817
3	1963	0.86 (0.63 to 1.19)	0.363
8	1965	0.86 (0.65 to 1.15)	0.303
22	1965	0.89 (0.67 to 1.18)	0.411
10	1968	0.92 (0.72 to 1.17)	0.474
17	1968	0.91 (0.73 to 1.13)	0.385
5	1969	0.90 (0.72 to 1.12)	0.337
16	1969	0.92 (0.77 to 1.09)	0.339
7	1970	0.89 (0.75 to 1.04)	0.143
11	1971	0.85 (0.73 to 0.99)	0.037
18	1971	0.86 (0.74 to 0.99)	0.034
13	1972	0.87 (0.75 to 1.00)	0.050
34	1973	0.87 (0.75 to 1.00)	0.049
15	1975	0.72 (0.66 to 0.79)	<0.001
20	1978	0.72 (0.66 to 0.79)	<0.001
21	1978	0.74 (0.68 to 0.80)	<0.001
31	1978	0.80 (0.74 to 0.87)	<0.001
4	1981	0.80 (0.74 to 0.87)	<0.001
23	1984	0.80 (0.74 to 0.87)	<0.001
28	1984	0.81 (0.75 to 0.87)	<0.001
26	1987	0.81 (0.75 to 0.87)	<0.001
29	1987	0.81 (0.76 to 0.88)	<0.001
6	1988	0.81 (0.75 to 0.87)	<0.001
9	1989	0.82 (0.76 to 0.88)	<0.001
24	1989	0.85 (0.80 to 0.91)	<0.001
25	1990	0.85 (0.79 to 0.90)	<0.001
32	1990	0.85 (0.79 to 0.90)	<0.001
33	1990	0.85 (0.79 to 0.90)	<0.001
19	1991	0.85 (0.79 to 0.90)	<0.001
30	1991	0.85 (0.80 to 0.91)	<0.001
1	1992	0.84 (0.79 to 0.90)	<0.001
14	1992	0.84 (0.79 to 0.90)	<0.001
27	1993	0.85 (0.79 to 0.90)	<0.001

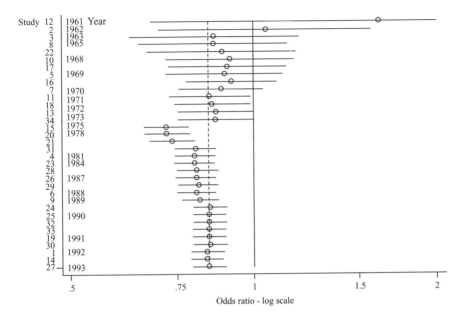

Figure 19.1 Cumulative meta-analysis for total mortality in the cholesterol lowering trials – ordering by year of publication.

The total mortality outcome for these trials is combined sequentially, in ascending publication date order using the fixed effect inverse-variance weighted method. Figure 19.1 displays the pooled estimate and 95% confidence interval for the analyses performed after the addition of each trial (the top point estimate and confidence interval relates to study 2 (published in 1961) alone, the second to studies 12 and 2 (published in 1961 and 1962) combined, and so on). The first point of interest is the general reduction in width of the confidence interval as the number of studies increases, due to inclusion of increasing evidence as each trial is sequentially added. The plot shows that the treatment effect estimates first became formally statistically significant at the 5% level (using a fixed effect model), after inclusion of study 11, published in 1971. However, it changed to non-significance again briefly before reverting to and retaining its significance with the addition of study 15, published in 1975. Study 15 is clearly very influential as the pooled result changes dramatically after it is included in the analysis.

It should be noted that a fixed effect model was used in this analysis, although it was shown in Chapter 3 that there was great heterogeneity between trials, and a random effect analysis is considered more appropriate (as carried out in Chapter 5). The reason for using the fixed effect model here is that it aids the identification of influential studies when this plot is being used in an exploratory manner (opposed to simply updating the meta-analysis when new trials are published). If a random effect model were used, it would have the effect of 'smoothing' the plot as more equal weights are given to each study, restricting the impact of individual studies. Additionally, when sources of heterogeneity have been identified, performing a cumulative meta-analysis on subgroups defined by those sources (in this example, baseline risk could have been used [5]) may be more informative.

19.3   Using study characteristics other than date of publication

In theory, the effect of any continuous or ordinal study-related variable can be assessed using cumulative meta-analysis. Lau *et al.* [6] consider the potential benefits of exploring specific ones in a clinical trial setting.

Cumulatively combining the trials ordered by study size may also be beneficial, as smaller studies results will generally vary greatest and be at greatest risk from publication bias. In the case of streptokinase for myocardial infarction, such an analysis showed that the pooled result became statistically significant with fewer patients when ordered by study size compared with publication year.

A cumulative meta-analysis could be carried out ordering by a study quality score (see Chapter 8). Hence, the variability in the pooled result obtained by including studies of decreasing quality can be ascertained. Similarly, if studies are ordered by control group event rate, with the highest control rates first, and the treatment effect systematically reduces as more trials are combined, then this suggests the treatment could be less effective in patients at lower levels of risk.

19.3.1   Example: Ordering the cholesterol trials by baseline risk in the control group

The influence of baseline risk in the control group on the cholesterol lowering trials was first assessed using a L'Abbé plot in Section 3.3.4, and then again using regression methods in Section 6.3.5. Here the trials are combined in decreasing baseline risk (defined by equation (6.10)). The resulting plot is provided in Figure 19.2. This is a curious shape; as in Figure 19.1, study 15 is highly influential, as it is large and suggests a large benefit in the treatment arm. The pooled estimates gradually moves back towards unity (no effect) as more studies are added. It is difficult to tell from this plot whether effectiveness does decrease with baseline risk, or simply whether the influence of study 15 has resulted in the observed shape of the plot. A more detailed examination would be to perform the analysis separately for the drug and the diet trials, as it is possible that the lack of effectiveness in low risk patients could be due to side effects of the drugs administered.

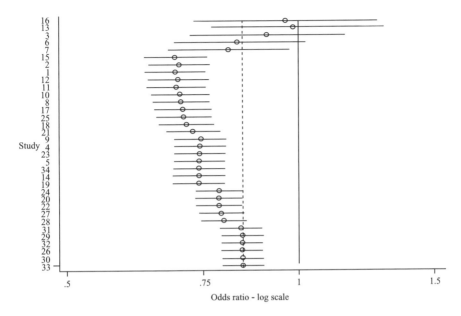

Figure 19.2 Cumulative meta-analysis for total mortality in the cholesterol lowering trials – ordering by baseline risk (descending).

19.4   Bayesian approaches

The 'revision in light of new information' nature of cumulative meta-analysis makes it naturally amenable to a Bayesian analysis [6, 7], since the sequential updating of the current state of knowledge concerning a particular intervention or risk factor mirrors that used in Bayes' Theorem. Lau *et al.* [6] and Schmid *et al.* [8] do not formally adopt a Bayesian methodology. Similarly, the link with decision theory is relevant, since using a cumulative or temporally sequential approach, a decision is ultimately made, either formally or informally, that there is sufficient evidence to warrant the adoption of a particular technology in routine health care practice. To date, however, this has yet to be developed (see Section 20.3.3) [9].

19.5   Issues regarding the uses of cumulative meta-analysis

Several problems exist with cumulative meta-analysis methodology. First, from a classical (frequentist) perspective, when the method is used to decide when an intervention effect first becomes statistically significant, a correction factor for multiple testing needs to be applied, as for stopping rules for clinical trials [10]. Berkey *et al.* [11] suggest the need for a stopping rule for cumulative meta-analyses, but conclude that no straightforward general method is available as each trial has its own sample size.

Indeed, Flather *et al.* [12] suggest the need more generally to validate cumulative application of meta-analysis as a research methodology. The merits of using cumulative meta-analysis as an exploratory tool, instead of other approaches such as regression methods, excluding each study sequentially, or simply plotting the pooled estimates from the individual studies in order of the variable under investigation still needs to be established.

For a discussion of *prospectively planned* cumulative meta-analysis see Section 20.3.1.

19.6   Summary/Discussion

This chapter considers cumulative meta-analysis, and two potential uses. First, it can be used to update meta-analyses as results of new studies are published. Performing such analyses provides answers regarding the effectiveness of a certain intervention at the earliest possible date in time.

Cumulative meta-analysis can also be used as an exploratory tool; however, in many instances the advantages of using this over other exploratory approaches described in this book are not obvious.

References

1. Whiting, G.W., Lau, J., Kupelnick, B., Chalmers, T.C. (1995). Trends in inflammatory bowel disease therapy: A meta-analytic approach. *Canadian J. Gastroenterology* 9: 405–11.

2. Antman, E.M., Lau, J., Kupelnick, B., Mosteller, F., Chalmers, T.C. (1992). A comparison of results of meta-analyses of randomized control trials and recommendations of clinical experts: Treatments for myocardial infarction. *J. Am. Med. Assoc.* 268: 240–8.

3. The Cochrane Database of Systematic Reviews. (1998). Oxford: Cochrane Collaboration.

4. Lau, J., Antman, E.M., Jimenez-Silva, J., Kupelink, B., Mosteller, S.F., Chalmers, T.C., JimenezSilva, J., Kupelnick, B., Mosteller, F. (1992). Cumulative meta-analysis of therapeutic trials for myocardial infarction. *New Engl. J. Med.* 327: 248–54.

5. Smith, G.D., Song, F., Sheldon, T.A., Song, F.J. (1993). Cholesterol lowering and mortality: The importance of considering initial level of risk. *Br. Med. J.* 306: 1367–73.

6. Lau, J., Schmid, C.H., Chalmers, T.C. (1995). Cumulative meta-analysis of clinical trials: Builds evidence for exemplary medical care. *J. Clin. Epidemiol.* **48**: 45–57.

7. Spiegelhalter, D.J., Myles, J.P., Jones, D.R., Abrams, K.R. (2000). Bayesian methods in health technology assessment. *Health Technol. Assess.* (to appear).

8. Schmid, C.H., Cappelleri, J.C., Lau, J. (1994). Applying bayesian meta-regression to the study of thrombolytic therapy. *Clin. Res.* **42**: A 290.

9. Petitti, D.B. (1994). *Meta-Analysis, Decision Analysis and Cost-Effectiveness Analysis.* New York: Oxford University Press.

10. Piantadosi, S. (1997). *Clinical Trials: A methodologic perspective.* New York: John Wiley & Sons. 1997.

11. Berkey, C.S., Mosteller, F., Lau, J., Antman, E.M. (1996). Uncertainty of the time of first significance in random effects cumulative meta-analysis. *Controlled Clin. Trials.* **17**: 357–71.

12. Flather, M.D., Farkouh, M.E., Yusuf, S. (1994). Meta-analysis in the evaluation of therapies. In: Julian, D., Braunwald, E., (editors). *Management of Acute Myocardial Infarction.* London: WB Saunders. 393–406.

CHAPTER 20

Miscellaneous and Developing Areas of Application in Meta-analysis

20.1 Introduction

This chapter outlines some alternative approaches to meta-analysis which have been suggested, and briefly considers some areas in which the use of meta-analysis is developing and may increase.

20.2 Alternatives to conventional meta-analysis

Three approaches to meta-analysis outlined below can be viewed as alternatives to the general approach of the weighted average of individual effect sizes idea, which underpins all the other methods described in this book. To-date, none of these three other methods has widespread use.

20.2.1 Estimating and extrapolating a response surface

In 1990 Rubin presented a new and provocative perspective for meta-analysis [1]. The crux of this method is that it aims to estimate the effect of an ideal study, rather than to calculate average effects over the studies being combined. Rubin questioned the aims of a meta-analysis, identifying two kinds: (i) those for literature synthesis; and (ii) those for understanding the underlying science [1]. He believed that the current view was to carry out literature synthesis, with the aim of summarizing all existing studies by estimating their average population effect. A conceptual framework for a contrasting method, 'building and extrapolating a response surface', is described, though no technical details of the model are given. Although there has been support for the idea [2], and it has been implemented [3], it has not, to our knowledge, been implemented in a health research setting, probably due to the requirement for a large number of studies to estimate a response surface successfully.

20.2.2   Odd man out method

The odd man out method [4] is a novel graphical approach to meta-analysis that also has not been widely adopted. Dickersin and Berlin give a concise explanation:

> The areas of overlap of confidence intervals from individual studies are used to construct summary "confidence regions". These regions are within the graphic display and include information about both the influence of individual studies and the overall results [5].

The probability that a true parameter value will lie in at least $N - 1$ 95% confidence intervals of N studies which estimate it, is approximately 95%, for N taking values between 5 and 10. Using this result, a summary 95% confidence bound can be constructed including all values included in at least $N - 1$ of the confidence intervals from the individual studies. This method has been used in some meta-analyses to explore heterogeneity; if the summary confidence interval is disjoint (i.e. is made up of two or more regions which are not connected) this may reflect heterogeneity between studies [4] There are questions about its validity, however, because it excludes trials on the basis of their results, not their design. In such circumstances, it may be more helpful to investigate why the results are different, rather than simply exclude studies.

20.2.3   Best evidence synthesis

Slavin has proposed an alternative method to meta-analysis for synthesizing results [6, 7]. He calls this method 'best evidence synthesis', and claims it combines the strengths of meta-analytic and traditional reviews. Put simply, this method does not combine all the studies available, but considers and combines only the best of them. Criteria for 'best' are defined *a priori*. This is linked with a more formal discussion of the studies/results. He criticizes traditional meta-analysis due to the incorporation of the biases of the primary studies, and notes that the reader is rarely able to form his or her own opinion as the studies are not described in any sufficient detail. This method incorporates many of the important contributions of meta-analysis, but also retains features of intelligent and insightful narrative reviews. Letzel [8] compares in detail the uses and characteristics of 'best-evidence synthesis' and meta-analysis, and concludes that both approaches supplement one another, rather than being considered as alternatives. In essence, this method is conceptually similar to a traditional meta-analysis using a quality scoring system with zero weight given to the poor quality studies.

20.3   Developing areas

20.3.1   Prospective meta-analysis

Margitic *et al.* [9] discuss a new approach they call *prospective* meta-analysis. It has been defined as 'a compilation of common data collected prospectively from trials testing related interventions' [9], and involves planning trials in the context of knowing they will be included in a meta-analysis. The method combines elements of a multicentre clinical trial with specific features of a retrospective meta-analysis. It thus reduces a number of the disadvantages inherent in a traditional retrospective meta-analysis. Disadvantages of standard meta-analysis overcome by this method include (a) possibility of biases (selection, trial-related, publication) (b) trial heterogeneity (different protocols and data, varying outcomes, and data quality). and (c) incomplete access to the databases of the individual trials comprising a particular meta-analysis. Using this method, selection and publication biases are minimized, and a complete pooled database is obtained with more consistent quality data across sites. It also allows a pre-trial statement of objectives, definition of the population of interest, and *a priori* hypothesis testing.

Friedenreich [10] discusses prospective meta-analysis designs for epidemiological studies. Planning data pooling during the design phase, facilitates combined analyses since the studies being combined will have similar designs and standardized methods. This approach has already been used by the International Agency for Research on Cancer for a number of cohort and case-control studies [11], but requires collaboration, often on an international scale.

A prospectively planned cumulative meta-analysis applied to a series of concurrent clinical trials

Whitehead [12] combines methods for sequential designs (specifically, the triangular test), and those for combining studies. This method was developed for use in the context of a series of studies, following broadly similar protocols, each comparing the same form of new treatment with a control treatment.

An example given is the assessment of a drug for the prevention of a serious side-effect resulting from chemotherapy treatment for patients with cancer. Different studies may deal with cancers at different sites, however, the primary efficacy variable (occurrence of a specific side effect) is the same in all studies. In this situation, individual fixed sample size studies may be designed for the primary efficacy variable, but the safety variable would be analysed according to a sequential design with stopping boundaries. Significant evidence demonstrating that the new treatment was harmful could then lead to the stopping of all the studies [12]. The author goes on to comment that one may wish to treat the design as a multi-centre trial, or stratified study, rather than planning each trial with enough power to detect, say, side effects individually. One would want to know as quickly as possible if there was a difference (i.e. there was a side effect), so a sequential design would be suitable. If a fixed effects model is used in this situation, no new methodology is necessary, but new methodology is needed if random effects are included in the analysis [12].

20.3.2 Economic evaluation through meta-analysis

Economic evaluations through meta-analysis are sometimes referred to as 'secondary economic evaluations', evaluations which use available economic data either alongside a review and meta-analysis of clinical trials or as a summary of self-standing evaluations [13]. Despite the number of studies available, the process of reviewing and summarizing economic evidence has been relatively little developed, with many fewer meta-analytic economic evaluations than meta-analyses of RCTs having been carried out [14]. It has been suggested that the latter appear to have greater scientific acceptance, possibly because RCTs use harder endpoints than costs [14]. The distinction between primary and secondary research in economic evaluation is particularly difficult, given the range of steps and diversity of sources of data required for a typical economic evaluation [14].

Jefferson et al. [14] detailed some of the problems of carrying out a secondary economic evaluation: (i) substantial methodological variations have been demonstrated between economic evaluations that are superficially comparable, raising questions about the reliability of cost-effectiveness 'league tables' and other comparative devices, (ii) doubts have been expressed about the theoretical basis for transferring results from the study setting to another; and (iii) it is not clear what technical options exist for summarizing and transferring results from a number of economic evaluations. There is also the problem that in most economic evaluations, the cost data are viewed as constant and not stochastic—thus making the combination of study results problematic. A study into the feasibility of such an analysis [14] concluded that progress may require a more coherent theoretical framework linked to cost and production function theory. Additionally, they suggest that inclusion criteria could be tied to increasingly explicit reporting guidelines now being urged for economic studies. They observe that the UK Medical Research Council and NHS National Research and Development Programme are placing increasing emphasis on the prospective study design, in which economic evaluations are performed alongside and as part of randomised controlled trials. Although this may be the best way forward, such prospective studies may themselves be costly, or be insufficiently powered to show significant cost differences.

20.3.3   Combining meta-analysis and decision analysis

When a meta-analysis is performed and the results reported, others are usually left to assess its implications for changes in policy or practice. However, Midgette *et al.* [15] have combined the synthesis and decision process by incorporating the treatment effect estimates, provided by a meta-analysis, into a decision model assessing when the treatment should be used. They assess the effectiveness of Intravenous Streptokinase (IVSK) on short term survival after suspected Acute Myocardial Infarction (AMI). Using a meta-analysis of the effects of streptokinese on short term mortality in patients with different locations of infarction, a simple decision tree was developed to compare streptokinese with conservative treatment for AMI. Short-term mortality, costs, and marginal cost-effectiveness ratios as a ratio of additional dollars per additional life saved were all predicted.

20.3.4   Net benefit model synthesizing disparate sources of information

Glasziou and Irwig [16, 17] consider generalizing randomized trial results to patient groups with specific levels of risk (in order to be able to better tailor treatments for individual patients) using additional trial information. They consider the following equation:

$$Net\ Benefit = (Risk\ Level \times Risk\ Reduction) - Harm. \tag{20.1}$$

This model suggests potential benefit increases with risk, but that harm will remain relatively fixed. Thus, at low levels of risk, the benefits may not outweigh the harm and we should refrain from intervening, but at higher levels the benefit may outweigh the harm. Completing the above equation for population subgroups generally requires several sources of data. The authors suggest that the estimate of relative risk reduction should come from (a meta-analysis of) randomized trials, the adverse event rates may come from both randomized trials and other epidemiological studies; and the level of risk will usually come from multivariate risk equations derived from large cohort studies.

References

1. Rubin, D. (1992). A new perspective. In: Wachter, K.W., Straf M.L. (editors). *The Future of Meta-analysis*. New York: Russell Sage Foundation. 155–65.

2. Lau, J., Ioannidis, J.P., Schmid, C.H. (1998). Summing up evidence: one answer is not always enough. *Lancet* **351**: 123–7.

3. Vanhonacker, W.R. (1996). Meta-analysis and response surface extrapolation: a least squares approach. *Am. Statistician* **50**: 294–9.

4. Walker, A.M., Martin-Moreno, J.M., Artalejo, F.R. (1988). Odd man out: a graphical approach to meta-analysis. *Am. J. Public Health* **78**: 961–6.

5. Dickersin, K., Berlin, J.A. (1992). Meta-analysis: state-of-the-science. *Epidemiol Rev.* **14**: 154–76.

6. Slavin, R.E. (1986). Best-evidence synthesis: An alternative to meta-analytic and traditional reviews. *Educ. Res.* **15**: 5–11.

7. Slavin, R.E. (1995). Best evidence synthesis: an intelligent alternative to meta-analysis. *J. Clin. Epidemiol.* **48**: 9–18.

8. Letzel, H. (1995). 'Best-evidence synthesis: an intelligent alternative to meta-analysis': discussion. A case of 'either-or' or 'as well'. *J. Clin. Epidemiol* **48**: 19–21.

9. Margitic, S.E., Morgan, T.M., Sager, M.A., Furberg, C.D. (1995). Lessons learned from a prospective meta-analysis. *J. Am. Geriatr Soc.*. **43**: 435–9.

10. Friedenreich, C.M. (1993). Methods for pooled analyses of epidemiologic studies. *Epidemiology* **4**: 295–302.

11. World Health Organization—International Agency for Research on Cancer. (1989). Biennial Report, 1988–1989. Lyon: Lyon: International Agency for Research on Cancer.

12. Whitehead, A. (1997). A prospectively planned cumulative meta-analysis applied to a series of concurrent clinical trials. *Stat. Med.* **16**: 2901–13.

13. Jefferson, T., DeMicheli V., Mugford, M. (1996). Current issues. In: *Elementary Economic Evaluation in Health Care*. London: BMJ Publishing Group.

14. Jefferson, T., Mugford, M., Gray, A., DeMicheli, V. (1996). An exercise in the feasibility of carrying out secondary economic analysis. *Health Economics* **5**: 155–65.

15. Midgette, A.S., Wong, J.B., Beshansky, J.R., Porath, A., Fleming, C., Pauker, S.G. (1994). Cost-effectiveness of streptokinase for acute myocardial-infarction – a combined metaanalysis and decision-analysis of the effects of infarct location and of likelihood of infarction. *Med. Decis. Making* **14**: 108–17.

16. Glasziou, P., Irwig, L. (1995). Generalizing randomized trial results using additional epidemiologic information. *Am. J. Epidemiol* **141**: S47.

17. Glasziou, P.P., Irwig, L.M. (1995). An evidence based approach to individualizing treatment. *Br. Med. J.* **311**: 1356–9.

APPENDIX I:

Software Used for the Examples in this Book

Although several stand-alone programs for performing meta-analysis exist [1–4] they generally have limitations in the range of analyses they perform, and the graphical displays they can produce. The available programs have been reviewed elsewhere [5–8] and a summary table outlining the capabilities of the most modern software is given in Sutton *et al.* [8]

Unfortunately, the majority of the large statistical packages such as SAS® [9] Splus [10], Stata [11], etc., do not have built in routines for performing meta-analysis. This can be attributed to the fact that these programs are orientated to dealing with individual subject, rather than study summary data, and due to the limited application of the statistical techniques used for meta-analysis. (However, a book describing how to perform meta-analysis using the SAS system has recently been published [12]). Fortunately, in recent years, several functions/macros have been written for various statistical packages which carry out many of the different analyses described in this book [13–24]. For a review of many of these, again see Sutton *et al.* [8]. In addition, the calculations required for many of the simpler methods described in the early chapters can easily be performed in a spreadsheet package such as Excel. [25].

In an effort to assist the reader who is interested in carrying out the methods described in this book the software used in each example is indicated below. Where possible, links to the code, or the code itself have been made available, via the website:

http://www.prw.le.ac.uk/epidemio/personal/ajs22/meta/book.html

This site also provides further details about software for meta-analysis generally (and may be updated as new developments occur). Additionally, the datasets used in this book are also available to download, where possible. Note that STATA (version 5.0) was used for many of the figures in the book. Editing of these figures was often required, and carried out with the companion package STAGE [26].

301

Chapter 3

3.2.3 The test for heterogeneity was carried out using the *meta* macro [15, 16] for STATA.

3.3.1 The calculations and histogram of normalized z-scores was plotted using SPSS [27].

3.3.2 The radial plot was created using the STATA macro *galbr* [22].

3.3.3 This basic forest plot was produced using the *meta* macro [15, 16] for STATA.

3.3.4 The L'Abbé plot was produced using the standard *graph* command in STATA.

Chapter 4

4.2.1 and 4.2.2 The inverse variance weighted analysis and corresponding forest plot were produced using the *meta* macro [15, 16] for STATA.

4.3.1 The Mantel–Haenszel method was calculated using the macro *metan* [17] (as well as by hand) for STATA.

4.3.2 The Peto method was calculated using the macro *metan* [17] (as well as by hand) for STATA.

Chapter 5

5.5 The random effect estimate and corresponding forest plot were calculated using the macro *meta* for STATA.

5.6.1 The method of Hardy and Thompson [28] can be implemented using an S-plus macro written by the authors, and available from the website. The method of Biggertaff and Tweedie [29] was implemented using SAS code provided in the original paper.

5.6.2 Although this exact approach to random effects meta-analysis of binary data is not illustrated, software written for Gauss by Van Houwelingen *et al.* [30] is available on request from the first author and website.

Chapter 6

6.2.1 and 6.2.2 Subgroup analyses were performed and forest plots created using the *meta* macro [15, 16] for STATA. (Note: Each subgroup analyses was performed individually, creating individual datasets for each subgroup.)

6.3.2 Fixed effect regression was performed using weighted regression analysis within SPSS.

6.3.4 The mixed effect regression model was fitted using the STATA macro *metareg* [18].

6.3.5 Mixed modelling using patient baseline risk as a covariate, adjusting for regression to the mean was performed using BUGS code in the original paper [31]. Figure 6.6 was produced using the graph command within STATA (see Sharp [18] for details).

Chapter 7

7.5.1 The Funnel plot in Figure 7.3 was custom-drawn; however, similar plots can be obtained using the *metan* [17] and *metabias* [21, 23] macros for STATA.

7.5.2 The rank correlation test was performed using the *metabias* [21, 23] macro for STATA.

7.5.3 The linear regression test and Figure 7.4 were carried out using the *metabias* [21, 23] macro for STATA. (Note: there was some confusion over the best weighting method to use in this test and this has changed in the second release, [23] which will give different answers from the first release [21]

7.6.2 No specialist software was used in the calculation of the file drawer number.

7.6.5 No software currently exists for performing 'Trim and Fill' (our example was kindly calculated by S. Duval and R. Tweedie, the original authors of the method). Software for this method may be accessible through our website in the future.

Chapter 8

8.3.1 The graphical plot was produced 'line by line' using S-plus.

8.3.3 The random effect regression model and plot were obtained using the *metareg* macro [18] for STATA.

8.3.4 The calculations required for incorporating study quality in the study weightings were done 'by hand'.

Chapter 9

9.2 Excluding each study individually from the analysis. Figure 10.1 and the corresponding analysis were produced using the STATA macro *metainf* [24].

9.3.1 The assessment of the impact of choice of study weighting plot was produced in SAS using the macros described by Kuss and Koch [13], and available from our website.

Chapter 10

The high quality, annotated forest plots in Figures 11.1 and 11.2 were produced using the S-plus macro ci.plot written and described by Higgins [32] (code and manual available from the website).

Chapter 11

11.2.3 Bayesian analysis of a single trial. No software was required for the computations. Splus using custom code was used to draw Figure 11.1.

11.3.1 Bayesian meta-analysis of normally distributed data. This analysis was performed using WinBUGS, which at time of press could be downloaded for free (link provided on website). An online manual is included with the software; additionally see Smith *et al.* [33].

11.4.1 Bayesian meta-analysis of binary data. This analysis was performed using the WinBUGS software using code described in Smith *et al.* [33]. Figures 11.2 and 11.3 were created in Splus using custom code.

11.6 Shrinkage example. Figure 11.3 was created in R [34] using custom code.

11.7.5 Trace plot. This was created using an Splus program for Bayesian meta-analysis written by DuMouchel [35] available from the website.

Chapter 14

14.4.1 Combining binary diagnostic test result data. The analysis, regression plot (Figure 14.3) and SROC plot (Figure 14.4) were all done using a stand-alone program called Meta-Test, [36] written specifically to perform meta-analyses of diagnostic test data (link provided on the website).

14.8.9 Combining *p*-values. This example was carried out using the STATA macro *metap* [20].

Chapter 15

15.5.6 Multiple outcomes models. Code corresponding to the fixed and random effect multiple outcome models of Sections 15.5.3 and 15.5.4, respectively, is available for SAS from the website.

Chapter 16

The oral contraceptive use and risk of breast cancer meta-analysis was carried out some time ago, before the advent of specially written software. Any software which carries out meta-regression analyses, such as the STATA macro *metareg* [18] could be used.

Chapter 17

17.7.2 Bayesian hierarchical model for cross-design synthesis. This was implemented using the WinBUGS software using code described in Prevost *et al.* [37].

Chapter 18

18.6.1 Iterative generalised least squares for meta-analysis of survival data at multiple times. Code is available from the website.

Chapter 19

19.2 Cumulative meta-analysis ordering by date. Analysis and plot carried out using the STATA macro *metacum* [19].

19.3.1 Cumulative meta-analysis ordering by covariate. Analysis and plot carried out using the STATA macro *metacum* [19].

References

1. Review Manager [computer program]. 7.00. (1996). Oxford: Update Software Ltd.

2. Cucherat, M., Boissel, J.P., Leizorovicz, A., Haugh, M.C. (1997). EasyMA: a program for the meta-analysis of clinical trials. *Comput Methods Programs Biomed* **53**: 187–90.

3. DuMouchel, W., Fram, D., Jin, Z., Normand, S.L., Snow, B., Taylor, S., Tweedie, R. (1997). Software for exploration and modeling of meta-analysis (abstract). *Controlled Clin. Trials* **18**: 181S.

4. Eddy, D.M., Hasselblad, V. (1992). FastPro: Software for MetaAnalysis by the Confidence Profile Method [computer program]. San Diego, California: Academic Press, 3.5–inch disk. IBM-PC.

5. Normand, S.L.T. (1995). Metaanalysis software – a comparative review – DSTAT, version 1.10. *Am. Statistician* **49**: 298–309.

6. Behar, D. (1992). FastPro – software for metaanalysis by the confidence profile method. *J. Am. Med. Assoc.* **268**: 2109.

7. Egger, M., Sterne, J.A.C., Davey Smith, G. (1998). Meta-analysis software. *Br. Med. J.* 316. Website only: http://www.bmj.com/archive/7126/7126ed9.htm

8. Sutton, A.J., Lambert, P.C., Hellmich, M. *et al.* (2000). Meta-analysis in practice: a critical review of available software. In: Berry, D.A., Stangl. D.K., editors. *Meta-Analysis in medicine and Health policy*. Marcel Dekker.

9. SAS Institute Inc. (1992). SAS Technical Report P-229, SAS/STAT Software: Changes and Enhancements. Release 6.07.

10. S-Plus [computer program]. 4.0. (1997). Statistical Sciences, Inc., 1700 Westlake Ave. North, Suite 500, Seattle, WA 98109.

11. STATA [computer program]. 702 (1985). University Drive East, Texas: Stata Press.

12. Wang, M.C., Bushman, B.J. (1999). Integrating results through meta-analytic review using SAS(R) software. Cary, NC, USA: SAS Institute Inc.

13. Kuss, O., Koch, A. (1996). Metaanalysis macros for SAS. *Computational Statistics & Data Anal* **22**: 325–33.

14. Higgins, J.P.T. ci.plot: (1999). Confidence interval plots using S-PLUS, Manual version 2. London: Institute of Child Health.

15. Sharp, S., Sterne, J. Meta-analysis. (1997). *Stata Tech. Bull.* 38(sbe16):9–14.

16. Sharp, S., Sterne, J. (1998). New syntax and output for meta-analysis command. *Stata Tech. Bull.* 42(sbe 16):6.

17. Bradburn, M.J., Deeks, J.J., Altman, D.G. (1998). metan – an alternative meta-analysis command. *Stata Tech. Bull.* STB 44(sbe24):4–15.

18. Sharp, S. (1998). Meta-analysis regression. *Stata Tech. Bull.* 42(sbe23):16–22.

19. Sterne, J. (1998). Cumulative meta-analysis. *Stata Tech. Bull.* 42(sbe22):13–16.

20. Tobias, A. (1999). Meta-analysis of p-values. *Stata Tech. Bull.* STB 49(sbe28):15–7.

21. Steichen, T.J. (1998). Tests for publication bias in meta-analysis. *Stata Tech. Bull.* 41(sbe20):9–15.

22. Tobias, A. (1998). Assessing heterogeneity in meta-analysis: the Galbraith plot. *Stata Tech. Bull.* 41(sbe20):15–7.

23. Steichen, T.J., Egger, M., Sterne, J. (1998). Modification of the metabias program. *Stata Tech. Bull.* STB 44(sbe19.1):3–4.

24. Tobias, A. (1999). Assessing the influence of a single study in the meta-analysis estimate. *Stata Tech. Bull.* STB 47(sbe26):15–17.

25. Excel 2000 [computer program]. (1999). Redmond, WA: Microsoft.

26. STAGE [computer program]. (1993). 702 University Drive East, Texas: Stata Corporation.

27. SPSS BASE [computer program]. (1999). 233 S. Wacker Drive, 11th Floor: SPSS Inc.

28. Hardy, R.J., Thompson, S.G. (1996). A likelihood approach to meta-analysis with random effects. *Stat. Med.* **15**: 619–29.

29. Biggerstaff, B.J., Tweedie, R.L. (1997). Incorporating variability in estimates of heterogeneity in the random effects model in meta-analysis. *Stat. Med.* **16**: 753–68.

30. Van Houwelingen, H.C., Zwinderman, K.H., Stijnen, T. (1993). A bivariate approach to meta-analysis. *Stat. Med.* **12**: 2273–84.

31. Thompson, S.G., Smith, T.C., Sharp, S.J. (1997). Investigation underlying risk as a source of heterogeneity in meta-analysis. *Stat. Med.* **16**: 2741–58.

32. Higgins, J.P.T. ci.plot: (1999). Confidence interval plots using S-PLUS, Manual version 2. London: Institute of Child Health.

33. Smith, T.C., Spiegelhalter, D.J., Thomas, A. (1995). Bayesian approaches to random-effects meta-analysis: A comparative study. *Stat. Med.* **14**: 2685–99.

34. Ihaka, R., Gentleman, R.R. (1996). A language for data analysis and graphics. *J. Computational and Graphical Stat.* **5**: 299–314.

35. DuMouchel, W. (1994). Predictive cross-validation in Bayesian meta-analysis. *Proceedings of Fifth Valencia International Meeting*, Valencia, Spain.

36. Meta-Test [computer program]. Lau J. .6. (1997). New England Medical Center, Boston.

37. Prevost, T.C., Abrams, K.R., Jones, D.R. Using hierachical models in generalized synthesis of evidence. An example based on studies of breast cancer screening. *Stat. Med.* (Forthcoming).

Index

a priori beliefs 168–9
absolute risk reduction (ARR) 25–6, 155
acute myocardial infarction (AMI) 299
adjustment factors 241
AIDS 19, 215–16, 223
all-cause mortality 173
alternative hypothesis 219
analysis of variance (ANOVA) 93
angiography 211
asymmetric studies 124
asymmetry measure 117
Asymptotic Relative Efficiency (ARE) 283
Attachment Level (AL) 233–6

Bacillus Calmette-Guérin (BCG) vaccine 95–7
baseline risk 102–4
Bayes' Theorem 164, 165
Bayesian analysis, conjugate model 167–9
Bayesian hierarchical models 176, 181, 215, 267–71
  combining evidence from different study types 269–71
  Larose and Dey 271
Bayesian methods 163–89
  advantages/disadvantages 166–7, 176–8
  binary data 171–5
  comparison with classical approaches 179
  comprehensive modelling 182
  cumulative meta-analysis 291

disadvantages 178–9
extensions 179–83
further developments 183
health research 163–9
missing data 203
model selection 180
normally distributed data 169–71
sensitivity analysis 181–2
specific areas of application 179–83
best evidence synthesis 296–7
between study heterogeneity 87–104
  see also heterogeneity
between study variance 99
  uncertainty induced by estimating 80–1
bias
  exclusion of studies 52
  misclassification 244
  observational studies 239–40
  publocation *see* publication bias
  subject selection 244
binary data
  Bayesian methods 171–5
  model formulations 172
  random effects meta-analysis 81–2
  scales of measurement 28
  summarizing 28
binary diagnostic test data, combining 211–15
binary outcome data reported on different scales 207
binary outcome measures, combining 173–5

binary outcomes, combining with continuous outcomes   207
binary test results, combining   209–15
biochemical evidence, combining with epidemiological evidence   272
bivariate normality   102
blinding   136
bone-marrow transplantation (BMT)   281–2
borrowing strength   176
breast cancer   240, 241, 248, 252, 270
BUGS   165, 266, 268

Campbell Collaboration   4
cancer studies, combining results   272
cancer treatments   126
case-control studies   253, 255, 262
case-control study   20
case mix, coronary artery bypass graft studies   139
CD4 count   215–16
chi-square distribution   39, 52
chi-square test   40
choice of study weighting   150
clinical trials
  multiple outcome model for   232
  prospectively planned cumulative meta-analysis   297–8
clozapine   171, 177
Cochrane Collaboration   4, 147
Cochrane Database of Systematic Reviews   137
Cochrane Working Group   193, 194
coefficient estimation from reports presenting only means   245–6
cohort studies, meta-analysis   173
combined log odds ratio   62
combined odds ratio   62
combining data
  ordinal data   205–6
  scales of measurement for   206–8
  see also specific cases
comparative binary outcomes   20–8
compliance rate   49
conditional maximum likelihood estimates   69

confidence intervals   19, 42, 59, 62, 63, 65, 67, 68, 75, 79, 84, 155, 164, 173, 177, 243–4
confidence profile method   265–6
confounder-exposure information   244
confounding   250
conjugate models   165
  Bayesian analysis   167–9
CONSORT statement   153
continuous data   28–9
Continuous Electronic Heart Rate Monitoring (CEHRM)   137
continuous exposure parameter estimates, combining   251–2
continuous outcome scale   62–3
continuous outcomes, combining with summaries of binary outcomes   207
continuous scales   28
continuous test results, combining   215
coronary artery bypass graft studies, case mix   139
coronary artery bypass graft study   138–9
coronary artery bypass graft surgery   139, 140
coronary heart disease (CHD)   93, 142, 173, 174
corrected standard error   94
correlated outcome measures   229–37
correlation coefficient   216
covariates   150, 241, 279–80
  inclusion of   180
  individual Patient Data (IPD)   194
  patient baseline risk as   102–4
  regression analysis   51–2
  study level missing data   201
credibility intervals   103, 164
cross-design synthesis   266–73
cross-over trials, combining   216–17
cross-sectional studies   240
cumulative meta-analysis   138, 287–93
  Bayesian methods   291
  ordering by date of publication   288–90
  using study characteristics other than date of publication   290

data extraction, data manipulation
    for 243–4
data manipulation for data
    extraction 243–4
data type effect measures, combining 207
data types, combining studies with
    differing 207
Database of Abstracts of Reviews of
    Effectiveness (DARE) 154–5
decision analysis, combining with meta-
    analysis 299
diagnostic test, accuracy 209
discussion requirements 155
dissemination 155
distribution of effect size estimates 158
dose-response data 272
    combining 251
duplex Doppler ultrasound 211

Early Breast Cancer Trialists' Collaborative
    Group 192, 194
ecologic studies 240
economic evaluation 298
effect size estimates 37, 230
    distribution 158
    missing 200–1
effect size model 123
EM algorithm 76
Empirical Bayes (EB) methods 166, 175–6
epidemiological evidence, combining with
    biochemical evidence 272
epidemiological studies 239–58
    influence analysis 248–9
    presentation of results 240
    sensitivity analysis 248–9
evidence
    combining different sources 259–76
    generalized synthesis 259–76
evidence-based decision making 8
evidence-based health care (EBHC) 3–4, 11
Evidence-Based Medicine (EBM) 4
evidence hierarchy 135
exact approach, random effects meta-
    analysis 81–2
exact methods, interval estimation 69
exchangeability 181

exclusion of studies 52, 142–3
    quality factor 143
expected number of deaths (O-E) 278–9
experimental studies 135–6
exposure coefficients, calculating 244–5
exposure variables, qualitative/
    categorical 244
extensions/alternative tests 39–40
external reference population 245

F statistic 38
failure-time data, combining 279–80
file drawer problem 120–2
fixed effects 247
fixed effects methods 57–72, 82
fixed effects models 78, 79, 83–4, 94, 141,
    236
fixed effects regression 93–4
fixed effects regression models
    alternative formulations 100
    generalization and extensions 100–1
follow-up length 50, 158
forest plot 37, 42–5, 61, 64, 79, 92, 155–6
    annotated 156, 157
    including extra information 156
funnel plot 113–16, 117, 120, 124, 157
    limitations 115
further research 155

Galbraith diagram 46–7, 118, 119
general fixed effect model 58–63
general inverse weighted variance model 74
Generalised Estimating Equations
    (GEE) 211
Gibbs sampling 165
graphical displays 155–8
graphical plot 137
Greenwood's formula 279
grouped random effects models 271

hazard ratio, see also (log) hazard ratio
health research, Bayesian methods 163–9
heterogeneity
    assessment 37–56
    underlying patient risk 101–2
    between study 87–104

heterogeneity assessment (*cont.*)
  dealing with 50–3, 248
  graphical informal tests/
    explorations 41–8
  impact of early stopping rules 50
  impact of size of intervention dose 49
  impact of underlying risk 49
  methods for investigating 50–3
  observational studies 246
  possible causes 48–50
  sources of 50–3
  specific factors causing 49–50
  testing for presence of 38–41
heterogeneous outcomes, pooling studies
  with 53
hierarchical models 101, 181
  see also Baysian hierarchical models
histogram of normalized (z) scores 41–2
historical controls 259–62
HIV 215–16
hypothesis tests, presence of
  heterogeneity 38–41

incidence rates 19
  ratio 28
inclusion criteria
  ambiguity 148
  changing 147–8
  sensitivity of results to 147–9
Individual Patient Data (IPD) 191–7
  checking data 193
  combining with summary data 195
  comparison with summary data 194–5
  costs 192
  covariates 194
  data collection 193
  guidelines 194
  issues involved in meta-analyses 193–4
  motivation for meta-analysis 191–2
  observational studies 250–1
  potential advantages and disadvantages
    194
  procedural methodology 193
  survival data 283
influence analysis, epidemiological
  studies 248–9

inputs and results 144
interactions 250
Intermittent Auscultation (IA) 137
internal reference group 245
interval estimation, exact methods 69
Inverse Gamma distribution 174
Inverse Gamma prior distribution 171
inverse variance weighted method 58–63,
  79
  combining odds ratios 59
iterative generalized least squares,
  survival data 280–2

Jeffreys' prior 166
joint posterior density 165
journals 126

Kaplan–Meier curves 278
Kaplan–Meier estimate 279
Kendall's tau 216

L'Abb plot 47–8
language bias, evidence 111–12
largest studies analysis 120
least squares methods 233
likelihood function 164
linear regression test 117–19
  application 118
(log) hazard ratio 278
log odds ratio 59, 168, 173, 205
log-rank odds ratio 278–9
log (relative risk), prognostic factors 282
logit method 220
lung cancer 272

Mann–Whitney statistic 207
Mantel–Haenszel chi squared statistic 263
Mantel–Haenszel method 64–6, 69, 70
Mantel–Haenszel test 283
marginal posterior densities 165
Markov Chain Monte Carlo (MCMC)
  methods 165, 170, 171, 174, 180,
  181, 183, 271
masking 136
matched and unmatched data,
  combining 262–3

matching variables 241
maximum likelihood (ML) estimate 68–9,
    74–6, 96
mean difference 29–31
  calculation 30–1
medulloblastoma 261
meta-analysis
  alternative approaches 295–7
  analytic 51
  combining with decision analysis 299
  developing areas 297–9
  development and uses 3–16
  exploratory 51
  guidelines for good practice 9
  impact of study quality 101
  miscellaneous applications 295–300
  novel applications using non-standard
      methods or data 223
  prospective 297–8
  role in systematic review 8–12
  study quality 137–43
  types 51
  uncertainty induced by estimating between
      study variance 81
  see also specific data types and
      applications
meta-regression 51–2, 248
meta-regression model 93–4
Meta-Test 304
method of moments 98
methodology quality 148
minimum p method 219
misclassification bias 244
missing data 199–204
  analytic methods 201–3
  Bayesian methods 203
  general analytic methods 201
  individual patient level 199
  methods specific to meta-analysis 202
  reasons for 200
  standard deviations 202–3
  study level 199, 200–1
  study level characteristics/covariates 201
  study level effect sizes 200–1
  whole studies 199
missing values 148

mixed effect models 53, 93, 97–9
  alternative methods 99–100
  extensions 99–104
  patient baseline risk as covariate 102–4
moment based estimator 81, 82
multi-level models 101, 181
multiple outcome measures 229–37
  reducing to single measure for each
      study 231
multiple outcome models
  clinical trials 232
  DuMouchel's model 233
  illustration of use 233–6
multiple treament arms, combining 263–5
Multivariate Maximum Likelihood (MML)
    methods 233
multivariate model development 231–6
  Gleser and Olkin model 232
  Raudenbush et al. model 231
myelogenous leukaemia 281–2

net benefit model synthesizing disparate
    sources of information 299
non-comparative binary outcomes 18–19
non-empirical evidence 111
non-parametric methods of combining
    different data type effect measures
    207
Normal prior distribution 171, 174
normalized (z) scores 41–2
normally distributed data, Bayesian
    approach 169–71
nuisance parameters 165
null hypothesis 41, 219
number needed to treat (NNT) 27, 155
  calculation 27

observational studies 136, 194, 239–58
  bias 239–40
  extraction and derivation of study
      estimates 240–6
  guidelines 239
  heterogeneity 246
  individual patient data (IPD) 250–1
  methods for combining estimates 247
  pooling of 254

observational studies (*cont.*)
  reporting   248
  scales of measurement   243
  sensitivity analysis   240
  study quality considerations   249–50
  unresolved issues   254–5
  weighting   247
occupational studies   142
odd man out method   296
odds   18–19
  calculation   19
odds/log odds ratio   21
odds ratio   20–3, 139, 168, 207, 210, 252,
    278
  calculation   21–4
odds ratios, combining   59, 63–70
oral contraceptives   240, 241, 248, 252
ordinal data, combining   205–6
ordinal outcomes   33
ordinal regression equations   215
outcome measures   17–35
outcome variable, change of scale   51
outcomes
  defined on original metric   29–31
  standardized mean difference   31–3
overall effect   170

p-values   76, 123, 244
  combining   218–21, 230
    example   221
  multiple   230
passive smoking   272
patient baseline risk as covariate   102–4
patient follow-up   136
patient preference outcome   216–17
patient risk   101–2
periodontal disease   233
periodontal trials   235
Peto method   66–9
point estimates   155
pool-first method   253
pooled estimates   84
pooled result, impact of individual
    studies   149
pooled standard deviation   29
pooled survival rates   279

pooling
  independent samples of survival data   283
  observational studies   254
  studies with heterogeneous outcomes   53
population effect   58, 171
population risk   102
posterior density function   164
posterior distribution   168
  estimation   179
precision, missing measures   201
prediction   178
predictive distribution   164
predictor variables   94
presence of heterogeneity, hypothesis
    tests   38–41
prior distribution   168, 169, 171, 174
  choice of   170
  sensitivity   179
  specification   178
prior probability   168
probability, direct statements   178
Probing Depth (PD)   233–6
profile likelihood method   81
prognostic factors, log (relative risk)   282
proportional-hazards model   280
prospective meta-analysis   297–8
prospective registration of trials   126
publication bias   109–32
  adjustment of results   119–20
  assessment   148
  assessment of presence of   115
  broader perspective solutions   126–7
  consequences of   112
  definition   109
  detection   119
  evidence   110–12
  identifying   112
  predictors   112
  reduction   127
  seriousness   112
  taking into account   119
publication process and journals   126–7
published trials versus registered
    trials   110

Q statistic   38–40, 52, 62, 76, 80, 139

Q-TWiST 282
qualitative/categorical exposure
      variables 244
quality adjustment score 139
quality of life adjusted survival data 282–3
Quality Of Reporting Of Meta-analyses
      (QUORUM) 153
quality score 138, 141
quantitative variables, risk associations
      of 246

radial plots 46–7, 61, 157
random component models 181
random effects 247
random effects meta-analysis
      binary data 81–2
      exact approach 81–2
random effects methods 73–86
random effects models 52, 73, 76–9, 83–4,
      141, 169, 265
      extensions 80–3
      miscellaneous 82–3
      weighted method 74–5
random effects multiple outcome regression
      model 232–3
random effects regression 97–8
random effects regression models
      alternative formulations 100
      generalization and extensions 100–1
Randomized Controlled Trials (RCTs) 4,
      17–24, 28, 30, 32, 41–2, 47–50, 52, 61,
      88, 90, 95, 115, 126, 134, 136, 166,
      194, 202, 239, 266–7
      cumulative meta-analysis 287–93
      reporting 153
      sources of evidence other than 178
rank correlation test 116–17
      application 117
rate comparisons 28
rate ratio 28
rate ratio/relative rate 23
Receiver Operating Characteristic (ROC)
      curves 210, 215
reference lists 155
registered studies, follow-up of cohorts 111
registered trials versus published trials 110

regression 246
regression analysis 213, 241
      covariates 51–2
regression coefficients 98
regression models 93–104, 138–40, 235
      types 93
regression parameters 236
relative risk 23–5, 207, 270
      adjusting an unadjusted 244
      calculation 25
      repeated estimate using broad exposure
          categories 245
reporting
      meta-analysis 153–9
      observational studies 248
      overview and structure 154–5
      research practices 200
research quality, assessing 133
research reporting practices 200
response surface, estimating and
          extrapolating 295–6
restricted maximum likelihood (REML)
          estimate 74–6
results and inputs 144
risk associations of quantitative
          variables 246
risk difference (RD) 25
      calculation 26
risk/odds estimates, combining 252
robustness of results 155
Rosenthal's 'file drawer' method 120–2
      application 122

sampling error 37
SAS system 301, 304
scales of measurement
      combining data 206–8
      observational studies 243
scattergram 213
schizophrenia 115–18, 120, 122, 125, 171,
      177
selection models 123
sensitivity, prior distributions 179
sensitivity analysis 52, 87, 147–52
      Bayesian methods 181–2
      definition 147

sensitivity analysis (*cont.*)
  epidemiological studies 248–9
  observational studies 240
sensitivity approach of Copas 125
sensitivity of results
  inclusion criteria 147–9
  meta-analytic methods 150–1
shrinkage 176, 177, 269
significance levels, combining 218–21
Simpson's paradox 57
simulation of extra trials 151
single-arm studies 259–62
single case research 253–4
single outcome measures using different
    treatment arms 263–5
software 301–5
sources of evidence other than RCT 178
S-plus 302, 303
STAGE 301
standard chi square test 38–9
standard deviation 29, 173
  missing data 202–3
standard error 63, 94
  calculation 243
standard random effect method 82
standardized effect measure 32
standardized effect size 116
standardized likelihood 166, 169
standardized mean difference 31–3
  calculation 32
  combining 62–3
standardized morbidity ratio (SMR) 245
Standardized Normal Distribution 65
standardized normal test statistic 244
STATA 301–5
statistical analysis 136
statistical approaches 150
statistical chi square test 40
Stouffer's formula 230
stratification 246
  patient characteristics 89–93
  study characteristics 89
stratified case-control studies 64
stratified log-rank test 283
stratified ordinal regression, combining
    information from disparate

toxicological studies using 272
study assessment, practical
    implementation 143–4
study estimates
  combining 73–86
  comparison of methods 76
study quality 133–46
  and study weighting 141–2
  incorporation 179–80
  meta-analysis 137–43
  methodological factors 134–6
study weighting, and study quality 141–2
subgroup analyses 88–93
subgroup effects 88
subgroups 52
subject selection bias 244
subjective probability 164
sum of logs method 220
sum of z's method 220
Summary Receiver Operating Characteristics
    (SROC) curve 210–14
surrogate markers 215–16
survey of authors 110
survival data 277–85
  individual patient data (IPD) 283
  iterative generalized least squares 280–2
  multiple times 280–2
  obtaining and using 283–4
  pooling independent samples of 283
survival rates, pooled 279
synthesizing studies with disparate
    designs 271
systematic review 5–8
  advantages 6–8
  characteristic features 6
  characteristics 7
  methodology 6–8

target audience 155
test statistic 62
toxocological studies 272–3
transforming scales, maintaining same data
    type 207
treatment difference variance 29
treatment effect 29, 38, 75, 102, 114
  pooled estimate 58

treatment estimates, combining   57–72
trials, registration   126–7
trim and fill method   123–5
   application   125
tuberculosis prevention   95–9
2×2 table   207
2×2-table, for binary test   209
Type I error   50, 230
Type II error   10

uncertainty   148
   allowance   178
   induced by estimating between study
      variance   80–1
   reduced   168
unified modelling approach   176
unpublished studies   127, 148
   estimating number of   122
   trial amnesty   127
Unweighted Least Squares (UWLS)   74
U.S. General Accounting Office
      (GAO)   267

vague prior distributions   166
variance estimates   65
variance formulae   58
variance-weighted fixed effect
      model   142
vote-counting methods   201, 217–18

weight functions   123
weighted distribution theory   123
weighted least squares (WLS) method   74,
      98
weighted mean difference   63
weighted non-interative random effects
      model   78
weighting   140–2, 150
   influence of   150–1
   observational studies   247
WinBUGS   165, 171, 174, 304

Yule's paradox   57

z-scores   41–2, 220

WILEY SERIES IN PROBABILITY AND STATISTICS
ESTABLISHED BY WALTER A. SHEWHART AND SAMUEL S. WILKS

Editors
Vic Barnett, Noel A. C. Cressie, Nicholas I. Fisher, Iain M. Johnstone,
J. B. Kadane, David W. Scott, Bernard W. Silverman,
Adrian F. M. Smith, Jozef L. Teugels, Ralph A. Bradley, Emeritus,
J. Stuart Hunter, Emeritus, David G. Kendall, Emeritus

Probability and Statistics Section

*ANDERSON · The Statistical Analysis of Time Series
ARNOLD, BALAKRISHNAN, and NAGARAJA · A First Course in Order Statistics
ARNOLD, BALAKRISHNAN, and NAGARAJA · Records
BACCELLI, COHEN, OLSDER, and QUADRAT · Synchronization and Linearity: An
  Algebra for Discrete Event Systems
BARNETT · Comparative Statistical Inference, *Third Edition*
BASILEVSKY · Statistical Factor Analysis and Related Methods: Theory and
  Applications
BERNARDO and SMITH · Bayesian Theory
BILLINGSLEY · Convergence of Probability Measures, *Second Edition*
BOROVKOV · Asymptotic Methods in Queuing Theory
BOROVKOV · Ergodicity and Stability of Stochastic Processes
BRANDT, FRANKEN, and LISEK · Stationary Stochastic Models
CAINES · Linear Stochastic Systems
CAIROLI and DALANG · Sequential Stochastic Optimization
CONSTANTINE · Combinatorial Theory and Statistical Design
COOK · Regression Graphics
COVER and THOMAS · Elements of Information Theory
CSÖRGŐ and HORVÁTH · Weighted Approximations in Probability Statistics
CSÖRGŐ and HORVÁTH · Limit Theorems in Change Point Analysis
*DANIEL · Fitting Equations to Data: Computer Analysis of Multifactor Data, *Second
  Edition*
DETTE and STUDDEN · The Theory of Canonical Moments with Applications in
  Statistics, Probability, and Analysis
DEY and MUKERJEE · Fractional Factional Plans
*DOOB · Stochastic Processes
DRYDEN and MARDIA · Statistical Shape Analysis
DUPUIS and ELLIS · A Weak Convergence Approach to the Theory of Large
  Deviations
ETHIER and KURTZ · Markov Processes: Characterization and Convergence
FELLER · An Introduction to Probability Theory and Its Applications, Volume 1, *Third
  Edition*, Revised; Volume II, *Second Edition*
FULLER · Introduction to Statistical Time Series, *Second Edition*
FULLER · Measurement Error Models

---

*Now available in a lower priced paperback edition in the Wiley Classics Library.

Probability and Statistics Section (Continued)

GHOSH, MUKHOPADHYAY, and SEN · Sequential Estimation

GIFI · Nonlinear Multivariate Analysis

GUTTORP · Statistical Inference for Branching Processes

HALL · Introduction to the Theory of Coverage Processes

HAMPEL · Robust Statistics: The Approach Based on Influence Functions

HANNAN and DEISTLER · The Statistical Theory of Linear Systems

HUBER · Robust Statistics

HUŠKOVÁ, BERAN, and DUPAČ · Collected Works of Jaroslav Hájek—With Commentary

IMAN and CONOVER · A Modern Approach to Statistics

JUREK and MASON · Operator-Limit Distributions in Probability Theory

KASS and VOS · The Geometrical Foundations of Asymptotic Inference

KAUFMAN and ROUSSEEUW · Finding Groups in Data: An Introduction to Cluster
    Analysis

KELLY · Probability, Statistics, and Optimization

KENDALL, BARDEN, CARNE, and LE · Shape and Shape Theory

LINDVALL · Lectures on the Coupling Method

McFADDEN · Management of Data in Clinical Trials

MANTON, WOODBURY, and TOLLEY · Statistical Applications Using Fuzzy Sets

MARDIA and JUPP · Directional Statistics

MORGENTHALER and TUKEY · Configural Polysampling: A Route to Practical
    Robustness

MUIRHEAD · Aspects of Multivariate Statistical Theory

OLIVER and SMITH · Influence Diagrams, Belief Nets, and Decision Analysis

*PARZEN · Modern Probability Theory and Its Applications

PRESS · Bayesian Statistics: Principles, Models, and Applications

PUKELSHEIM · Optimal Experimental Design

RAO · Asymptotic Theory of Statistical Inference

RAO · Linear Statistical Inference and Its Applications, *Second Edition*

RAO and SHANBHAG · Choquet-Deny Type Functional Equations with Applications
    to Stochastic Models

ROBERTSON, WRIGHT, and DYKSTRA · Order Restricted Statistical Inference

ROGERS and WILLIAMS · Diffusions, Markov Processes, and Martingales, Volume I:
    Foundations, *Second Edition*, Volume II: Itô Calculus

RUBINSTEIN and SHAPIRO · Discrete Event Systems: Sensitivity Analysis and
    Stochastic Optimization by the Score Function Method

RUZSA and SZEKELEY · Algebraic Probability Theory

SALTELLI, CHAN and SCOTT · Sensitivity Analysis

SCHEFFÉ · The Analysis of Variance

SEBER · Linear Regression Analysis

SEBER · Multivariate Observations

SEBER and WILD · Nonlinear Regression

SERFLING · Approximation Theorems of Mathematical Statistics

SHORACK and WELLNER · Empirical Processes with Applications to Statistics

SMALL and McLEISH · Hilbert Space Methods in Probability and Statistical Inference

STAPLETON · Linear Statistical Models

STAUDTE and SHEATHER · Robust Estimation and Testing

STOYANOV · Counterexamples in Probability

TANAKA · Time Series Analysis: Nonstationary and Noninvertible Distribution Theory

THOMPSON and SEBER · Adaptive Sampling

WELSH · Aspects of Statistical Inference

WHITTAKER · Graphical Models in Applied Multivariate Statistics

---

*Now available in a lower priced paperback edition in the Wiley Classics Library.

Probability and Statistics Section (Continued)
YANG · The Construction Theory of Denumerable Markov Processes

Applied Probability and Statistics Section

ABRAHAM and LEDOLTER · Statistical Methods for Forecasting
AGRESTI · Analysis of Ordinal Categorical Data
AGRESTI · Categorical Data Analysis
ANDERSON, AUQUIER, HAUCK, OAKES, VANDAELE, and WEISBERG ·
    Statistical Methods for Comparative Studies
ARMITAGE and DAVID (editors) · Advances in Biometry
*ARTHANARI and DODGE · Mathematical Programming in Statistics
ASMUSSEN · Applied Probability and Queues
*BAILEY · The Elements of Stochastic Processes with Applications to the Natural
    Sciences
BARNETT and LEWIS · Outliers in Statistical Data, *Third Edition*
BARTHOLOMEW, FORBES, and McLEAN · Statistical Techniques for Manpower
    Planning, *Second Edition*
BATES and WATTS · Nonlinear Regression Analysis and Its Applications
BECHHOFER, SANTNER, and GOLDSMAN · Design and Analysis of Experiments
    for Statistical Selection, Screening, and Multiple Comparisons
BELSLEY · Conditioning Diagnostics: Collinearity and Weak Data in Regression
BELSLEY, KUH, and WELSCH · Regression Diagnostics: Identifying Influential Data
    and Sources of Collinearity
BHAT · Elements of Applied Stochastic Processes, *Second Edition*
BHATTACHARYA and WAYMIRE · Stochastic Processes with Applications
BIRKES and DODGE · Alternative Methods of Regression
BLISCHKE and PRABHAKAR MURTHY · Reliability: Modeling, Prediction, and
    Optimization
BLOOMFIELD · Fourier Analysis of Time Series: An Introduction, *Second Edition*
BOLLEN · Structural Equations with Latent Variables
BOULEAU · Numerical Methods for Stochastic Processes
BOX · Bayesian Inference in Statistical Analysis
BOX and DRAPER · Empirical Model-Building and Response Surfaces
*BOX and DRAPER · Evolutionary Operation: A Statistical Method for Process Improvement
BUCKLEW · Large Deviation Techniques in Decision, Simulation, and Estimation
BUNKE and BUNKE · Nonlinear Regression, Functional Relations, and Robust Methods:
    Statistical Methods of Model Building
CHATTERJEE and HADI · Sensitivity Analysis in Linear Regression
CHERNICK · Bootstrap Methods: A Practitioner's Guide
CHILÈS and DELFINER · Geostatistics: Modeling Spatial Uncertainty
CHOW and LIU · Design and Analysis of Clinical Trials: Concepts and Methodologies
CLARKE and DISNEY · Probability and Random Processes: A First Course with
    Applications, *Second Edition*
*COCHRAN and COX · Experimental Designs, *Second Edition*
CONOVER · Practical Nonparametric Statistics, *Second Edition*
CORNELL · Experiments with Mixtures, Designs, Models, and the Analysis of Mixture
    Data, *Second Edition*
*COX · Planning of Experiments
CRESSIE · Statistics for Spatial Data, *Revised Edition*
DANIEL · Applications of Statistics to Industrial Experimentation

---

Applied Probability and Statistics (Continued)

DANIEL · Biostatistics: A Foundation for Analysis in the Health Sciences, *Sixth Edition*

DAVID · Order Statistics, *Second Edition*

*DEGROOT, FIENBERG, and KADANE · Statistics and the Law

DODGE · Alternative Methods of Regression

DOWDY and WEARDEN · Statistics for Research, *Second Edition*

DUNN and CLARK · Applied Statistics: Analysis of Variance and Regression, *Second Edition*

*ELANDT-JOHNSON and JOHNSON · Survival Models and Data Analysis

*FLEISS · The Design and Analysis of Clinical Experiments

FLEISS · Statistical Methods for Rates and Proportions, *Second Edition*

FLEMING and HARRINGTON · Counting Processes and Survival Analysis

GALLANT · Nonlinear Statistical Models

GLASSERMAN and YAO · Monotone Structure in Discrete-Event Systems

GNANADESIKAN · Methods for Statistical Data Analysis of Multivariate Observations. *Second Edition*

GOLDSTEIN and LEWIS · Assessment: Problems, Development, and Statistical Issues

GREENWOOD and NIKULIN · A Guide to Chi-squared Testing

*HAHN · Statistical Models in Engineering

HAHN and MEEKER · Statistical Intervals: A Guide for Practitioners

HAND · Construction and Assessment of Classification Rules

HAND · Discrimination and Classification

HEDAYAT and SINHA · Design and Inference in Finite Population Sampling

HEIBERGER · Computation for the Analysis of Designed Experiments

HINKELMAN and KEMPTHORNE · Design and Analysis of Experiments, Volume 1: Introduction to Experimental Design

HOAGLIN, MOSTELLER, and TUKEY · Exploratory Approach to Analysis of Variance

HOAGLIN, MOSTELLER, and TUKEY · Exploring Data Tables, Trends and Shapes

HOAGLIN, MOSTELLER, and TUKEY · Understanding Robust and Exploratory Data Analysis

HOCHBERG and TAMHANE · Multiple Comparison Procedures

HOCKING · Methods and Applications of Linear Models: Regression and the Analysis of Variables

HOGG and KLUGMAN · Loss Distributions

HOSMER and LEMESHOW · Applied Logistic Regression

HØYLAND and RAUSAND · System Reliability Theory: Models and Statistical Methods

HUBERTY · Applied Discriminant Analysis

HUNT and KENNEDY · Financial Derivatives in Theory and Practice

JACKSON · A User's Guide to Principal Components

JOHN · Statistical Methods in Engineering and Quality Assurance

JOHNSON · Multivariate Statistical Simulation

JOHNSON & KOTZ · Distributions in Statistics

JOHNSON, KOTZ, and BALAKRISHNAN · Continuous Univariate Distributions, Volume 1, *Second Edition*

JOHNSON, KOTZ, and BALAKRISHNAN · Continuous Univariate Distributions, Volume 2, *Second Edition*

JOHNSON, KOTZ, and BALAKRISHNAN · Discrete Multivariate Distributions

JOHNSON, KOTZ, and KEMP · Univariate Discrete Distributions, *Second Edition*

JUREČKOVÁ and SEN · Robust Statistical Procedures: Asymptotics and Interrelations

KADANE · Bayesian Methods and Ethics in a Clinical Trial Design

KADANE and SCHUM · A Probabilistic Analysis of the Sacco and Vanzetti Evidence

---

*Now available in a lower priced paperback edition in the Wiley Classics Library.

Applied Probability and Statistics (Continued)

KALBFLEISCH and PRENTICE · The Statistical Analysis of Failure Time Data

KELLY · Reversibility and Stochastic Networks

KHURI, MATHEW, and SINHA · Statistical Tests for Mixed Linear Models

KLUGMAN, PANJER, and WILLMOT · Loss Models: From Data to Decisions

KLUGMAN, PANJER, and WILLMOT · Solutions Manual to Accompany Loss Models: From Data to Decisions

KOTZ, BALAKRISHNAN and JOHNSON · Continuous Multivariate Distributions, Volume 1, *Second Edition*

KOVALENKO, KUZNETZOV, and PEGG · Mathematical Theory of Reliability of Time-Dependent Systems with Practical Applications

LACHIN · Biostatistical Methods: The Assessment of Relative Risks

LAD · Operational Subjective Statistical Methods: A Mathematical, Philosophical, and Historical Introduction

LANGE, RYAN, BILLARD, BRILLINGER, CONQUEST, and GREENHOUSE · Case Studies in Biometry

LAWLESS · Statistical Models and Methods for Lifetime Data

LEE · Statistical Methods for Survival Data Analysis, *Second Edition*

LEPAGE and BILLARD · Exploring the Limits of Bootstrap

LEYLAND and GOLDSTEIN · Multilevel Modelling of Health Statistics

LINHART and ZUCCHINI · Model Selection

LITTLE and RUBIN · Statistical Analysis with Missing Data

LLOYD · The Statistical Analysis of Categorical Data

MAGNUS and NEUDECKER · Matrix Differential Calculus with Applications in Statistics and Econometrics, *Revised Edition*

MALLER and ZHOU · Survival Analysis with Long Term Survivors

MANN, SCHAFER, and SINGPURWALLA · Methods for Statistical Analysis of Reliability and Life Data

McLACHLAN and KRISHNAN · The EM Algorithm and Extensions

McLACHLAN · Discriminant Analysis and Statistical Pattern Recognition

McNEIL · Epidemiological Research Methods

MEEKER and ESCOBAR · Statistical Methods for Reliability Data

*MILLER · Survival Analysis, *Second Edition*

MONTGOMERY and PECK · Introduction to Linear Regression Analysis, *Second Edition*

MYERS and MONTGOMERY · Response Surface Methodology: Process and Product in Optimization Using Designed Experiments

NELSON · Accelerated Testing, Statistical Models, Test Plans, and Data Analyses

NELSON · Applied Life Data Analysis

OCHI · Applied Probability and Stochastic Processes in Engineering and Physical Sciences

OKABE, BOOTS, CHUI, and SUGIHARA · Spatial Tesselations: Concepts and Applications of Voronoi diagrams, *Second Edition*

PANKRATZ · Forecasting with Dynamic Regression Models

PANKRATZ · Forecasting with Univariate Box–Jenkins Models: Concepts and Cases

PIANTADOSI · Clinical Trials: A Methodologic Perspective

PORT · Theoretical Probability for Applications

PUTERMAN · Markov Decision Processes: Discrete Stochastic Dynamic Programming

RACHEV · Probability Metrics and the Stability of Stochastic Models

RÉNYI · A Diary on Information Theory

RIGDON and BASU · Statistical Methods for the Reliability of Repairable Systems

RIPLEY · Spatial Statistics

RIPLEY · Stochastic Simulation

*Now available in a lower priced paperback edition in the Wiley Classics Library.

Applied Probability and Statistics (Continued)

ROLSKI, SCHMIDLI, SCHMIDT, and TEUGELS · Stochastic Processes for Insurance and Finance

ROUSSEEUW and LEROY · Robust Regression and Outlier Detection

RUBIN · Multiple Imputation for Nonresponse in Surveys

RUBINSTEIN · Simulation and the Monte Carlo Method

RUBINSTEIN and MELAMED · Modern Simulation and Modeling

RYAN · Statistical Methods for Quality Improvement, *Second Edition*

SCHUSS · Theory and Applications of Stochastic Differential Equations

SCOTT · Multivariate Density Estimation: Theory, Practice, and Visualization

*SEARLE · Linear Models

SEARLE · Linear Models for Unbalanced Data

SEARLE, CASELLA, and McCULLOCH · Variance Components

STOYAN, KENDALL, and MECKE · Stochastic Geometry and Its Applications, *Second Edition*

STOYAN and STOYAN · Fractals, Random Shapes, and Point Fields: Methods of Geometrical Statistics

SUTTON, ABRAMS, JONES, SHELDON and SONG · Methods for Meta-analysis in Medical Research

THOMPSON · Empirical Model Building

THOMPSON · Sampling

THOMPSON · Simulation: A Modeler's Approach

TIJMS · Stochastic Modeling and Analysis: A Computational Approach

TIJMS · Stochastic Models: An Algorithmic Approach

TITTERINGTON, SMITH, and MARKOV · Statistical Analysis of Finite Mixture Distributions

UPTON and FINGLETON · Spatial Data Analysis by Example, Volume 1: Point Pattern and Quantitative Data

UPTON and FINGLETON · Spatial Data Analysis by Example, Volume II: Categorical and Directional Data

VAN RIJCKEVORSEL and DE LEEUW · Component and Correspondence Analysis

VIDAKOVIC · Statistical Modeling by Wavelets

WEISBERG · Applied Linear Regression, *Second Edition*

WESTFALL and YOUNG · Resampling-Based Multiple Testing: Examples and Methods for p-Value Adjustment

WHITTLE · Systems in Stochastic Equilibrium

WINKER · Optimization Heuristics in Econometrics

WOODING · Planning Pharmaceutical Clinical Trials: Basic Statistical Principles

WOOLSON · Statistical Methods for the Analysis of Biomedical Data

*ZELLNER · An Introduction to Bayesian Inference in Econometrics

Texts and References Section

AGRESTI · An Introduction to Categorical Data Analysis

ANDERSON · An Introduction to Multivariate Statistical Analysis, *Second Edition*

ANDERSON and LOYNES · The Teaching of Practical Statistics

ARMITAGE and COLTON · Encyclopedia of Biostatistics. 6 Volume set

BARTOSZYNSKI and NIEWIADOMSKA-BUGAJ · Probability and Statistical Inference

BENDAT and PIERSOL · Random Data: Analysis and Measurement Procedures, *Third Edition*

---

*Now available in a lower priced paperback edition in the Wiley Classics Library.

Texts and References Section (Continued)

BERRY, CHALONER, and GEWEKE · Bayesian Analysis in Statistics and Econometrics: Essays in Honor of Arnold Zellner

BHATTACHARYA and JOHNSON · Statistical Concepts and Methods

BILLINGSLEY · Probability and Measure, *Second Edition*

BOX · R. A. Fisher, the Life of a Scientist

BOX, HUNTER, and HUNTER · Statistics for Experimenters: An Introduction to Design, Data Analysis, and Model Building

BOX and LUCEÑO · Statistical Control by Monitoring and Feedback Adjustment

BROWN and HOLLANDER · Statistics: A Biomedical Introduction

CHATTERJEE and PRICE · Regression Analysis by Example, *Third Edition*

COOK and WEISBERG · An Introduction to Regression Graphics

COOK and WEISBERG · Applied Regression Including Computing and Graphics

COX · A Handbook of Introductory Statistical Methods

DILLON and GOLDSTEIN · Multivariate Analysis: Methods and Applications

*DODGE and ROMIG · Sampling Inspection Tables, *Second Edition*

DRAPER and SMITH · Applied Regression Analysis, *Third Edition*

DUDEWICZ and MISHRA · Modern Mathematical Statistics

DUNN · Basic Statistics: A Primer for the Biomedical Sciences, *Second Edition*

EVANS, HASTINGS and PEACOCK · Statistical Distributions, *Third Edition*

FISHER and VAN BELLE · Biostatistics: A Methodology for the Health Sciences

FREEMAN and SMITH · Aspects of Uncertainty: A Tribute to D. V. Lindley

GROSS and HARRIS · Fundamentals of Queueing Theory, *Third Edition*

HALD · A History of Probability and Statistics and their Applications Before 1750

HALD · A History of Mathematical Statistics from 1750 to 1930

HELLER · MACSYMA for Statisticians

HOEL · Introduction to Mathematical Statistics, *Fifth Edition*

HOLLANDER and WOLFE · Nonparametric Statistical Methods, *Second Edition*

HOSMER and LEMESHOW · Applied Logistic Recession, *Second Edition*

HOSMER and LEMESHOW · Applied Survival Analysis: Regression Modeling of Time to Event Data

JOHNSON and BALAKRISHNAN · Advances in the Theory and Practice of Statistics: A Volume in Honor of Samuel Kotz

JOHNSON and KOTZ (editors) · Leading Personalities in Statistical Sciences: From the Seventeenth Century to the Present

JUDGE, GRIFFITHS, HILL, LÜTKEPOHL, and LEE · The Theory and Practice of Econometrics, *Second Edition*

KHURI · Advanced Calculus with Applications in Statistics

KOTZ and JOHNSON (editors) · Encyclopedia of Statistical Sciences. Volumes 1 to 9 with Index

KOTZ and JOHNSON (editors) · Encyclopedia of Statistical Sciences: Supplement Volume

KOTZ, REED, and BANKS (editors) · Encyclopedia of Statistical Sciences: Update Volume 1

KOTZ, REED, and BANKS (editors) · Encyclopedia of Statistical Sciences: Update Volume 2

LAMPERTI · Probability: A Survey of the Mathematical Theory, *Second Edition*

LARSON · Introduction to Probability Theory and Statistical Inference, *Third Edition*

LE · Applied Categorical Data Analysis

LE · Applied Survival Analysis

MALLOWS · Design, Data, and Analysis by Some Friends of Cuthbert Daniel

MARDIA · The Art of Statistical Science: A Tribute to G. S. Watson

*Now available in a lower priced paperback edition in the Wiley Classics Library.

Texts and References Section (Continued)

MASON, GUNST, and HESS · Statistical Design and Analysis of Experiments with Applications to Engineering and Science

MURRAY · X-STAT 2.0 Statistical Experimentation, Design Data Analysis, and Nonlinear Optimization

PURI, VILAPLANA, and WERTZ · New Perspectives in Theoretical and Applied Statistics

RENCHER · Methods of Multivariate Analysis

RENCHER · Linear Models in Statistics

RENCHER · Multivariate Statistical Inference with Applications

ROBINSON · Practical Strategies for Experimenting

ROSS · Introduction to Probability and Statistics for Engineers and Scientists

ROHATGI · An Introduction to Probability Theory and Mathematical Statistics

RYAN · Modern Regression Methods

SCHOTT · Matrix Analysis for Statistics

SEARLE · Matrix Algebra Useful for Statistics

STYAN · The Collected Papers of T. W. Anderson: 1943–1985

TIERNEY · LISP-STAT: An Object-Oriented Environment for Statistical Computing and Dynamic Graphics

WONNACOTT and WONNACOTT · Econometrics, *Second Edition*

WU and HAMADA · Experiments: Planning, Analysis, and Parameter Design Optimization

WILEY SERIES IN PROBABILITY AND STATISTICS

ESTABLISHED BY WALTER A. SHEWHART AND SAMUEL S. WILKS

Editors
Robert M. Groves, Graham Kalton, J. N. K. Rao, Norbert Schwarz, Christopher Skinner

Survey Methodology Section

BIEMER, GROVES, LYBERG, MATHIOWETZ, and SUDMAN · Measurement Errors in Surveys

COCHRAN · Sampling Techniques, *Third Edition*

COUPER, BAKER, BETHLEHEM, CLARK, MARTIN, NICHOLLS and O'REILLY (editors) · Computer Assisted Survey Information Collection

COX, BINDER, CHINNAPPA, CHRISTIANSON, COLLEDGE, and KOTT (editors) · Business Survey Methods

*DEMING · Sample Design in Business Research

DILLMAN · Mail and Telephone Surveys: The Total Design Method

GROVES · Survey Errors and Survey Costs

GROVES and COUPER · Nonresponse in Household Interview Surveys

GROVES, BIEMER, LYBERG, MASSEY, NICHOLLS, and WAKSBERG · Telephone Survey Methodology

*Now available in a lower priced paperback edition in the Wiley Classics Library.

Survey Methodology Section (Continued)

*HANSEN, HURWITZ, and MADOW · Sample Survey Methods and Theory, Volume I: Methods and Applications

*HANSEN, HURWITZ, and MADOW · Sample Survey Methods and Theory, Volume II: Theory

KISH · Statistical Design for Research

*KISH · Survey Sampling

KORN and GRAUBARD · Analysis of Health Surveys

LESSLER and KALSBEEK · Nonsampling Error in Surveys

LEVY and LEMESHOW · Sampling of Populations: Methods and Applications

LYBERG, BIEMER, COLLINS, de LEEUW, DIPPO, SCHWARZ, TREWIN (editors) · Survey Measurement and Process Quality

SIRKEN, HERRMANN, SCHECHTER, SCHWARZ, TANUR and TOURNANGEAU (editors) · Cognition and Survey Research

*Now available in a lower priced paperback edition in the Wiley Classics Library.